S0-CAS-514

ACS SYMPOSIUM SERIES 569

Modeling the Hydrogen Bond

Douglas A. Smith, EDITOR
University of Toledo

Developed from a symposium sponsored
by the Division of Computers in Chemistry
at the 206th National Meeting
of the American Chemical Society,
Chicago, Illinois,
August 22–27, 1993

American Chemical Society, Washington, DC 1994

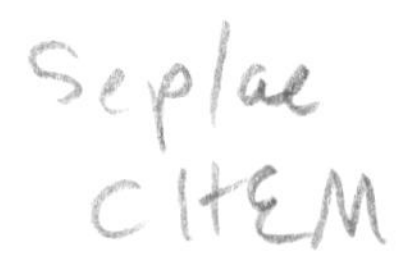

Library of Congress Cataloging-in-Publication Data

Modeling the hydrogen bond / Douglas A. Smith, editor

p. cm.—(ACS symposium series, ISSN 0097–6156; 569)

"Developed from a symposium sponsored by the Division of Computers in Chemistry at the 206th National Meeting of the American Chemical Society, Chicago, Illinois, August 22–27, 1993."

Includes bibliographical references and indexes.

ISBN 0–8412–2981–3

1. Hydrogen bonding—Simulation methods—Congresses.

I. Smith, Douglas A., 1959– . II. American Chemical Society. Division of Computers in Chemistry. III. Series.

QD461.M5964 1994
541.2′.24—dc20 94–31645
CIP

The paper used in this publication meets the minimum requirements of American National Standard for Information Sciences—Permanence of Paper for Printed Library Materials, ANSI Z39.48–1984. ♾

PRINTED IN THE UNITED STATES OF AMERICA

Foreword

THE ACS SYMPOSIUM SERIES was first published in 1974 to provide a mechanism for publishing symposia quickly in book form. The purpose of this series is to publish comprehensive books developed from symposia, which are usually "snapshots in time" of the current research being done on a topic, plus some review material on the topic. For this reason, it is necessary that the papers be published as quickly as possible.

Before a symposium-based book is put under contract, the proposed table of contents is reviewed for appropriateness to the topic and for comprehensiveness of the collection. Some papers are excluded at this point, and others are added to round out the scope of the volume. In addition, a draft of each paper is peer-reviewed prior to final acceptance or rejection. This anonymous review process is supervised by the organizer(s) of the symposium, who become the editor(s) of the book. The authors then revise their papers according to the recommendations of both the reviewers and the editors, prepare camera-ready copy, and submit the final papers to the editors, who check that all necessary revisions have been made.

As a rule, only original research papers and original review papers are included in the volumes. Verbatim reproductions of previously published papers are not accepted.

M. Joan Comstock
Series Editor

Contents

APPLICATIONS TO MOLECULES AND POLYMERS

INDEXES

Preface

THE CONCEPT OF HYDROGEN BONDING is one of the simplest yet most important in all of chemistry. It is a constant subject for research and discussion in the literature. Even so, our understanding of what hydrogen bonding is; how it is defined, measured, and quantified; and why it is so important and ubiquitous is far from circumscribed. Like many other phenomena in chemistry, hydrogen bonding is often invoked to account for what cannot be explained by other factors.

Recent research has led to the intentional inclusion of hydrogen bonding as a design criteria in molecules of interest, from biopolymers to synthetic polymers, from non-natural amino acids and molecular clefts to drugs, from crystals to liquid crystals, even to Lewis acid catalysts. Experimental and theoretical papers on hydrogen bonding abound. This symposium and book address a more fundamental question: How can the hydrogen bond be modeled in a way that is tractable yet lends added depth to our understanding of basic physical and chemical processes? The answers to this question (because surely there are many answers) can be applied to many experimental, theoretical, practical, and academic problems.

This book is divided into three sections. Chapters 2 through 5 focus on local or microscopic effects on and of hydrogen bonding. Chapters 6 through 13 discuss the numerical and visual methods used to study hydrogen bonding. Chapters 14 through 18 consider the use of modeling in understanding hydrogen bonding several classes of compounds.

Acknowledgments

There are too many people to thank than can be easily recounted. Most of the work for this volume was done by the people who presented their research at the American Chemical Society symposium and prepared their material for inclusion as chapters in this book. Thanks are due to the members of my research group, in particular Chuck Ulmer, Peter Nagy, and Joe Bitar, for their efforts helping me with this project, and to Sundararajan Vijaykumar. Thanks to the Computers in Chemistry Division officers, especially Angelo Rossi, Tom Pierce, and Phil Kutzenco, who were instrumental in getting me involved with the division and active in the program. Special thanks are due to George Famini for the friendly rivalry we established in trying to get the biggest and best symposium: both were fantastic; and to ACS editor Barbara Pralle, who put up with me during the long and sometimes difficult road to publication. Finally, as always, I must thank my wife Carolyn and my daughters Rachel and Rebecca for allowing me the time to focus on work and for always being there for me when I needed them.

DOUGLAS A. SMITH
Concurrent Technologies Corporation
1450 Scalp Avenue
Johnstown, PA 15904

August 17, 1994

Chapter 1

A Brief History of the Hydrogen Bond

Douglas A. Smith[1]

Department of Chemistry, University of Toledo, 2801 West Bancroft Street, Toledo, OH 43606–3390

Between the beginning of the twentieth century and the late 1930's many descriptions of interactions between hydrogen atoms in molecules and electronegative atoms to which these hydrogens were not covalently bound appeared in the literature. This was part of the general explosion of physical, structural and quantum chemistry that was providing a descriptive and explanatory underpinning for chemical phenomena as a whole. In 1920, Latimer and Rodebush, working on the structure and properties of water with G. N. Lewis at UC Berkeley, proposed:

> "[A] free pair of electrons on one water molecule might be able to exert sufficient force on a hydrogen held by a pair of electrons on another water molecule to bind the two molecules together.... Such an explanation amounts to saying that the hydrogen nucleus held between 2 octets constitutes a weak 'bond' (*1*)."

This description, based on the Lewis dot formalism, is the first to truly call this interaction a bond. It was in 1939, however, with the publication of the first edition of Pauling's "The Nature of the Chemical Bond" (*2*) and its seminal chapter *The Hydrogen Bond*, that this phenomenon really reached acceptance in the main stream. In this treatise was the following description:

> "[U]nder certain conditions an atom of hydrogen is attracted by rather strong forces to two atoms, instead of only one, so that it may beconsidered to be acting as a bond between them. This is called the *hydrogen bond*." (pp. 449, Third Edition)

Furthermore, Pauling put the hydrogen bond on firm quantum mechanical grounds:

> "[A] hydrogen atom, with only one stable orbital, cannot form more than one pure covalent bond and that the attraction of two atoms observed in hydrogen-bond formation must be due largely to ionic forces." (pp. 450-1, Third Edition)

[1]Current address: Concurrent Technologies Corporation, 1450 Scalp Avenue, Johnstown, PA 15904

0097–6156/94/0569–0001$08.00/0

As an ionic interaction, hydrogen bonding was limited to occurring only with the most electronegative atoms. It remained for Pimentel and McClellan provided a broader, operational definition in their 1960 book, "The Hydrogen Bond" (*3*):

> "A H Bond exists between a functional group A-H and an atom or a group of atoms B in the same or a different molecule when (a) there is evidence of bond formation (association or chelation), (b) there is evidence that this new bond linking A-H and B specifically involves the hydrogen atom already bonded to A."

This rather broad definition leaves open the possibility of hydrogen bonding to groups of atoms (or the electrons associated with those atoms) which are not themselves highly electronegative, or the involvement of hydrogen atoms covalently bound to non-electronegative atoms such as carbon. As both experiment and theory have shown, this is indeed a more correct definition. In current practice, the terms acceptor and donor refer to those atoms that either accept or donate the hydrogen of the hydrogen bond, respectively. Typical donors include but are not limited to electronegative atoms with at least one attached hydrogen, such as oxygen, nitrogen, sulfur and the halogens. Even carbon, silicon and phosphorus, however, can behave as donors. More generally, donors are simply Bronsted acids. Acceptor atoms or groups are Lewis bases which range from simple electronegative atoms with lone electron pairs to the π electrons of an unsaturated or aromatic system, as well as ionic species such as oxides and halides. Other attached atoms or functional groups influence the capabilities of both the donor and acceptor.

Even though a description of intramolecular hydrogen bonding appeared as early as 1906 in azo compounds (*4*) and 1913 for the interaction between a hydroxyl hydrogen and a carbonyl oxygen (*5*), hydrogen bonds were considered a linear phenomenon based primarily on experimental data from x-ray and neutron diffraction data (*2*). Recent experimental and theoretical data support this viewpoint, at least in some systems. For example, only linear hydrogen bonding is observed experimentally for water (*6*), and theoretical calculations predict a linear angle in the case of water dimers (*7*). Such evidence has led, unfortunately, to a common misconception that hydrogen bonds cannot occur unless the angle of association is close to 180°. There is, however, a significant body of evidence to the contrary.

A recent survey of 1357 crystallographically observed intermolecular N-H···O=C bonds indicates that the donor angle in about 28% of the cases falls in the range 170° -180°, and 38% fall in the range of 160°-170°. The mean angle is 161.2°. For O-H···O bonds, the distribution is similar and the mean angle is 163.1° (*8*). A similar survey of 152 crystallographically observed intramolecular N-H···O=C hydrogen bonds show that they are significantly less linear, with a mean angle of 132.5° (*9*).

An Historical Interlude: Ljubljana, Yugoslavia, 1957

From 29 July through 3 August 1957, some of the most important and influential names in bonding and theory, including Pauling, Pople, Pimentel, and Coulson, among others, met in Ljubljana at the Symposium on Hydrogen Bonding. Perhaps the most important event of that meeting was the paper presented by Coulson, simply titled "The Hydrogen Bond" (*10*). His paper has become a landmark in this field. Subtitled "A review of the present interpretations and an introduction to the theoretical papers presented at the Congress," it was a critical review of the status of theory in its attempts to understand and explain hydrogen bonding.

In his paper, Coulson examined the electrostatic model of hydrogen bonding, which owed much to the work of Lennard-Jones and Pople (*11*) and discounted this model as insufficient to account for all of the phenomena associated with hydrogen bonding. Instead, Coulson divided the Coulombic attraction energy of the hydrogen bond into four causative factors: electrostatic, covalent, repulsive and dispersion contributions. Almost twenty years later, Kitaura and Morokuma developed a scheme within the framework of ab initio SCF theory in which they decompose the interaction energy of two molecules into five components: electrostatic (ES), polarization (PL), exchange repulsion (EX), charge transfer (CT), and coupling (MIX) (*12,13*). Application of this method to the origin of the hydrogen bond suggested that hydrogen bonds between neutral donors and acceptors were "*strongly ES in nature, with a small but significant contribution of CT* (*14*)." Furthermore, the uniqueness of hydrogen bonding was simply and unassumingly that "*it always involves a moderately polar, short, and strong [H-A] bond as the proton donor* (*14*)." The major objection to the Morokuma method, that two of the intermediate wave functions used to calculate the ES and PL components violate the Pauli exclusion principle (*13*), has recently been addressed by an energy decomposition analysis based on the natural bond orbital (NBO) method (*15,16*). This natural energy decomposition analysis (NEDA) indicates that both ES and CT make significant and similarly strong contributions to the hydrogen bonding interaction energy, at least in the water dimer (*17*). The important molecular orbital interaction in hydrogen bonded systems is the two electron two orbital interaction between the antibonding σ^* orbital corresponding to the donor H-A bond and the highest occupied molecular orbital (HOMO) of the acceptor (*18*). This is the predominant contribution to the charge transfer component of the interaction energy (*14,16*) and corresponds to the dominant component of van der Waals forces at short range. Even though the maximum orbital overlap should occur for a colinear arrangement of A-H···B, the local symmetry of the σ^* orbital is essentially spherical ($1s_H$). Thus, the directionality of the hydrogen bonding interaction is not critical.

Since 1957, a significant amount of theoretical work, compute cycles, and personnel time has been expended to examine the phenomenon of hydrogen bonding. Many empirical, semiempirical and ab initio molecular orbital studies have looked at hydrogen bonding systems and the nature of the hydrogen bond itself. Theoreticians have been examining many aspects of computational methods as applied to the study and accurate representation of hydrogen bonding. Recently, for example, Del Bene has published extensively on the basis set effects (*19*) and Del Bene (*20*) and Davidson (*21*) have studied the electron correlation contribution to the computed hydrogen bond energies of many species. Parameterization of molecular mechanics force fields (*22*), inclusion of more accurate charges in molecular mechanics calculations (*23*) have been examined, and calibration of hydrogen bonding potentials based on organic crystal structures have appeared (*24*). Applications to specific molecular systems of interest, such as oxygen hydrogen bonding to hydrogen attached to carbon (*25*) and DNA base pairs (*26*), are becoming pervasive. Several reviews have appeared, including excellent but now dated ones by Rao and Murthy (*27*) and Kollman and Allen (*28*). A recent book by Jeffry and Saenger, "Hydrogen Bonding in Biological Molecules" (*29*) gives an excellent compilation of more recent data, although it concentrates almost exclusively on crystallographic data. However, between 1957 and 1993, there has been no major symposium on modeling and theoretical methods for studying hydrogen bonding. The recent symposium on which this book is based, Modeling the Hydrogen Bond, sponsored by the Computers in Chemistry Division of the American Chemical Society, was put together in the hope that it would address many of the issues that have arisen since Ljubljana.

Literature Cited

1. Latimer and Rodebush, *J. Am. Chem. Soc.* **1920**, *42*, 1419.
2. Pauling, L. "The Nature of the Chemical Bond and the Structure of Molecules and Crystals: An Introduction to Modern Structural Chemistry," 3rd ed., Cornell University Press, Ithaca, NY, 1960, pp. 449 - 504.
3. Pimentel, G. C. and McClellan, A. L. "The Hydrogen Bond" W. H. Freeman and Co., San Francisco, 1960.
4. Oddo, G.; Puxeddu, E. *Gazz.* **1906,** *36*, 1.
5. Pfeiffer, P. *Ann.*, **1913**, *398*, 137.
6. Dyke, T. R.; Muenter, J. S. *J. Chem. Phys.* **1974**, *60*, 2929.
7. Harrell, S. A.; McDaniel, D. H. *J. Am. Chem. Soc.* **1964**, *86*, 4497. Morokuma, K.; Pedersen, L. *J. Chem. Phys.* **1969**, *51*, 3286.
8. Vinogradov, S. N.; Linnell, R. H. "Hydrogen Bonding," Van Nostrand Reinhold, New York, 1971.
9. Taylor, R.; Kennard, O.; Versichel, W. *Acta Crystallogr. B* **1984**, *B40*, 280.
10. Coulson, C. A. "The Hydrogen Bond. A review of the present interpretations and an introduction to the theoretical papers presented at the Congress" in *Hydrogen Bonding*, D. Hadzi and H. W. Thompson, Eds., Pergamon Press, London, 1959, pp. 339.
11. Lennard-Jones, J.; Pople, J. A. *Proc. Roy. Soc. (London)* **1951**, *205A*, 155, 163.
12. Morokuma, K. *J. Chem. Phys.* **1971**, *55*, 1236. Kitaura, K.; Morokuma, K. *Int. J. Quantum Chem.* **1976**, *10*, 325.
13. Morokuma, K. *Acc. Chem. Res.*, **1977**, *10*, 294.
14. Umeyama, H.; Morokuma, K. *J. Am. Chem. Soc.* **1977**, *99*, 1316.
15. Foster, J. P.; Weinhold, F. *J. Am. Chem. Soc.* **1980**, *102*, 7211.
16. Reed, A. E.; Weinstock, R. B.; Weinhold, F. J. Chem. Phys. **1985**, *83*, 735.
17. Glendening, E. D.; Streitwieser, A. *J. Chem. Phys.* **1994**, *100*, 2900.
18. Rauk, A. "Orbital Interaction Theory of Organic Chemistry," John Wiley & Sons, Inc., New York, 1994, pp. 160-2.
19. Del Bene, J. E. *Int. J. Quanum Chem.: Quantum Biol. Symp.* **1987,** *14*, 27.
20. Del Bene, J. E.; Shavitt, I. *Int. J. Quanum Chem.: Quantum Biol. Symp.* **1989,** *23*, 445.
21. Racine, S. C.; Davidson, E. R. *J. Phys. Chem.* **1993**, *97*, 6367.
22. See, for example, Damewood, J. R., Jr.; Kumpf, R. A.; Muhlbauer, W. C. F.; Urban, J. J.; Eksterowicz, J. E. *J. Phys. Chem.* **1990**, *94*, 6619.

23. See, for example, Cornell, W. D.; Cieplak, P.; Bayly, C. I.; Kollman, P. A. *J. Am. Chem. Soc.* **1993**, *115*, 9620.
24. Gavezzotti, A.; Filippini, G. *J. Phys. Chem.* **1994**, *98*, 4831.
25. See, for example, Turi, L.; Dannenberg, J. J. *J. Phys. Chem.* **1993**, *97*, 7899.
26. Gould, I. R.; Kollman, P. A. *J. Am. Chem. Soc.* **1994**, *116*, 2493. Del Bene, J. E. *Int. J. Quanum Chem.: Quantum Biol. Symp.* **1988,** *15*, 119.
27. Murthy, A. S. N.; Rao, C. N. R. *J. Mol. Struct.* **1970**, *6*, 253.
28. Kollman, P. A.; Allen, L. C. *Chem. Rev.* **1972**, *72*, 283.
29. Jeffrey, G. A.; Saenger, W. "Hydrogen Boding in Biological Structures," Springer-Verlag, Berlin, 1991.

RECEIVED July 29, 1994

SOLVENT AND MOLECULAR ELECTROSTATIC POTENTIAL EFFECTS

Chapter 2

Computation of Intermolecular Interactions with a Combined Quantum Mechanical and Classical Approach

Jiali Gao

Department of Chemistry, State University of New York at Buffalo, Buffalo, NY 14214

Hydrogen-bonded complexes for a series of organic compounds with water have been investigated using a combined quantum mechanical and molecular mechanical potential. The performance of this hybrid QM/MM approach is verified via comparison with ab initio 6-31G(d) calculations and with the empirical OPLS optimizations. Good accord in the predicted interaction energies between these methods was obtained.

The ability to model hydrogen-bonding interactions between solute and solvent is critically important in computer simulation of condensed-phase systems (*1*). Obviously, ab initio quantum mechanical methods can, in principle, provide a systematic and reliable treatment of intermolecular interactions and yield the necessary structural and energetic data. However, the computer time required, even with supercomputers, for these calculations rapidly becomes prohibitive as the number of atoms increases. Consequently, molecular mechanics force fields are typically used in fluid simulations. Despite its crude approximations, the method can provide valuable and meaningful insights on intermolecular interactions in solution (*2*). However, the empirical potential functions are generally not suitable for chemical processes involving electronic structural reorganizations, and are usually difficult to parameterize due to the lack of sufficient experimental data (*3*). An alternative approach is to combine the quantum mechanical (QM) and molecular mechanical (MM) methods such that only a small portion of the system, the solute, is represented quantum-mechanically, while the surrounding solvent molecules are approximated by molecular mechanics. Thus, it has the advantage of both the flexibility of the QM method and computational efficiency of the MM approach.

The combined QM/MM approach has been pioneered by several groups. In a seminal paper (*4*), Warshel and co-workers have laid out the basic

0097–6156/94/0569–0008$08.00/0

algorithms of the method and subsequently investigated numerous chemical and biological systems using, in particular, the empirical valence bond (EVB) theory (*5*). Karplus and co-workers employed the semiempirical AM1 and MNDO theory to carry out a molecular dynamics simulation of the S_N2 reaction, Cl^- + CH_3Cl → $ClCH_3$ + Cl^-, in water (*6*). Recently, we have extented the combined AM1/classical approach to determine the solvent polarization effects and free energies of hydration of organic compounds (*7*). Encouraging results have been obtained in these studies. In addition, a number of groups have combined ab initio method with molecular mechanics programs (*8*). However, only energy minimization studies have been carried out using such potentials.

In this paper, the performance of the combined QM-AM1/MM potential is verified through computation of bimolecular, hydrogen-bonded complexes of organic molecules with water. The results are compared with those from ab initio molecular orbital calculations and with the predictions using the empirical OPLS potential function (*9*). Similar studies have been carried out by Field et al. for many systems (*10*). The present investigation is an extension of that work. Below, computational details are first given followed by results and discussion.

Computational Details

The combined QM/MM calculations have been performed for a series of bimolecular hydrogen-bonded complexes with water using the MCQUB/BOSS program (*11*), in which QM energies are evaluated using the MOPAC program (*12*). We choose to investigate hydrogen-bonded complexes with water in order to evaluate the trends for these interactions.

In the present study, the organic species are treated by the semiempirical AM1 theory introduced by Dewar and co-workers (*13*), and the water molecule is treated by the three-site TIP3P model (*14*). Thus, the effective Hamiltonian is as follows (*4,10,15*):

$$\hat{H}_{eff} = \hat{H}^{o}_{qm} + \hat{H}_{qm/mm} \tag{1}$$

where $\hat{H}^{o}_{qm}$ is the AM1 Hamiltonian for the isolated QM molecule and $\hat{H}_{qm/mm}$ is the QM/MM interaction Hamiltonian describing "solute"-water interactions. $\hat{H}_{qm/mm}$ contains electrostatic and van der Waals terms (equation 2). The latter

$$\hat{H}_{qm/mm} = \hat{H}^{el}_{qm/mm} + \hat{H}^{vdW}_{qm/mm} =$$

$$\left(-\sum_{s=1}^{3}\sum_{i=1}^{2N}\frac{eq_s}{r_{si}} + \sum_{s=1}^{3}\sum_{m=1}^{M}\frac{q_s Z_m}{R_{sm}}\right) + \sum_{m=1}^{M} 4\epsilon_{Om}\left[\left(\frac{\sigma_{Om}}{R_{Om}}\right)^{12} - \left(\frac{\sigma_{Om}}{R_{Om}}\right)^{6}\right] \tag{2}$$

accounts for the QM/MM dispersion interactions that are omitted in the density matrix, and contains the only adjustable parameters, which are optimized by comparing with ab initio results for the bimolecular complexes considered here. These parameters are listed in Table I.

Table I. Lennard-Jones Parameters used in the AM1/TIP3P Model

atom	*σ,Å*	*ε,kcal/mol*
C	3.5000	0.0800
H (on C)	2.0000	0.0100
H (on heteroatoms)	0.8000	0.1000
O (sp^2)	2.9500	0.2000
O (sp^3)	2.1000	0.1500
N	2.5000	0.1400
N (in NH and NH_2)	2.8000	0.1500
N (in ammonium ions)	3.1000	0.1600
Cl	4.1964	0.1119
	Water	
O	3.1506	0.1521
H	0.0	0.0

In equation 2, e is the charge of electrons, q_s and Z_m are charges on water atoms and the solute nuclei, respectively, M is the total number of QM nuclear sites, and r_{si} and R_{sm} are the distances of the solute electrons and nuclei from water atoms. For use of the combined QM/MM potential in fluid simulations, an additional term describing solvent-solvent interactions is included in equation 1 (*7*).

Ab initio molecular orbital calculations were also executed for systems where previous studies are not available in the literature. In these calculations, the split-valence 6-31G(d) basis set was used with Gaussian 90 (*16*). Clearly, electron correlation and basis set superposition error corrections are important for determining the energetics of these hydrogen bonding complexes. However, the 6-31G(d) basis set is known to perform particularly well for hydrogen bonding interactions, perhaps due to fortuitous error cancellations (*17*). For comparison, similar computations were repeated using the empirical OPLS potential function (*9*). In this case, all species are represented by interaction sites, while methyl groups are typically treated as a single unit.

All partial geometry optimizations were executed with fixed monomer geometries at the corresponding level of theory for the "QM solute" molecules. The experimental configuration was adopted for water throughout (*14*).

Results and Discussion

Amide/Water Complexes. The formamide/water and N,N-dimethylformamide (DMF)/water complexes were examined (*10,18*), along with the **syn** and **anti** transition states (TS) in DMF. For each system, four structures were investigated, whose geometrical arrangements are shown in Figures 1 and 2 along with the optimized hydrogen bond parameters obtained from 6-31G(d) and AM1/TIP3P optimizations. In formamide/water complexes (**1-4**), both ab initio 6-31G(d) and AM1/TIP3P optimizations predict that the carbonyl group is a better hydrogen bond acceptor by about 1 kcal/mol than donor by the amide hydrogens (Table II). The OPLS functions showed little discrimination on the

Table II. Computed Hydrogen-Bond Energies for Amide/Water Complexes[a]

species	*6-31G(d)*	*AM1/TIP3P*	*OPLS[b]*
1	-6.2	-5.9	-6.2
2	-7.2	-6.6	-8.9
3	-5.9	-4.7	-7.2
4	-5.5	-4.9	-7.8
5	-6.5	-5.9	-6.2
6	-6.3	-6.2	-7.3
7	-1.5	-1.2	-2.1
8	-2.6	-2.8	-2.4
9	-4.7	-4.7	
10	-5.6	-5.4	
11	-4.4	-4.0	
12	-1.8	-2.1	
13	-5.0	-4.7	
14	-5.1	-4.5	
15	-6.4	-6.3	
16	-2.1	-2.5	

[a]Energies are given in kcal/mol.
[b]TIP3P model was used for water throughout.

energy differences, though it correctly predicted that **2** has the largest interaction energy, which is in accord with the quantum mechanical calculations. The calculated binding energies based on the AM1/TIP3P potential are uniformly smaller than the 6-31G(d) results by about 0.3 to 1 kcal/mol, whereas the OPLS function tends to over-estimate the binding energies.

For the ground state DMF, **5** is predicted to form the strongest hydrogen bond with water at the 6-31G(d) level, whereas **6** is found to be the best using the AM1/TIP3P model (Table II). This reversal of hydrogen-bond strengths has been noticed with the OPLS function for dimethylacetamide (DMA) and is

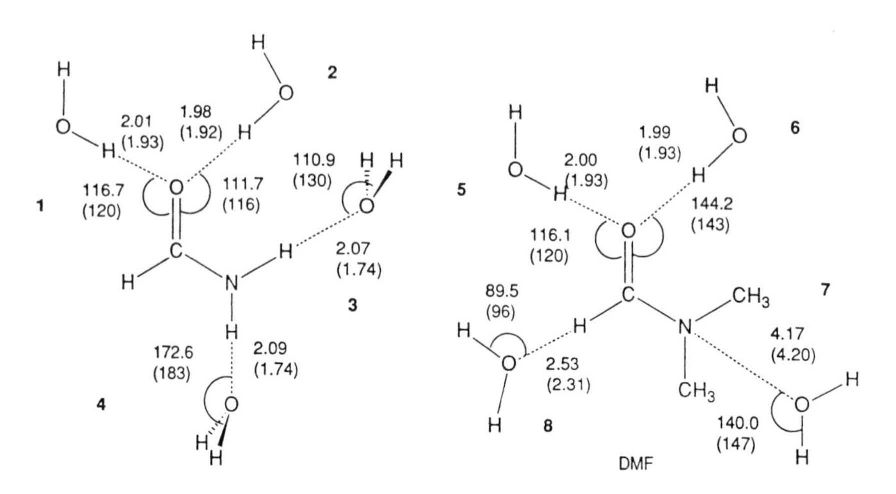

Figure 1. Formamide/water and DMF/water bimolecular complexes. Results of 6-31G(d) optimization are given, while those from the AM1/TIP3P model are shown in parentheses. Distances are given in angstroms and angles in degrees. This convention is used throughout.

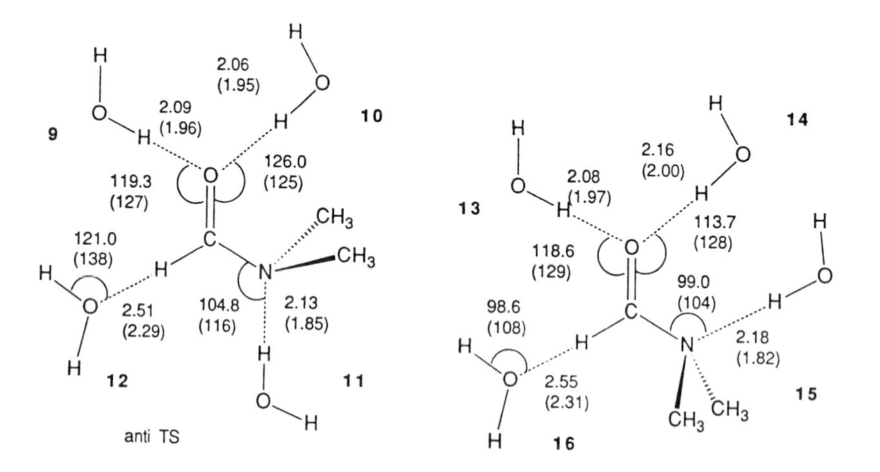

Figure 2. DMF/water bimolecular complexes for the anti and syn transition states.

confirmed for DMF here (*19*). Duffy et al. indicated that it was not possible to reproduce the ab initio ordering by adjusting potential function parameters (*19*). Therefore, it seems to be desirable to further verify the ab initio results by geometry optimizations including correlation corrections. In the transition state DMF/water complexes (Figure 2), hydrogen bonding to the pyramidal nitrogen becomes significant. In fact, structure **15** is predicted to be the most stable bimolecular complex for the syn TS. OPLS parameters for the DMF transition states are not available, though we note that the TS's of DMA have been investigated (*19*). It is not clear if these parameters are transferrable to DMF.

Geometries from the AM1/TIP3P calculations are also in reasonable accord with the ab initio data (Figures 1 and 2). The error ranges are similar to the difference between the OPLS and ab initio 6-31G(d) results. As has been noted previously, the OPLS functions tends to underestimate the hydrogen bond distances in order to yield good results in liquid simulations (*9*). In the AM1/TIP3P treatment, the agreement is much improved for hydrogen bonds with QM acceptors; however, the hydrogen-bond distances for complexes containing QM donors are typically about 0.1 Å shorter than the OPLS results, or about 0.3-0.4 Å shorter than the 6-31G(d) values. It seems to be more important for the hybrid AM1/TIP3P potential to yield a reasonably good prediction of the energetics for intermolecular interactions, with some sacrifice of the geometrical parameters involving QM hydrogen-bond donors, than to provide marginally improved hydrogen-bond lengths, but poor interaction energies. The present choice of the van der Waals parameters (Table I) is the result of a compromise of these considerations. The difference in the angles between AM1/TIP3P and 6-31G(d) calculations is similar to those between OPLS and ab initio results.

In an effort to elucidate the solvent effects on the barrier to amide isomerization, the AM1/TIP3P potential has been used to determine the potential of mean force for the dihedral rotation about the peptide bond in DMF in water (*20*), chloroform, and carbon tetrachloride (Gao, J. *Proc. Ind. Acad. Sci.*, in press.) The issue is important for the understanding of protein folding and the catalytical mechanism in peptidyl prolyl isomerases. It was found that the syn and anti TS's are destabilized by 0.5 and 2.7 kcal/mol, respectively, relative to the ground state in water, by 0.13 and 0.43 kcal/mol in CCl_4, and by -0.08 and 1.05 kcal/mol in $CHCl_3$ (*20*). These findings are consistent with the relevant experimental and theoretical results (*19,21*). The good perfomance of the AM1/TIP3P model in these simulations is a reflection of the agreement for the bimolecular complexes considered here.

Carboxylic Acid/Water Complexes. A total of eight structures (**17-24**) have been investigated for acetic acid/water complexes (*22*). Two conformers in acetic acid were examined, corresponding to the anti and syn orientations between the hydroxyl and carbonyl groups. Geometrical parameters optimized using the 6-31G(d) basis set and the AM1/TIP3P potential are given in Figure 3. As in the previous system, the agreement between the two calculations is good, particularly for QM-acceptor/MM-donor complexes. The accord between the AM1/TIP3P and ab initio hydrogen-bond distances is poor for structures **19** and

24, with an error up to 0.4 Å. The OPLS function gives hydrogen-bond distances for **19** and **24** that are about 0.2 Å better than the AM1/TIP3P values. However, OPLS optimizations yield hydrogen-bond distances about 0.1 to 0.2 Å shorter than the QM/MM and ab initio calculations for other complexes.

Energetically, the agreement between the AM1/TIP3P and ab initio calculations is excellent (Table III), particularly in view of the trends on

Table III. Computed Hydrogen-Bond Energies for Carboxylic Acid/Water Complexes (kcal/mol)

species	*6-31G(d)*	*AM1/TIP3P*	*OPLS*[a]
17	-5.4	-5.7	-5.4
18	-5.5	-5.5	-7.0
19	-8.5 (-7.4)[b]	-7.1	-9.6
20	-2.4	-3.6	-3.2
21	-5.9	-5.7	-6.4
22	-5.2	-5.4	-5.8
23	-4.9	-6.2	-6.5
24	-8.5 (-7.8)[b]	-7.7	-9.2

[a]TIP3P model was used for water throughout.
[b]Computed with the 6-31+G(d) basis set.

interaction energies for the syn and anti conformers with water. The OPLS function generally gives stronger interaction energies than both the ab initio and AM1/TIP3P calculations. Structures **19** and **24** are predicted to be the lowest energy form in either syn or anti acetic acid by all three methods. The OPLS function, however, overestimates the interaction energy by 1-2 kcal/mol (*22*).

The free energy difference between the syn and anti conformations in acetic acid in water has been computed via statistical perturbation theory with the combined QM/MM Monte Carlo method (*23*). The interest in this quantity stems from its relationship to the basicity difference of the carboxylate syn and

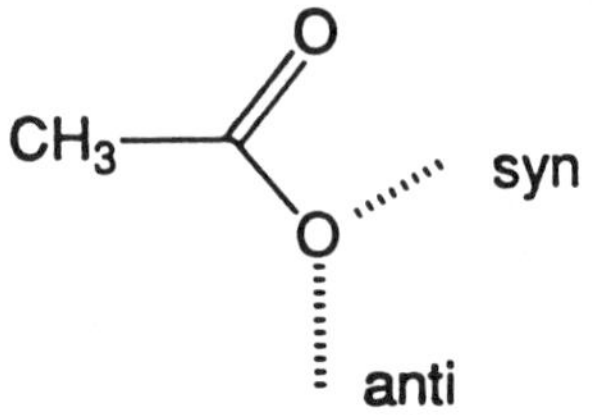

anti lone pairs. The syn lone pair has been found to predominate the His-Asp interaction in enzymes, which has been attributed to the larger basicity of the syn lone pair relative to the anti (by 10^5 times in the gas phase) (*24*). Our computation and others suggest that the difference is smeared out owing to a more favorable hydration of the anti form over that of the syn conformation (*25*). This finding is in excellent accord with the experimental results based on synthetically designed host-guest systems (*26*). Thus, it might be argued that the observed syn-preference in the His-Asp dyad is due to efficient packing in proteins.

Ion/Water Complexes. Nine simple organic ions were included in the present consideration (Figure 4 and Table IV). Ab initio calculations for bimolecular

Table IV. Computed Hydrogen-Bond Energies for Ion/Water Complexes (kcal/mol)

species	*6-31G(d)*	*AM1/TIP3P*	*OPLS*[a]
25	-14.3	-13.5	-13.6
26	-25.3	-30.1	-19.3
27	-19.1	-17.6	-19.9
28	-16.3	-15.2	-16.5
29	-21.8 (-18.2)[b]	-19.3	-19.8
30	-14.6	-12.8	-14.3
31	-18.2	-15.1	-17.8
32	-13.8	-13.7	-15.8
33	-16.1	-16.3	-18.1

[a]TIP3P model was used for water throughout.
[b]Computed with the 6-31+G(d) basis set.

complexes with water have been carried out previously in deriving the OPLS function (*10,27*); these results are compared with the predictions using the combined AM1/TIP3P treatment. Overall, the RMS deviation between the 6-31G(d) and AM1/TIP3P interaction energies is the same as that between the ab initio and the OPLS results. However, the AM1/TIP3P calculations tend to underestimate the binding energies, while the empirical potential function yields stronger hydrogen bonds in a few cases than those predicted by the 6-31G(d) calculation. The need for an overestimate in bimolecular complexes with the empirical function is due to the use of an effective potential that must take into account of the many-body polarization effect in the parameters. For the combined QM/MM approach, on the other hand, the polarization effect on the solute electronic structure is specifically treated. Thus, the interaction energies for hydrogen-bonded complexes are expected to increase in solution simulations because of the medium polarization contribution.

The only exception in the combined QM/MM treatment of the ion-water complexes is methoxide ion, which has a large deviation from the ab initio number. The AM1/TIP3P optimization predicted an interaction energy of -30.1

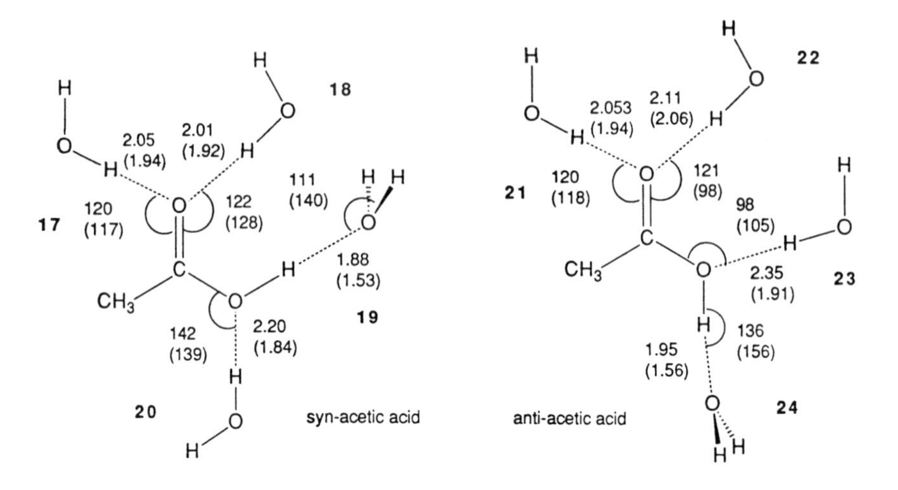

Figure 3. Acetic acid/water bimolecular complexes.

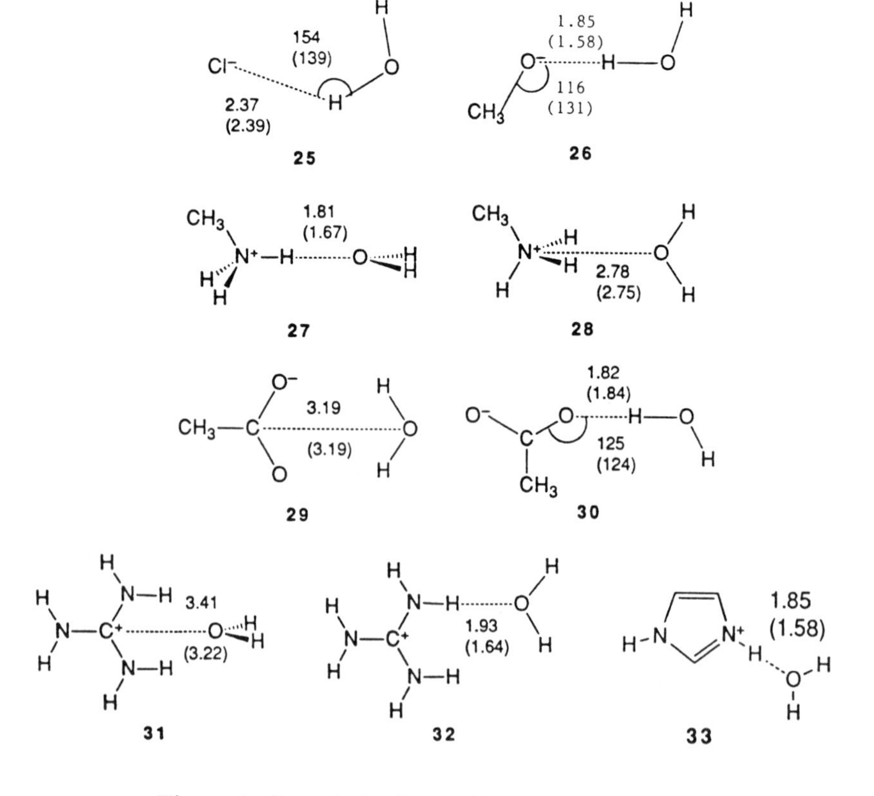

Figure 4. Organic ion/water bimolecular complexes.

kcal/mol, which may be compared with the 6-31G(d) and OPLS values of -25.3 and -19.3 kcal/mol, respectively. Two experimental measurements are available; the initial work of Caldwell et al. predicted a value of -19.9 kcal/mol, while the high-pressure mass spectrometry experiments of Meot-Ner and Sieck yielded an energy of -23.9 kcal/mol (*28*). Thus, the AM1/TIP3P model appears to overestimate the hydrogen-bonding energy by about 6 kcal/mol. Although increase the σ value of the Lennard-Jones parameters for oxygen may improve the agreement, it is not clear whether this is desirable in view of the simplicity in these parameters.

Other Organic Compounds/Water Complexes. Finally, some additional results are presented in Table V for complexes containing other functionalities (*17*).

Table V. Computed Hydrogen-Bond Energies for Organic Compound/Water Complexes (kcal/mol)

species[a]	*6-31G(d)*	*AM1/TIP3P*	*OPLS*[b]
$(CH_3)_2CO...HOH$	-5.8	-5.8	-5.3
$CH_3OH...OH_2$	-5.6	-5.3	-6.8
$CH_3OH...HOH$	-5.7	-5.4	-6.3
$(CH_3)_2O...HOH$	-5.7	-5.4	-5.4
$CH_3NH_2...OH_2$	-2.7	-2.1	-3.6
$CH_3NH_2...HOH$	-6.5	-4.6	-5.9
$CH_3CN...HOH$	-4.4	-3.0	-4.7
imidazole...OH_2	-5.7	-6.2	-8.7
imidazole...HOH	-6.3	-5.5	-5.3
urea...OH_2	-7.1	-4.6	-7.3
urea......HOH	-9.4	-7.9	-8.8
benzene...HOH	-3.1	-2.5	-3.8

[a]HOH represents water as a hydrogen-bond donor, and OH_2 represents water as an acceptor.
[b]TIP3P model was used for water throughout.

Among the carbonyl groups, hydrogen-bonding strength increases in the order, acetone, acetic acid > amides > urea, according to the 6-31G(d) and AM1/TIP3P optimizations (Tables II to V). For methanol, it is computed to be almost isoenergetic as a hydrogen-bond acceptor or a donor, while the OPLS function predicts that it is a better donor by about 0.5 kcal/mol. This contradicts the previous finding using the TIP4P model for water, where methanol was found to be a better hydrogen-bond acceptor than donor by 0.8 kcal/mol (*1*). Thus, the use of different water models in conjuction with a particular force field may alter the order of interaction energies that was originally developed in the parametrization. Methylamine is certainly a better hydrogen-bond acceptor than donor from the results of all computational methods; however, the AM1/TIP3P model underestimates the interaction energy by about 2 kcal/mol in comparison with the ab initio data. It should also be noted that benzene is predicted to form a weak "hydrogen-bonding" complex with water with an interaction energy of -2.5 kcal/mol using the AM1/TIP3P model (*29*). Hartree-Fock 6-31G(d) and MP2/6-31G(d,p) optimizations yielded complexation energies of -3.1 and -4.2 kcal/mol (*30*), respectively, whereas the interaction energy is reduced to -1.8 kcal/mol at the MP2 level with the correction of basis set superposition error (*30b*). Experimentally, binding energies of -1.6 to -2.8 (*30c*), and -1.9 kcal/mol have been reported (*30b*). The OPLS potential gives an interaction energy of -3.8 kcal/mol for this system.

The present results are also illustrated in Figures 5 and 6. The accord for the interaction energies between the AM1/TIP3P model and ab initio 6-31G(d) calculation is good with an RMS deviation of 1.2 kcal/mol for a total of 45 complexes. The OPLS/TIP3P values are somewhat not as good in agreement with the ab initio data as the AM1/TIP3P model. The computed RMS deviation in this case is 1.6 kcal/mol for 37 complexes, excluding the transition state structures that do not have the OPLS parameters. However, the seemingly poorer agreement might be due to the parametrization of the OPLS function that has to consider the medium polarization effect.

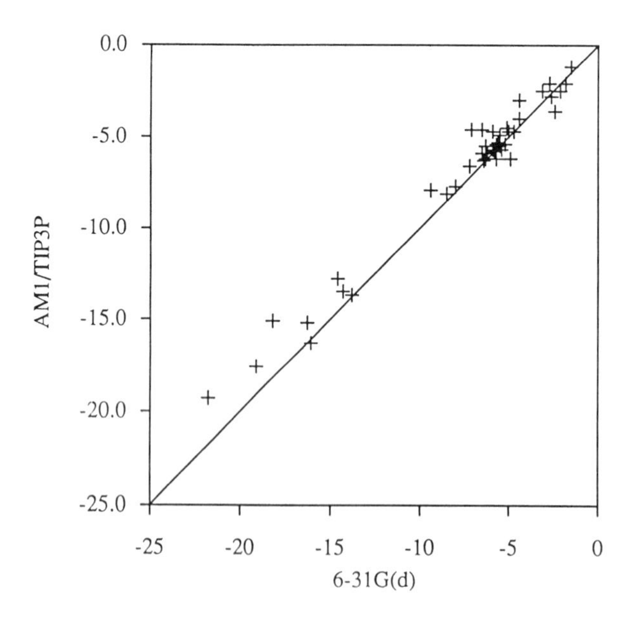

Figure 5. Comparison between ab initio 6-31G(d) and AM1/TIP3P interaction energies (kcal/mol).

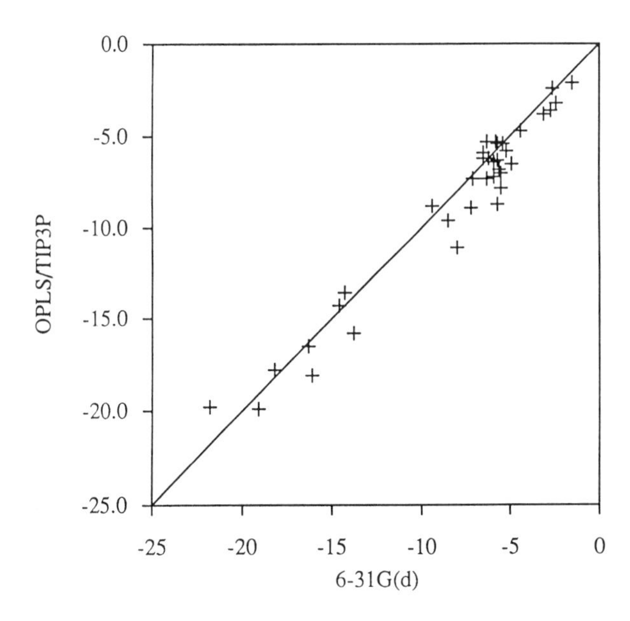

Figure 6. Comparison between ab initio 6-31G(d) and OPLS/TIP3P interaction energies (kcal/mol).

Conclusions

The validity on the use of the combined quantum mechanical and classical approach with the AM1/TIP3P model for liquid simulations has been investigated through comparison with ab initio 6-31G(d) results for hydrogen-bonded complexes of organic compounds with water. The accord between the AM1/TIP3P and 6-31G(d) interaction energies is encouraging with an RMS deviation of 1.2 kcal/mol for 45 complexes covering an energy range of -1 to -25 kcal/mol. The performance of the empirical OPLS potential for these complexes was also investigated. The agreement between the OPLS and 6-31G(d) results has an RMS deviation of 1.6 kcal/mol for 37 complexes. The results demonstrated here indicate that the combined QM/MM AM1/TIP3P potential can provide results that are as reliable as other approaches in simulating chemical systems in solution. However, the method has the advantage of providing additional insights into the solute electronic relaxations in solution (*7*).

Acknowledgement. Gratitude is expressed to the NSF Supercomputing Center, the Pittsburgh Supercomputing Center and to the NIH for support of our research.

Literature Cited

1. Jorgensen, W. L. *Chemtracts-Org. Chem.* **1991**, *4*, 91.
2. (a) Karplus, M.; Petsko, G. A. *Nature* **1990**, *347*, 631. (b) van Gunsteren, W. F.; Berendsen, H. J. C. *Angew. Chem., Int. Ed. Engl.* **1990**, *29*, 992.
3. McCammon, J. A.; Harvey, S. C. *Dynamics of Proteins and Nucleic Acids*; Cambridge University Press: Cambridge, 1987.
4. Warshel, A.; Levitt, M. *J. Mol. Biol.* **1976**, *103*, 227.
5. Warshel, A. *Computer Modeling of Chemical Reactions in Enzymes and Solutions*; Wiley: NY, 1991.
6. Bash, P. A.; Field, M. J.; Karplus, M. *J. Am. Chem. Soc.* **1987**, *109*, 8192.
7. Gao, J.; Xia, X. *Science* **1992**, *258*, 631.
8. (a) Weiner, S. J.; Seibel, G. L.; Kollman, P. A. *Proc. Natl. Acad. Sci.* **1986**, *83*, 649. (b) Rullmann, J. A. C.; Bellido, M. N.; van Duijnen, P. T. *J. Mol. Biol.* **1989**, *206*, 101. (c) Waszkowycz, B.; Hillier, I. H.; Gensmantel, N.; Payling, D. W. *J. Chem. Soc. Perkin Trans. 2* **1991**, 225.
9. Jorgensen, W. L.; Tirado-Rives, J. *J. Am. Chem. Soc.* **1988**, *110*, 1657.
10. Field, M. J.; Bash, P. A.; Karplus, M. *J. Comput. Chem.* **1990**, *11*, 700.
11. (a) Gao, J. *MCQUB*; SUNY at Buffalo: Buffalo, NY, 1992. (b) Jorgensen, W. L. *BOSS*, Version 2.9; Yale University: New Haven, CT, 1990.
12. Stewart, J. J. P. *MOPAC*, Version 5; QCPE 455, **1986**, Vol. 6, p 391.
13. Dewar, M. J. S.; Zoebisch, E. G.; Healy, E. F.; Stewart, J. J. P. *J. Am. Chem. Soc.* **1985**, *107*, 3902.
14. Jorgensen, W. L.; Chandrasekhar, J.; Madura, J. D.; Impey, R. W.; Klein, M. L. *J. Chem. Phys.* **1983**, *79*, 926.
15. Gao, J. *J. Phys. Chem.* **1992**, *96*, 537.
16. Frisch, M. J.; Head-Gordon, M.; Trucks, G. W.; Foresman, J. B.; Schlegel, H. B.; Raghavachari, K.; Robb, M.; Binkley, J. S.; Gonzalez, C.; Defrees, D. J.; Fox, D. J.; Whiteside, R. A.; Seeger, R.; Melius, C. F.; Baker, J.; Martin, R. L.; Kahn, L. R.; Stewart, J. J. P.; Topiol, S.; Pople, J. A. *GAUSSIAN 90*, Gaussian Inc.: Pittsburgh, PA, 1990.
17. (a) Hehre, W. J.; Radom, L.; Schleyer, P. v. R.; Pople, J. A. *Ab Initio Molecular Orbital Theory*; Wiley: NY, 1986. (b) Frisch, M. J., et al. *J. Chem. Phys.* **1986**, *84*, 2279. (c) Dill, J. D.; Allen, L. C.; Topp, W. C.; Pople, J. A. *J. Am. Chem. Soc.* **1975**, *97*, 7220.
18. (a) Lovas, F. J.; Suenram, R. D.; Fraser, G. T.; Gillies, C. W.; Zozom, J. *J. Chem. Phys.* **1987**, *88*, 722. (b) Jaisen, P. G.; Stevens, W. J. *J. Chem. Phys.* **1986**, *84*, 3271.
19. Duffy, E. M.; Severance, D. L.; Jorgensen, W. L. *J. Am. Chem. Soc.* **1992**, *114*, 7535.
20. Gao, J. *J. Am. Chem. Soc.* **1993**, *115*, 2930.
21. Drakenberg, T.; Dahlqvist, K.-I.; Forsen, S. *J. Phys. Chem.* **1972**, *76*, 2178.
22. Gao, J. *J. Phys. Chem.* **1992**, *96*, 6432.
23. Gao, J.; Pavelites, J. J. *J. Am. Chem. Soc.* **1992**, *114*, 1912.
24. Gandour, R. D. *Bioorg. Chem.* **1981**, *10*, 169.
25. Pranata, J. *J. Comput. Chem.* **1993**, *14*, 685.

26. (a) Zimmerman, S. C.; Cramer, K. D. *J. Am. Chem. Soc.* **1988**, *110*, 5906. (b) Tadayoni, B. M.; Rebek, J., Jr. *Bioorg. Med. Chem. Lett.* **1991**, *1*, 13.
27. Gao, J.; Garner, D. S.; Jorgensen, W. L. *J. Am. Chem. Soc.* **1986**, *108*, 4784.
28. (a) Caldwell, G.; Rozeboom, M. D.; Kiplinger, J. P.; Bartmess, J. E. *J. Am. Chem. Soc.* **1984**, *106*, 4660. (b) Neot-Ner, M.; Sieck, L. W. *J. Am. Chem. Soc.* **1986**, *108*, 7525.
29. Gao, J.; Chou, L. W.; Auerbach, A. *Biophys. J.* **1993**, *65*, 43.
30. (a) Jorgensen, W. L.; Severance, D. L. *J. Am. Chem. Soc.* **1990**, *112*, 4768. (b) Suzuki, S.; Green, P. G.; Bumgarner, R. E.; Dasgupta, S.; Goddard, W. A.; Blake, G. A. *Science* **1992**, *257*, 942. (c) Gotch, A. J.; Zwier, T. S. *J. Chem. Phys.* **1992**, *96*, 3388.

RECEIVED June 2, 1994

Chapter 3

Inclusion of Explicit Solvent Molecules in a Self-Consistent-Reaction Field Model of Solvation

T. A. Keith and M. J. Frisch

Lorentzian, Incorporated, 140 Washington Avenue, North Haven, CT 06473

Preliminary results from the inclusion of one to several explicit solvent water molecules in a self-consistent reaction field model of solvation are presented for the aqueous solvation free energies of simple solutes. Standard free energies of aqueous solvation for the first three alkali metal cations, calculated using a varying number of explicit water molecules together with an extension of the SCRF model of Tomasi et al, are compared with experiment, with SCRF results obtained without the inclusion of explicit water molecules and with gas-phase hydration free energies.

There are essentially two extreme approaches to the prediction of solvation phenomena in quantum chemistry. One is the "super-molecule" (SM) approach (*1 - 4*), in which conventional ab-initio calculations are performed on a model solution consisting of a solute molecule complexed with one or more solvent molecules. The other is the self-consistent reaction field (SCRF) approach (*5 - 11*), in which ab-initio calculations are performed on a solute molecule embedded in a cavity and allowed to interact electrostatically with the surrounding solvent, which is modeled as a polarizable dielectric continuum. Each extreme has advantages and disadvantages relative to the other. The SM approach has the advantage of treating all solute-solvent interactions exactly, in principle. In practice, of course, only a limited number of neighbouring solvent molecules and configurations can be considered, and thus long-range solute-solvent interactions as well as dynamical effects are inevitably neglected. The SCRF approach has the advantage of including long-range solute-solvent interactions and, to some extent, dynamical effects, but specific, short-range solute-solvent interactions are neglected.

Given the opposing nature of the SM and SCRF approaches, it is natural to consider employing them in combination, whereby a complex consisting of the solute and one or more neighboring solvent molecules is treated by the SCRF approach. This "semi-continuum" approach, hereafter abbreviated as SC, is beginning to be investigated by researchers. For example, Tunon et al (*12*) have calculated the

0097–6156/94/0569–0022$08.00/0

standard free-energy of solvation of the proton in water by embedding various protonated water clusters in a cavity of an aqueous polarizable continuum and employing the SCRF model of Rivail et al. (*7 - 9*), hereafter abbreviated as SCRFR. Rinaldi et al (*13*) have also used the SCRFR model to calculate standard free energies of solvation for several alkaline-earth cations by embedding hydrates of the ions in a cavity surrounded by an aqueous polarizable continuum.

The SCRFR model used in (*12*) and (*13*) is one in which the solute cavity is constrained to be spherical (in general, the SCRFR model allows for ellipsoidal cavities) and the electrostatic interaction between the solute and the polarized solvent is approximated in terms of a sixth-order multipole expansion of the electrostatic potential produced by the solute charge distribution (*7 - 10*). In principle, the SCRFR multipole expansion can be carried out to arbitrary order.

An alternative SCRF approach, first formulated by Tomasi et al (*10 - 11*) and summarized below, has recently been extended and implemented (*14*) in Gaussian 92 (*15*). Foresman et al (*16*) have shown that, in general, the extended Tomasi SCRF model, hereafter abbreviated as SCRFT, predicts more reliable solvation properties than the SCRFR model. This is due to the explicit solute-dependent cavity shape employed in the SCRFT model as well as the use of the full electrostatic potential produced by the solute charge distribution to determine the solvent polarization distribution and the resulting solute-solvent electrostatic interaction energies.

It is the purpose of this work to begin a systematic investigation of the SM model and the SCRFT model, as well as the corresponding SC model in order to determine the feasibility of using the SC approach as a more reliable method of applying quantum chemistry to solvation phenomena. Preliminary results of this investigation, employing relatively simple solutes in water, are described here.

Theoretical Solvation Models

The Supermolecule Model (SM). Perhaps the most conceptually straightforward approach to predicting solvation properties in quantum chemistry is simply to treat the solution just as one does a gas-phase system of interacting molecules, whereby all of the particles of the solute and solvent are included in the system's Hamiltonian and no phenomenological interaction operators are introduced to account for the solute-solvent interactions. In this "super-molecule" (SM) approach (*1 - 4*), the properties of a solute molecule M complexed with several solvent molecules m are calculated using conventional quantum chemistry techniques and compared with similar results for the isolated solute molecule, the solute molecule complexed with other types of solvent molecules or similar solutes complexed with the the same type of solvent molecules.

One problem with the SM approach is to relate the measurable gas-to-solution solvation free energy, $\Delta G[M,m]$, to the calculated free energy of formation $\Delta G[M(m)_n,g]$ of the gaseous supermolecule $M(m)_n$ from the solute M(g) and n solvent molecules m(g). The simple procedure of approximating $\Delta G[M,m]$ by $\Delta G[M(m)_n,g]$ amounts to assuming that the pure solvent consists of non-interacting molecules and that long-range interactions between solute and solvent are negligible. The former assumption can be partially removed by adding n times the vaporization free energy of the pure solvent to $\Delta G[M(m)_n,g]$. Solvation properties which are measurable without reference to the gas-phase or pure solvent phase and are expressible as energy differences or derivatives in solution, such as conformational equilibria, vibrational frequencies, NMR shieldings, relative solvation energies of similar solutes, etc., present less conceptual difficulty in relating SM results to experiment.

The SCRF Models. In the SCRF models (*5 - 11*), the solution is modeled by assuming that the solute molecule, treated quantum mechanically in the sense that all of its particles appear explictly in the model system's Hamiltonian, is contained within a solvent-free cavity of the otherwise uniform dielectric solvent. An interaction operator V_σ representing the interaction between the solute charge distribution and a solute-induced polarization distribution **P** in the solvent is introduced into the solute Hamiltonian and the desired form of quantum chemical calculation is performed based on this Hamiltonian. The interaction operator V_σ depends on the solute charge distribution, and thus a proper application of a reaction field model leads to self-consistency (SCRF) between the solute charge distribution and **P**. Since the known bulk dielectric properties of the pure solvent are explicitly included in an SCRF model, calculated solvation energies are, in one sense, more closely related to measured solvation energies than in the SM method. Like the SM method, however, an SCRF method can also be expected to predict energy differences and derivatives in solution more accurately than absolute solvation energies.

The Extended Tomasi SCRF model, SCRFT. The SCRF model employed here, SCRFT, is an extended version of the model originally formulated by Tomasi et al (*10 - 11*) and has been implemented (*14*) into Gaussian 92 (*15*). Within the SCRFT model the solute-solvent interaction operator V_σ is determined by the potential $v_\sigma(\mathbf{r})$ of a polarization charge distribution $\sigma(\mathbf{r}_s)$ induced on the surface S of the solute cavity

$$v_\sigma(\mathbf{r}) = \int_S d\mathbf{r}_s \sigma(\mathbf{r}_s)|\mathbf{r} - \mathbf{r}_s|^{-1} \tag{1}$$

where $\sigma(\mathbf{r}_s)$ is the polarization charge density at point $\mathbf{r}_s$ on the cavity surface, which is determined by the normal component of the total electric field $\mathbf{E}(\mathbf{r}_s)$ at the point $\mathbf{r}_s$

$$\begin{aligned}\sigma(\mathbf{r}_s) &= -(\varepsilon - 1)/4\pi\varepsilon)[\mathbf{E}(\mathbf{r}_s)].\mathbf{n}(\mathbf{r}_s) = -(\varepsilon - 1)/4\pi\varepsilon)[\mathbf{E}_0(\mathbf{r}_s) + \mathbf{E}_\sigma(\mathbf{r}_s)].\mathbf{n}(\mathbf{r}_s) \\ &= -(\varepsilon - 1)/4\pi\varepsilon)[\,\mathbf{E}_0(\mathbf{r}_s).\mathbf{n}(\mathbf{r}_s) - 2\pi\sigma(\mathbf{r}_s) \\ &\quad + \int_S d\mathbf{r}_s'\sigma(\mathbf{r}_s')|\mathbf{r}_s'-\mathbf{r}_s|^{-3}(\mathbf{r}_s'-\mathbf{r}_s).\mathbf{n}(\mathbf{r}_s)]\end{aligned} \tag{2}$$

In equation 2, $\mathbf{E}_0(\mathbf{r}_s)$ is the electric field due to the solute charge distribution while $\mathbf{E}_\sigma(\mathbf{r}_s)$ is the electric field due to the surface polarization charge distribution σ, including the contribution from $\sigma(\mathbf{r}_s)$ itself. The symbol $\mathbf{n}(\mathbf{r}_s)$ is the outward normal vector at the point $\mathbf{r}_s$. The quantity ε is the bulk dielectric constant of the pure solvent (ε = 78.3 for water at standard conditions). The Hamiltonian H for the solute within the SCRFT model is then

$$H = H^{(0)} + \sum_{i=1}^{n} v_\sigma(\mathbf{r}_i) + \sum_{A=1}^{M} v_\sigma(\mathbf{R}_A) = H^{(0)} + V_\sigma \qquad (3)$$

where the operator $H^{(0)}$ is the Hamiltonian of the isolated solute molecule, the second term is the interaction operator for the n electrons of the solute and the solvent polarization charge distribution σ, while the third term is the interaction operator for the M nuclei of the solute and σ.

In contrast to other SCRF models based on equations 1 - 3 , which invariably employ interlocking spheres to define S, in the SCRFT model S is defined explicitly in terms of a solute property, an isosurface of the solute's electron density distribution. There are both practical and physical reasons for such a definition, all of which will be detailed in a separate paper (*14*). For the results presented in this paper, the gas-phase isosurface with value 0.0004 au has been used. Though this choice is by no means "optimal", we have used it because the molar volumes defined by such isosurfaces have been found to give consistently good agreement with experimental molar volumes for a variety of molecules (*17*).

The surface integrals given in equations 1 and 2 cannot, in general, be calculated analytically, but must be determined numerically. Thus, the surface integrals become a finite summation over numerical integration points $\mathbf{r}_s^j$ on the surface with corresponding numerical integration weights w_s^j

$$v_\sigma(\mathbf{r}) = \sum_j w_s^j\, \sigma(\mathbf{r}_s^j)|\mathbf{r}-\mathbf{r}_s^j|^{-1} = \sum_j q(\mathbf{r}_s^j)|\mathbf{r}-\mathbf{r}_s^j|^{-1} \qquad (4)$$

$$\sigma(\mathbf{r}_s^i) = -(\varepsilon-1)/4\pi\varepsilon)[\, E_0(\mathbf{r}_s^i)\cdot\mathbf{n}(\mathbf{r}_s^i) - 2\pi\sigma(\mathbf{r}_s^i)$$
$$+ \sum_j q(\mathbf{r}_s^j)|\mathbf{r}_s^j - \mathbf{r}_s^i|^{-3}(\mathbf{r}_s^j - \mathbf{r}_s^i)\cdot\mathbf{n}(\mathbf{r}_s^i)] \qquad (5)$$

The evaulation of the i=j term in equation 5 requires special consideration which, together with the rest of the numerical integration procedure, will be fully described in a separate paper (*14*).

The formulation of the SCRFT model is based on the assumption that the solvent is homogenous, and thus that the solute charge distribution vanishes outside of the solute cavity, which is of finite extent. In practice, of course, the solute charge distribution does not vanish outside of the solute cavity. Following Tomasi et al (*10 - 11*), we apply a scaling of the surface charge distribution σ such that the total surface polarization charge Q_σ corresponds to the solute charge distribution being localized entirely in the solute cavity. It follows from Gauss' law that the relationship between Q_σ and the total charge of the solute cavity Q_{cav} is as follows

$$Q_\sigma = -(1 - 1/\varepsilon)Q_{cav} = \sum_j q(\mathbf{r}_s^j)^+ + \sum_j q(\mathbf{r}_s^j)^- = Q_\sigma^+ + Q_\sigma^- \qquad (6)$$

where the first summation is over all positive surface charges $q(\mathbf{r}_s^j)^+$, whose sum is Q_σ^+, while the second summation is over all negative surface charges $q(\mathbf{r}_s^j)^-$, whose sum is Q_σ^-. If the solute charge distribution were entirely localized in the the cavity, then Q_{cav} would equal Q_{sol}, the net charge of the solute in the gas phase. Thus, we apply a scaling of the surface charges as follows

$$Q_\sigma' = -(1 - 1/\varepsilon)Q_{sol} = \sum_j (1 + b)\, q(\mathbf{r}_s^j)^+ + \sum_j (1 - b)\, q(\mathbf{r}_s^j)^-$$

$$= (1 + b)Q_\sigma^+ + (1 - b)Q_\sigma^- \qquad (7)$$

where the scaling factor b approaches zero when Q_{cav} approaches Q_{sol} and is given by

$$b = [-(1 - 1/\varepsilon)Q_{sol} - Q_\sigma]/[Q_\sigma^+ + Q_\sigma^-] \qquad (8)$$

This scaling procedure, somewhat different than that employed by Tomasi et al (*10 -11*), produces a surface charge distribution σ' which corresponds to a solute charge distribution localized within the solute cavity, but preserves the structure of the unscaled surface charge distribution σ.

With a means of calculating the surface charge distribution σ and the corresponding operator V_σ accurately, it is a straightforward process to modify the desired ab-initio procedure to include the effects of the solvent polarization.

Our basic computational approach to solving the Hartree Fock SCRFT problem is similar to that recommended by Tomasi et al (*10 - 11*) and is as follows:

i) After choosing a basis set, geometry etc., solve the Hartree-Fock problem for the gas-phase solute molecule.

ii) Using equation 5, calculate the contribution to the surface charge distribution from the solute electric field $\mathbf{E}_0$.

iii) Keeping the contribution to σ from $\mathbf{E}_0$ fixed, determine the contribution from the electric field of the surface charge distribution $\mathbf{E}_\sigma$ self-consistently using equation 5.

iv) Construct the operator V_σ from the surface charge distribution determined in step iii) using equations 3 and 4 and solve the Hartree-Fock problem for the solute molecule, adding the operator V_σ to the core Hamiltonian operator.

v) Repeat steps ii) - iv) until the reaction field operator V_σ, and hence the total energy, has converged.

The total SCRFT energy is

$$E^{(SCRFT)} = \langle\psi^{(SCRFT)}|H^{(0)} + V_\sigma|\psi^{(SCRFT)}\rangle$$

$$+ (1/2)\int_S \int_S d\mathbf{r}_s d\mathbf{r}_s'\, \sigma(\mathbf{r}_s')\sigma(\mathbf{r}_s)|\mathbf{r}_s' - \mathbf{r}_s|^{-1} \qquad (9)$$

where $\psi^{(SCRFT)}$ is the solute wavefunction in the presence of the solvent reaction field and the second term in equation 9 represents the self-interaction energy of the surface charge distribution, ie. the energy required to polarize the solvent. When the energy $E^{(SCRFT)}$ is stationary with respect to variations in $\psi^{(SCRFT)}$ then the following is true

$$<\psi^{(SCRFT)}|V_\sigma|\psi^{(SCRFT)}> = -\int_S \int_S d\mathbf{r}_s d\mathbf{r}_s{}' \sigma(\mathbf{r}_s{}')\sigma(\mathbf{r}_s)|\mathbf{r}_s{}' - \mathbf{r}_s|^{-1} \quad (10)$$

Thus, $E^{(SCRFT)}$ can be expressed more simply as

$$E^{(SCRFT)} = <\psi^{(SCRFT)}|H_0 + (1/2)V_\sigma|\psi^{(SCRFT)}> \quad (11)$$

The difference between $E^{(SCRFT)}$ and the energy of the gas phase solute molecule $E^{(0)}$ is the electrostatic contribution to the free energy of solvation of the solute (M) in the solvent (m), within the SCRFT model.

$$\Delta G_{es}[M,m] = E^{(SCRFT)} - E^{(0)} \quad (12)$$

Within all SCRF models, another significant, but less well-defined, contribution to the solvation free energy of M in m is the so-called “cavitation free energy” $\Delta G_c[M,m]$. According to the model of Sinangolu (*18*), $\Delta G_c[M,m]$ can be expressed as

$$\Delta G_c[M,m] = k(M,m)S_M\gamma_m \quad (13)$$

where γ_m is the surface tension of the pure solvent, (γ_m = 71.96 N/m at stadnard conditions for water), S_M is the surface area of the solute cavity and k(M,m) is a scaling factor for γ_m, which is dependent on the solvent and the ratio of the molar volumes of the solute V_M and solvent V_m. The factor k(M,m) is given by (*18*)

$$k = 1 + (V_M/V_m)^{-2/3}[k_1(m) - 1] \quad (14)$$

where $k_1(m) = 1.302$ for water at standard conditions.

Inlcuding the cavitation contribution given in equation (13), the free energy of solvation within the SCRFT model is then

$$\Delta G[M,m] = \Delta G_{es}[M,m] + \Delta G_c[M,m] = E^{(SCRFT)} - E^{(0)} + k(M,m)S_M\gamma_m \quad (15)$$

There are, of course, other contributions to the solvation free energy which one can include within the SCRFT model, such as the effects of geometry relaxation, but these are expected to be much smaller than $\Delta G_{es}[M,m]$ and $\Delta G_c[M,m]$ and are neglected here.

An SCRFT-SC Model. Within the SC model derived from the SCRFT model, the solvation free energy of a solute M in water $\Delta G[M,aq]$ is calculated here using the following formula

$$\Delta G[M,aq] = \Delta G[M(H_2O)_n,g] + \Delta G[M(H_2O)_n,aq] - n\Delta G[H_2O,aq] \quad (16)$$

where $\Delta G[M(H_2O)_n,g]$ is the free energy change accompanying the formation of the complex $M(H_2O)_n$ in the gas-phase from M(g) and $n(H_2O)(g)$, while $\Delta G[M(H_2O)_n,aq]$ is the calculated solvation free energy of the gaseous $M(H_2O)_n$ complex in water and $-n\Delta G[H_2O,aq]$ is the calculated vaporization free energy of n water molecules from pure water. The latter term is necessary to close the thermodynamic cycle.

There are certainly other thermodynamic cycles one can devise to calculate $\Delta G[M,aq]$ within the SCRFT model. For example, it is arguably unjustified to use the a pure SCRF model for the vaporization free energy of water in evaluating an SC model of a solute in water, as we have done here. Perhaps a more self-consistent procedure would be to vaporize the water complex $(H_2O)_n$ using an SCRF model and, instead of using $-n\Delta G[H_2O,aq]$ in equation 16, include the vaporization free energy of the complex, $-\Delta G[(H_2O)_n,aq]$, together with its gas-phase dissociation to n water molecules. We will be investigating such alternative SC cycles in future work.

Results

Gas-Phase Free Energies of Hydration for Alkali Metal Cations. Using mass-spectrometric techniques, Kebarle (*19*) has determined standard free energies of hydration for several ion-water complexes $X(H_2O)_n^{m+}$ in the gas-phase. A number of studies, both experimental and theoretical, exist on the solvation of small ions in water (*13,20 - 22*). Consequently, we have chosen to examine the solvation free energies of the alkali-metal cations Li^+, Na^+ and K^+ in water as a first comparison of the SC model with the supermolecule model and the corresponding "straight" SCRFT model.

Using Gaussian 92 (*15*), gas-phase geometry optimizations and harmonic frequency calculations for the complexes $X(H_2O)_n^+$ of lithium, sodium and potassium cations with 1 - 4 water molecules have been carried out at the Hartree-Fock level using the standard 6-311++G(2d,2p) basis set for all centers except potassium, for which a [11s7p2d] basis set was used. Some geometric features of the complexes are summarized in Table I. In all cases, the n symmetrically equivalent water molecules are "dipole oriented", with the dipole axis of each water molecule parallel to its X-O bond. In each of the three series of complexes, the only significant geometric change which occurs in $X(H_2O)_{n-1}^+$ upon the addition of a water molecule is a lengthening of the X-O bond and, of course, the change in the O-X-O bond angles. Comparing complexes involving different X and a given n, one sees the expected lengthening of O-X bond in going down the periodic table, with other geometric features remaining essentially constant.

Table I. Bond Lengths and Bond Angles in $X(H_2O)_n^+$ Complexes.[a]

Molecule	Symmetry	X-O	O-H	O-X-O	H-O-X	H-O-H
H_2O	C_{2v}		0.940		106.3	
$Li(H_2O)^+$	C_{2v}	1.83	0.946		126.7	106.6
$Li(H_2O)_2^+$	D_{2d}	1.86	0.945	180.0	126.7	106.7
$Li(H_2O)_3^+$	D_3	1.90	0.943	120.0	126.6	106.8
$Li(H_2O)_4^+$	S_4	1.96	0.942	107.8	127.8	106.9
				110.3	125.1	
$Na(H_2O)^+$	C_{2v}	2.28	0.944		127.0	105.9
$Na(H_2O)_2^+$	D_{2d}	2.26	0.944	180.0	127.0	106.0
$Na(H_2O)_3^+$	D_3	2.29	0.943	120.0	126.9	106.2
$Na(H_2O)_4^+$	S_4	2.32	0.942	106.6	127.4	106.3
				111.2	126.0	
$K+(H_2O)^+$	C_{2v}	2.68	0.944		127.2	105.6
$K+(H_2O)_2^+$	D_{2d}	2.72	0.943	180.0	127.2	105.6
$K+(H_2O)_3^+$	D_3	2.74	0.942	120.0	127.1	105.8
$K+(H_2O)_4^+$	S_4	2.77	0.942	106.6	127.5	105.9
				110.9	126.6	

[a] Bond lengths are in Angstroms and bond angles are in degrees.

The calculated gas-phase standard free energies of complexation $\Delta G[X(H_2O)_n^+,g]$, including zero-point vibrations, thermal, $P\Delta V$ and entropic contributions are shown in Table II along with Kebarle's experimental results (*19*). The agreement is quite good, considering that the HF level of theory was used throughout. Counterpoise corrections for the $X(H_2O)_2^+$ complexes, also shown in Table II, indicate that basis set superposition error is not a serious problem with the relatively large basis sets used here.

It is noteworthy that the change in free energy upon the addition of H_2O to $X(H_2O)_{n-1}^+$ approaches, as it should, the free energy of solvation of water in water (-6.7 kcal/mol) as n increases.

Solvation Free Energies for Alkali Metal Cations. Given the good agreement between the level of theory used and experiment for the free energies of hydration in the gas-phase, SCRFT calculations at the same level have been carried out on the both $X(H_2O)_n^+$ complexes as well as a single water molecule, the latter being necessary for the SC model. The calculated solvation free energy results are shown in Table III. The first column, labeled $\Delta G_{es}[X(H_2O)_n^+,aq]$ lists the electrostatic solvation free energies of the complexes $X(H_2O)_n^+$ in water. One sees that as more water explicit water molecules are added to each of the complexes, the electrostatic solvation free energy decreases. Indeed, $\Delta G_{es}[X(H_2O)_n^+,aq]$ approaches zero with increasing n. The second column in Table III, labeled $\Delta G[X(H_2O)_n^+,aq]$, lists the calculated free energies of aqueous solvation including the cavitation free energy. This quantity approaches infinity with increasing n due to the cavitation free energy. The third column in Table III, labeled $\Delta G_{es}[X^+,aq]$, lists the SC electrostatic free energy of solvation for the X^+ cations in water, as calculated using equation 15 without cavitation free energies. One sees that this quantity appears to be converging, but to a value significantly below the experimental result. The final column in Table III, labeled $\Delta G[X^+,aq]$, lists the "total" solvation free energy for the X+ cations in water as calculated using equation 15 and including the cavitation free energy terms. Ideally, this quantity would, of course, converge to the experimental value with increasing n, and indeed it appears to do so for all three cations at n=4. This result must be considered fortuitous, however, given that the SCRFT free energy of solvation of water in water is calculated to be positive (+2.2 kcal/mol compared to the experimental value of -6.7 kcal/mol) when the cavitation term is included. As mentioned above, one of the ambiguities associated with the use of any SC model is how the vaporization free energies of the solvent molecules are calculated.

More meaningful than the results given in Table III are the relative solvation free energies, $\Delta G[X^+,aq]$ - $\Delta G[X^+,aq]$, shown in Table IV. Here the neglected contributions to $\Delta G[X^+,aq]$ should largely cancel and the somewhat ambiguous term $-n\Delta G[H_2O,aq]$, totally cancels. One sees that the straight SCRFT model poorly predicts relative solvation energies, but that the SC model improves the predictions significantly, giving fairly good agreement with experiment at n=4. Also shown in Table 4 are the relative solvation free energies of the cations predicted by the pure SM approach. One sees that for relative solvation free energies, the SM approach

Table II. Gas-Phase Hydration Free Energies for $X(H_2O)_n^+$ Complexes.[a]

Molecule	$\Delta G(X(H_2O)_n^+,g)_{calc}$ [b]	$\Delta\Delta G_{calc}$ [c]	$\Delta G(X(H_2O)_n^+,g)_{expt}$ [d]	$\Delta\Delta G_{expt}$ [e]
$Li+(H_2O)$	-27.5		-27	
$Li+(H_2O)_2$	-48.8 (-48.2[f])	-21.3	-47	-20
$Li+(H_2O)_3^+$	-61.5	-12.7	-60	-13
$Li+(H_2O)_4^+$	-68.9	-7.4	-68	-6
$Na(H_2O)^+$	-16.8		-17	
$Na(H_2O)_2^+$	-30.4(-30.0[f])	-13.6	-31	-14
$Na(H_2O)_3^+$	-39.6	-9.2	-40	-9
$Na(H_2O)_4^+$	-46.2	-6.6	-47	-7
$K+(H_2O)^+$	-10.3		-11	
$K+(H_2O)_2^+$	-17.9 (-17.7[f])	-7.6	-20	-9
$K+(H_2O)_3^+$	-23.6	-5.7	-26	-6
$K+(H_2O)_4^+$	-29.9	-6.3	-31	-5

[a] Free energies are in kcal/mol.
[b] $\Delta G(X(H_2O)_n^+,g)_{calc}$ is the calculated standard free energy change for the formation of $X(H_2O)_n^+(g)$ from $X^+(g)$ and $nH_2O(g)$.
[c] $\Delta\Delta G_{calc}$ is the difference between $\Delta G(X(H_2O)_n^+,g)_{calc}$ and $\Delta G(X(H_2O)_{n-1}^+,g)_{calc}$, i.e., the calculated free energy change for the addition of H_2O to $X(H_2O)_{n-1}^+$ in the gas phase.
[d] $\Delta G(X(H_2O)_n^+,g)_{expt}$ is the experimental standard free energy change for the formation of $X(H_2O)_n^+(g)$ from $X^+(g)$ and $nH_2O(g)$. Reference *19*.
[e] $\Delta\Delta G_{expt}$ is the difference between $\Delta G(X(H_2O)_n^+,g)_{expt}$ and $\Delta G(X(H_2O)_{n-1}^+,g)_{expt}$, i.e., the experimental free energy change for the addition of H_2O to $X(H_2O)_{n-1}^+$ in the gas phase.
[f] Includes counterpoise correction for basis set superposition error.

Table III. Standard Free Energies of Aqueous Solvation for $X+(H_2O)_n^+$ and X^+.[a]

Molecule	$\Delta G_{es}[X(H_2O)_n^+,aq]$[b]	$\Delta G[X(H_2O)_n^+,aq]$[c]	$\Delta G_{es}[X^+,aq]$[d]	$\Delta G[X^+,aq]$[e]
H_2O	-5.0	2.2		
Li^+	-154.4	-151.3	-154.4	-151.3
$Li(H_2O)^+$	-112.2	-104.7	-134.7	-134.4
$Li(H_2O)_2^+$	-76.8	-64.7	-115.5	-117.9
$Li(H_2O)_3^+$	-60.7	-45.5	-107.1	-113.5
$Li(H_2O)_4^+$	-55.1	-37.4	-103.9	-115.0
				-115.0 (expt)[f]
Na^+	-114.8	-110.5	-114.8	-110.5
$Na(H_2O)^+$	-93.2	-84.7	-105.0	-103.7
$Na(H_2O)_2^+$	-75.9	-63.0	-96.2	-97.7
$Na(H_2O)_3^+$	-62.8	-45.8	-87.3	-91.9
$Na(H_2O)_4^+$	-54.6	-34.2	-80.7	-89.1
				-89.6(expt)[f]
K^+	-85.6	-79.2	-85.6	-79.2
$K(H_2O)^+$	-74.4	-64.0	-79.7	-76.5
$K(H_2O)_2^+$	-65.6	-51.1	-73.4	-73.3
$K(H_2O)_3^+$	-58.7	-40.1	-67.2	-70.2
$K(H_2O)_4^+$	-53.8	-31.1	-63.6	-69.7
				-72.7(expt)[f]

[a] All free energies are given in kcal/mol.
[b] The calculated free energy of aqueous solvation for $X(H_2O)_n^+$, without the cavitation term.
[c] The calculated free energy of aqueous solvation for $X(H_2O)_n^+$, including the cavitation term.
[d] The calculated free energy of aqueous solvation for X^+ using equation 13, without the cavitation terms.
[e] The calculated free energy of aqueous solvation for X^+ using equation 13, including the cavitation terms.
[f] Reference *21*.

Table IV. Relative Free Energies of Aqueous Solvation for Alkali metal Cations.[a]

Molecule	[$\Delta G(Li^+,aq)-\Delta G(Na^+,aq)$]	[$\Delta G(Li^+,aq)-\Delta G(K^+,aq)$]	[$\Delta G(Na^+,aq)-\Delta G(K^+,aq)$]
SCRFT-SC[b]			
X^+	-39.6 (-40.8)	-68.8 (-72.1)	-29.2 (-31.3)
$X(H_2O)^+$	-29.7 (-30.7)	-55.0 (-57.9)	-25.3 (-27.2)
$X(H_2O)_2^+$	-19.3 (-20.2)	-42.1 (-44.6)	-22.8 (-24.4)
$X(H_2O)_3^+$	-19.8 (-21.6)	-39.9 (-43.3)	-20.1 (-21.7)
$X(H_2O)_4^+$	-23.2 (-25.9)	-40.3 (-45.3)	-17.1 (-19.4)
SM[c]			
$X(H_2O)^+$	-10.7	-17.2	-6.5
$X(H_2O)_2^+$	-18.4	-30.9	-12.5
$X(H_2O)_3^+$	-21.9	-37.9	-16.0
$X(H_2O)_4^+$	-22.7	-39.0	-16.3
Expt.[d]	-25.4	-42.3	-16.9

[a] All free energies are given in kcal/mol.

[b] Difference in SCRFT free energies of aqueous solvation calculated using equation 13. Quantities in parentheses include the cavitation free energy.

[c] Difference in gas-phase hydration free energies.

[d] Reference *21*.

with only a few waters appears to be adequate, i.e. the "continuum" contribution to the solvation energies of the complexes of a given n (> 2) are roughly the same.

Conclusions

The results presented here show that when the solute cavity is defined consistently in terms of an isodensity surface of a particular value (0.0004 au), a value chosen to correspond to the solute "size", the straight SCRFT model is not able to accurately predict either absolute or relative solvation free energies of the alkali metal cations in water. The use of the "supermolecule" method, on the other hand, predicts relative solvation free energies for the cations in relatively good agreement with experiment with only a few explicit water molecules. The use of the SC model yields reasonable absolute and relative solvation free energies for the cations, but the absolute free energy calculations within the SC model are somewhat ambiguous and should be viewed with caution. Overall, the preliminary results given here indicate that, within the SCRFT model for even the simplest of solutes in water, the inclusion of short-range, specific interactions between solvent and solute, which are neglected in the straight SCRFT model, appear to be necessary to predict both absolute and relative solvation free energies.

Literature Cited

1. Kollman, P.A.; Kuntz, I.D. *J. Amer. Chem. Soc.* **1972**, *94*, 9236.
2. Kistenmacher, H.; Popkie, H.; Clementi E. *J. Chem. Phys.* **1973**, *58*, 5627.
3. Kistenmacher, H.; Popkie, H.; Clementi E. *J. Chem. Phys.* **1974**, *61*, 799.
4. Pullman, A.; Armbruster, A.M. *Int. J. Quantum Chem Symp.* **1974**, 8, 169.
5. Angyan, J.G. *J. Math. Phys.* **1992**, *10*, 93.
6. Tapia, O. *J. Math. Chemistry* **1992**, *10*, 139.
7. Rivail, J.L.; Rinaldi, D. *Chem. Phys.* **1976**, *18*, 233.
8. Rinaldi, D.; Ruiz-Lopez, M.F.; Rivail, J.L. *J. Chem. Phys.* **1983**, *78*, 834.
9. Rivail, J.L.; Terryn, B.; Rinaldi, D.; Ruiz-Lopez, M.F. *J. Mol. Struct. (Theochem)* **1985**, *120*, 387.
10. Miertus, S.; Scrocco, E.; Tomasi, J. *Chem. Phys.* **1981**, *55*, 117.
11. Miertus, S.; Tomasi, *J. Chem. Phys.* **1981**, *65*, 239.
12. Tunon, I.; Silla, E.; Bertran, S. *J. Phys. Chem.* **1993**, *97*, 5547.
13. Marcos, E.S.; Pappalardo, R.R.; Rinaldi, D. *J. Phys. Chem.* **1991**, *95*, 8928.
14. Keith, T.A.; Foresman, J.B.; Frisch, M.J.; to be published.
15. *Gaussian 92*, M.J. Frisch, G.W. Trucks, M. Head-Gordon, P.M.W. Gill, J. B. Foresman, M.W. Wong, B.G. Johnson, H.B. Schlegel, M.A. Robb, E.S. Replogle, R. Gomperts, J.L. Andres, K. Raghavachari, J.S. Binkley, C. Gonzalez, R.L. Martin, D.J. Fox, D.J. Defrees, J. Baker, J.J.P. Stweart, and J.A. Pople, Gaussian Inc., Pittsburgh PA, 1992.
16. Foresman, J.B.; Keith, T.A.; Frisch, M.J.; Wiberg, K.B; to be published.
17. Wong, M.W.; Wiberg, K.B.; Frisch, M.J.; Keith, T.A.; to be published.

18. Sinanoglu, O. In *Molecular Interactions*; Ratajczak, H. and Orville-Thomas, W.J., Eds.; John Wiley and Sons, Ltd.: 1982, Vol. 3; pp. 283-342.
19. Kebarle, P. *Ann. Rev. Phys. Chem.* **1977**, *28*, 445.
20. Conway, B.E. *Ionic hydration in Chemistry and Biophysics*; Elsevier Scientific Publishing Co.: 1981.
21. Marcus, Y. *Ion Solvation*; John Wiley and Sons, Ltd: 1985.
22. Krestov, G.A. *Thermodynamics of Solvation: Solution and Dissolution Ions and Solvent Structure and Energetics*; Ellis Horwood: 1991.

RECEIVED July 29, 1994

Chapter 4

Effect of Hydrogen Bonding on Molecular Electrostatic Potential

M. Dominic Ryan

Department of Physical and Structural Chemistry, SmithKline Beecham Pharmaceuticals, King of Prussia, PA 19406

Electrostatic potential difference maps were calculated for 16 hydrogen bonded complexes typical of hydrogen bonds in proteins. The molecular electrostatic potentials of the complexes and the monomers were characterized by fitting atom centered charges to the molecular electrostatic potential. These charge sets were used to generate the difference maps which were then contoured at arbitrary values of potential energy. The difference maps permit the visualization of the perturbation of the electrostatic potential by hydrogen bond formation. N-methyl acetamide (nma) perturbs a complex more as a donor than acceptor; methyl imidazole results in greater perturbative effects and methanol lesser. The energetic effect of complex formation on interaction with a third molecule was investigated. Hydrogen bond formation stabilizes the electrostatic interaction with a third molecule, nma, in the geometries studied, except when the initial complex is an ion pair.

A large body of work has been published over the years on the quantum mechanical investigation of complex formation (*1*), particularly with respect to hydrogen bonding (*2*). The published work has concentrated on the search for local minima in the complex hypersurface and the structural and energetic characterizations of these minima. When the interaction energy of such complexes is subjected to Morokuma analysis (*3*) it becomes apparent that most of the energy arises from Coulomb terms. This is reflected well in recent work where the interaction energy was evaluated classically from a set of distributed multipoles on each monomer (*4,5*). The multipole sets were fitted to ab initio wavefunctions of the respective isolated monomers. The electrostatic interaction energy so calculated was found to closely parallel the supermolecule complexation energies.

Thus the molecular electrostatic potentials (MESP) of the monomers, represented by distributed multipoles, give a good indication of the interaction potential. However, once the complex has formed there is no reason to expect that the molecular electrostatic potential of the complex will remain unperturbed relative to the simple sum of the monomers. This question will be explored by calculating a difference potential:

0097–6156/94/0569–0036$08.72/0

$$\text{Difference Potential} = \text{MESP}(\text{complex}_{AB}) - \text{MESP}(\text{reference})$$

$$\text{MESP}(\text{reference}) = [\text{MESP}(A) + \text{MESP}(B)]$$

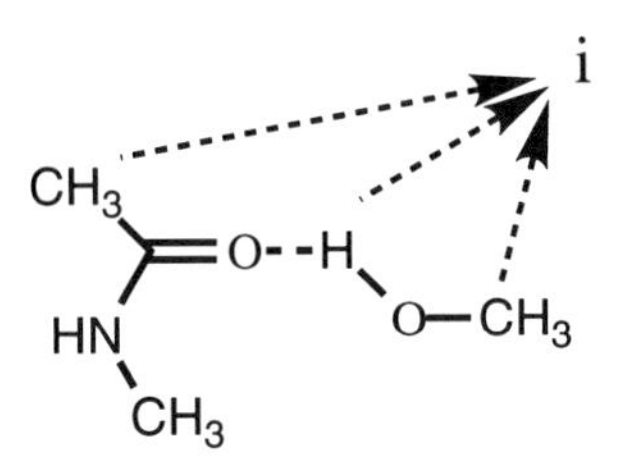

The molecular electrostatic potential (MEP) is approximated by the potential at points distributed around a molecule outside the VDW envelope of the molecule. It is the energy of interaction of a proton with the molecule at each point i. The difference potential is then calculated at a point i by subtracting a reference potential from the potential of the complex at that point. The reference potential is derived using the same geometry as the complex but with charges taken from the isolated monomers.

The potential can be represented in various forms. Most force fields in current usage (AMBER (*6*), CHARMM (*7*), Discover (*8*)) represent the electrostatic potential around a molecule with atom centered monopoles, i.e. point charges. Non-atom centered charges are also used, particularly for water models. (*9*, *10*, *11*). A more detailed description of the electrostatic potential may be obtained from a set of distributed multipoles. Finally, the electrostatic potential at each point in space may be evaluated directly from a wavefunction.

Due to the computational demand of obtaining a wavefunction for large systems, or systems requiring large numbers of configurations such as molecular dynamics, atom centered point charges are usually chosen to represent the electrostatic potentials. While this is a compromise, the accuracy with which a charge set represents a quantum mechanical electrostatic potential may be maximized by fitting the charges to values of electrostatic potential, calculated from a wavefunction, at points evenly distributed around the molecule. The set of atom-centered charges obtained in this manner is often referred to as ESP-fitted charges and the methodology has been the subject of much work and several reviews (*12*,*13*,*14*,*15*)

The need to limit the representation of the molecular electrostatic potential to atom centered monopoles is particularly acute in biomolecular systems such as protein-ligand complexes.

A force-field assumes the transferability of properties; i.e., that a set of charges determined for a molecule will be valid when the molecule is in a new environment. The severity of this approximation has been examined (*16*, *17*, *18*) from the point of view of conformational dependence of fitted charges where there can be important fluctuations in the molecular electrostatic potential as a function of intramolecular torsions. However there has been little consideration of the impact of the proximity of other molecules, particularly forming hydrogen bonds, on the electrostatic potential surrounding the system. Protein secondary structure elements such as helix and sheet have strong hydrogen bonds to the amide backbone which perturb the electrostatic potentials. The magnitude and anisotropy of this effect will be examined within the context of a fitted charge representation.

Methodology

Eight monomers were chosen to represent selected amino acid side-chains and the amide backbone. These are N-methyl acetamide, methanol, acetate, N-methyl guanidinium cation, 2-methyl imidazole, methyl ammonium cation, acetic acid and methyl amine.

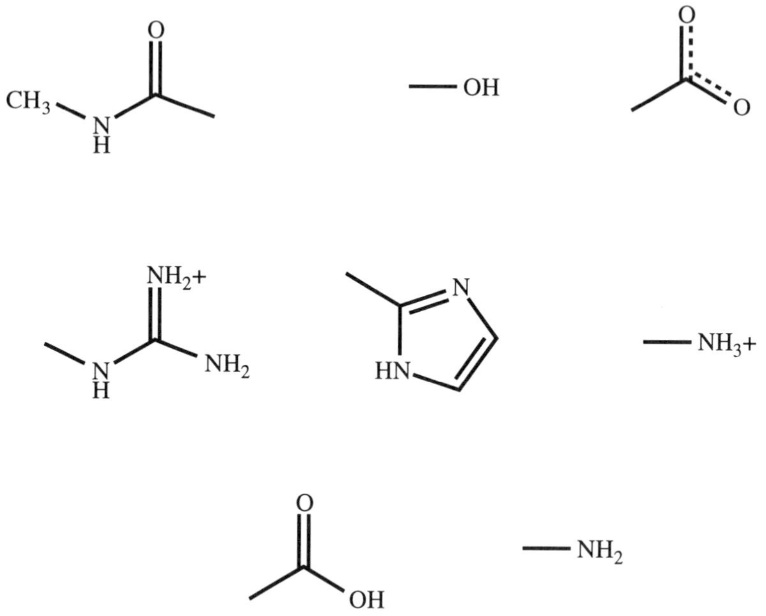

Gaussian 92-rev D2 (*19*) was used for all quantum mechanical calculations with the following basis sets (*20*:) 6-31G*//6-31G* for all neutral and cationic complexes and monomers, 6-31+G*//6-31G* for all complexes containing an anionic component and constituent monomers of those complexes.

Geometries of complexes were fully optimized to the default criteria of G92 for rms deviation and gradient without performing a vibrational analysis to determine the presence of a stationary point. Fitted charges of the monomers, the reference charges, were obtained using fixed monomer geometries from the optimized complexes. This ensures that the perturbation of the potentials arises from the intermolecular effects with no intramolecular contributions. In cases where non-minimum energy complexes were used they were minima with respect to all but a constraining coordinate. The complexation energies are not corrected for basis set superposition error and are computed using the energies of monomers fully optimized in the basis set used for the wavefunction of the complex.

Fitted charges were obtained with the Merz-Kollman (*21*) option of G92. This places the points at which electrostatic potential are evaluated on four shells around each atom in increasing distance from each nucleus at 1.2x, 1.4x, 1.8x and 2.0 times the VDW radius of that nucleus. Points are placed at a uniform spacing of 1 pt /Bohr on each surface. The quality of the fit to the molecular electrostatic potential is expressed in rms and relative rms (rrms) error functions, where the rrms is a percentage of the absolute value and more useful than the absolute rms for comparison purposes.

$$\text{rms: } \sigma = \left[N^{-1} \sum_i \left(V_i^{QM} - V_i^{C} \right)^2 \right]^{1/2}$$

$$\text{relative rms: } s = \sigma \left[\mathrm{N}^{-1} \sum_i \left(V_i^{QM} \right)^2 \right]^{-1/2}$$

V_i = Electrostatic potential at point i. QM = Quantum mechanical, C = classical

The selection of the MK point distribution scheme is based upon the observation that low values of relative rms are obtained with a lesser number of points than other schemes (*22*) based on a grid or random sampling. The atom-centered charges are fitted to the values of electrostatic potential with a Lagrange undetermined multiplier matrix solution that permits constraints to be applied to the system. The only constraint used was that the sum of the atomic charges equals the molecular charge.

Difference Potentials. The difference potentials were calculated by regenerating a regular three dimensional grid of points around the molecule using the Sybyl (*23*) POTENTIAL command. As with the original evaluation of electrostatic potential energy (EP) energy, the energy is evaluated between a proton probe and molecule at each point as shown in the equation below. At each point i the value of electrostatic potential is calculated from the charges at centers n. Molecular dynamics simulations of proteins commonly use a distance-dependent dielectric (*24*). Therefore the reconstructed potentials were also calculated with a distance-dependent dielectric.

$$\mathrm{ESP_i} = 332 \sum_{\mathrm{n}} \frac{q_{\mathrm{n}}}{\varepsilon d(\mathrm{ni})} \qquad \varepsilon = d(\mathrm{ni})$$

A second set of potentials was then calculated with the complex in the same coordinate frame using the reference charge set. The potentials were stored in Sybyl 'density' files and then subtracted from each other to give a difference potential which was written to a new 'density' file. This new file was contoured at arbitrary values of electrostatic potential energy and viewed in 3-D.

The only effect of complexation considered here, and referred to below, is the effect on EP energy. For a constant value of energy level contoured, the volume enclosed by the contour characterizes the extent of the perturbation. If the volume is zero the perturbation is less than the threshold set for the contour. In comparing contour volumes a larger volume indicates that the perturbation of EP energy at the level contoured extends farther into space and interpreted as a larger impact of complexation. Assessments of volume are qualitative in this work, serving to rank the complexes in their mutual perturbations of the original MEP.

Electrostatic complexation energies between bimolecular complexes and N-methyl acetamide were calculated within Sybyl using the energy command. They were calculated as the difference between the total energy of the trimers and the sum of the energies of the complex and N-methyl acetamide. Only the electrostatic component of the force-field was used.

Results

Sixteen complexes were studied in this manner. In each difference plot a decrease in potential energy is displayed in black contour volumes, and an increase is displayed in gray. An increase in potential energy indicates energy becoming either less negative or more positive while a decrease indicates energy becoming either more negative or less positive. The donor molecule generally exhibits an decrease in electrostatic potential energy surrounding it, rationalized by the partial sharing of a proton (a positive charge) with the acceptor, resulting in a more negative potential. Similarly the acceptor, in gaining some fraction of a proton, experiences an increase in electrostatic potential around the molecule. These changes are not uniform, however, and depend on the specifics of the hydrogen bonding interaction. The complexes are discussed in approximately the order of the magnitude of the impact of complexation: neutral pairs, ion-neutral pairs and ion-ion pairs. The contour level varies from 2.0 Kcal/mol to 8 Kcal/mol such that the sizes of the contours are approximately similar from one complex to the next, small enough to visualize around the complexes and still be close to the intermolecular binding energies that range from about 4 to 10 Kcal/mol.

The results are discussed first in terms of the qualitative perturbations that result from hydrogen bond formation and second in terms of the energetic consequences to the electrostatic interaction energy of the complex with a third molecule. The former provides some insight into the possible perturbations that may result to the electrostatic potential environment of a receptor site.

Qualitative Characterization of Perturbations

Methanol / N-methyl acetamide. Figures 1 - 4 show methanol / N-methyl acetamide complexes, with methanol as hydrogen bond donor. The difference potentials are contoured at 2 Kcal/mol. The structure in Figure 1 is a local energy minimum. The largest perturbation on N-methyl acetamide occurs around the methyl α to the carbonyl group with a very small positive volume outside the NH proton. When the hydrogen bonding distance is increased by 0.5Å, destabilizing the complex by 1.85 Kcal, the effects are attenuated. This is evident in Figure 2 with smaller volumes in the contours at the same value of electrostatic potential energy.

In Figures 1 and 2 the hydrogen bond geometry corresponds to one of the lone-pairs of electrons on oxygen being the hydrogen bond acceptor. When the molecular electrostatic potential of an amide carbonyl group is contoured, two minima in potential are located at the positions of lone-pairs of electrons (*25*). Such lone-pairs are also apparent as electron density in low temperature high resolution crystal structures (*26*). Computed hydrogen bond geometries also display a preference for an orientation that suggests hydrogen bonding to lone-pairs, but this preference is relatively small.

Figure 3 shows contours for a complex geometry 2.16 Kcal/mol higher in energy, with the CO:H angle of 180°. The difference potential shows a slight increase in the volume of the contour surrounding the amide methyl group, an increase in the volume above and below the amide plane and a small decrease in the plane of the NH proton. When the hydrogen bond donor is in the plane of the carbonyl group but below the plane of the amide group, in Figure 4, the volume around the amide methyl group again changes little, but the volume at the NH proton returns and is larger than in the minimum energy geometry. The volume around the amide N-methyl also increases.

An amide bond is often described as a partially delocalized double bond due in part to the barrier to rotation (*27*) about the amide bond of 15 to 23 Kcal/mol.

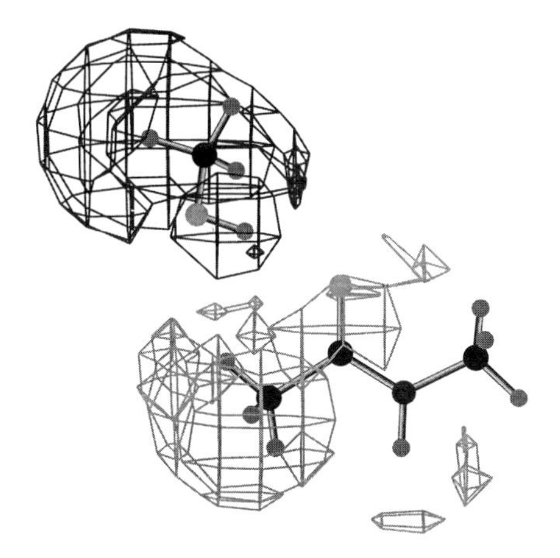

Figure 1. Methanol (OH donor) / N-methyl acetamide (carbonyl oxygen acceptor). Minimum energy conformation, d(H..O)=1.98Å. Electrostatic potential energy difference contoured at +/- 2 Kcal/mol.

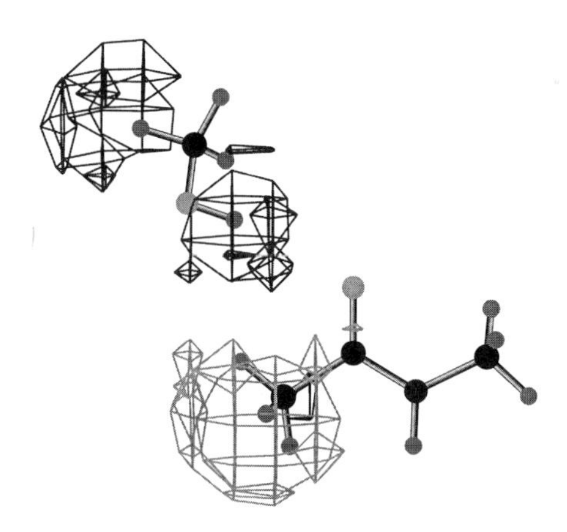

Figure 2. Methanol (OH donor) / N-methyl acetamide (carbonyl oxygen acceptor). d(H..O)=2.5Å. Electrostatic potential energy difference contoured at +/- 2 Kcal/mol.

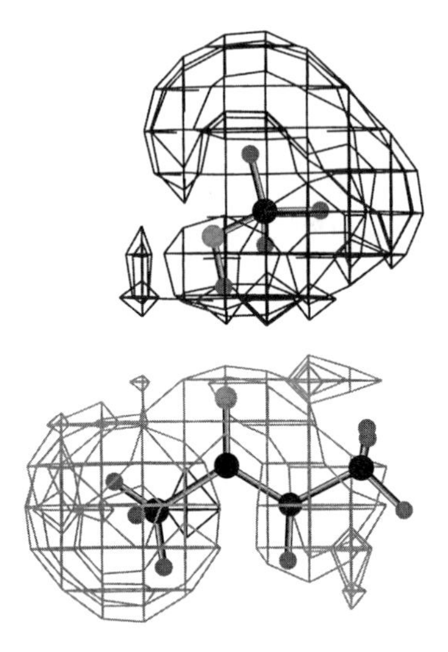

Figure 3. Methanol (OH donor) / N-methyl acetamide (carbonyl oxygen acceptor). C=O:H angle=180°. Electrostatic potential energy difference contoured at +/- 2 Kcal/mol.

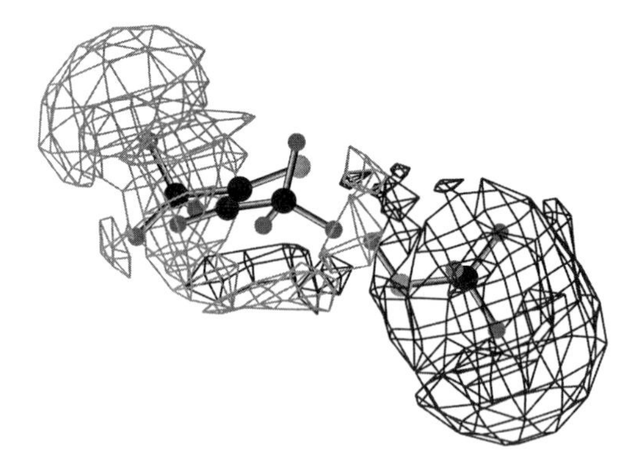

Figure 4. Methanol (OH donor) / N-methyl acetamide (carbonyl oxygen acceptor). C=O:H angle=118°, OH hydrogen in line with carbonyl, below plane of amide. Electrostatic potential energy difference contoured at +/- 2 Kcal/mol.

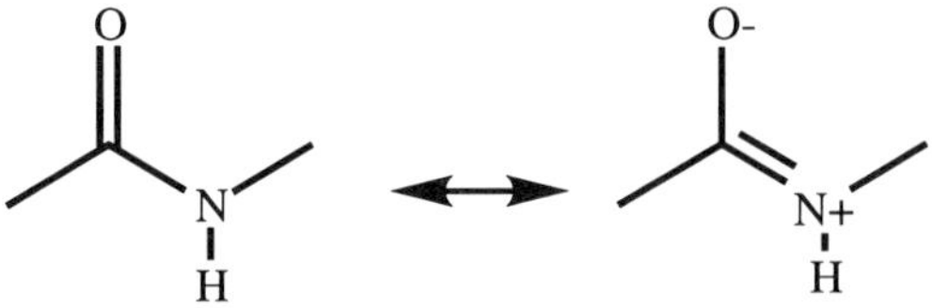

The importance of the zwitterionic resonance description should be increased by hydrogen bonding, leading to a difference potential volume in the vicinity of the NH proton. While there is some evidence for this, the perturbation of the electrostatic potential at the NH proton is much smaller than the perturbation around the carbonyl methyl group. However, it does increases slightly when the hydrogen bond geometry does not permit hydrogen bonding to the lone-pairs of an sp^2 model as in Figure 3. The slight increase suggests that the importance of the zwitterionic resonance contributor may be greater in the distorted structure. This may be due to greater flexibility in accommodating non-ideal hydrogen bond geometries in the zwitterionic form. The keto form of the resonance description localizes lone-pairs, the preferred acceptor region, at 120° relative to the carbonyl axis. An acceptor able to adopt this geometry may reinforce the keto resonance form.

Figure 5 shows the same pair of monomers with N-methyl acetamide NH as donor. At the same level of energy contour the volumes are larger on both donor and acceptor. The impact on the carbonyl group oxygen appears to be larger here than the impact on the NH proton was when N-methyl acetamide was the acceptor in Figure 1, reflecting a directional inequality across the amide π-system. Complexation with methanol also exerts a larger effect on the methyl group of methanol when N-methyl acetamide is a donor than when it is an acceptor. Thus the effect of hydrogen bonding on the MEP of N-methyl acetamide is larger when N-methyl acetamide is a donor than when it is an acceptor, suggesting a greater importance of the zwitterionic resonance form when it is a donor.

Methyl imidazole / N-methyl acetamide. Figures 6 and 7, contoured at 2 Kcal/mol, show methyl imidazole / N-methyl acetamide complexes with N-methyl acetamide as acceptor and donor respectively. Figure 6a shows a larger perturbation on N-methyl acetamide than in Figure 1, in particular over the NH group. The complexation energy (ΔEscf) of -8.01 Kcal is slightly greater than ΔEscf of Figure 1. Figure 6b, in the plane of imidazole, shows contours covering the entire edge of imidazole distal to the complex interface. The region of energy increase around the acceptor extends to the NH group of imidazole.

As with N-methyl acetamide / methanol, the contour volumes around N-methyl acetamide are larger when N-methyl acetamide is the donor (Figure 7a) than when it is the acceptor (Figure 6a). Imidazole shows a non-uniform perturbation, particularly in Figure 7b, where it is apparent that there are both regions of increase in electrostatic potential energy and regions of decrease centered over the ring carbons.

The contour volumes of N-methyl acetamide in N-methyl acetamide / imidazole are larger than in corresponding N-methyl acetamide / methanol pairs (Figure 1 vs. Figure 6, and particularly Figure 5 vs. Figure 7a) at the same value of potential energy, indicating that the perturbation of the electrostatic potential of N-methyl acetamide by imidazole is larger than the perturbation by methanol.

Methanol / Methyl imidazole. Methanol / imidazole complexes are shown in Figures 8 (methanol donor) and 9 (methanol acceptor). Figure 8 shows a smaller perturbation on methyl imidazole than seen in Figure 7b. The perturbation on imidazole as donor is also slightly smaller than in the corresponding N-methyl

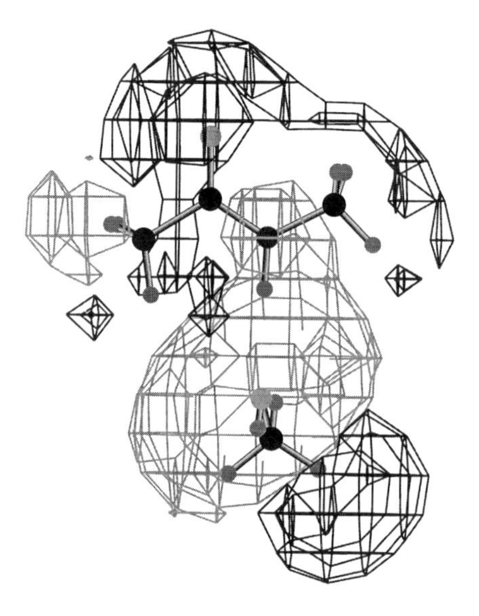

Figure 5. N-methyl acetamide (NH donor) / methanol (OH acceptor). N-H::O angle=180°. Electrostatic potential energy difference contoured at +/- 2 Kcal/mol.

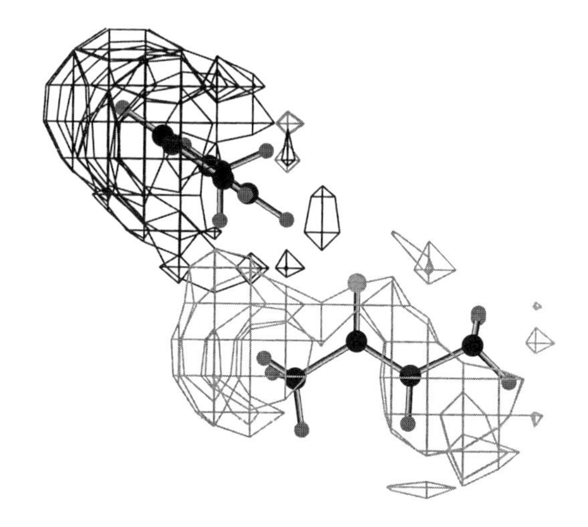

Figure 6a. Methyl imidazole (NH donor) / N-methyl acetamide (carbonyl oxygen acceptor). Minimum energy conformation. View in plane of amide. Electrostatic potential energy difference contoured at +/- 2 Kcal/mol.

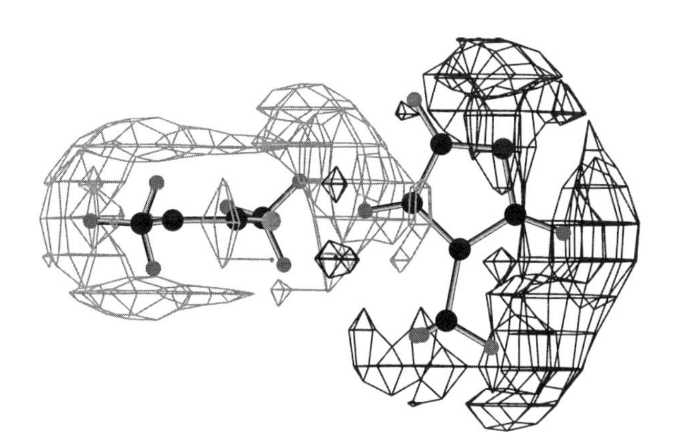

Figure 6b. Methyl imidazole (NH donor) / N-methyl acetamide (carbonyl oxygen acceptor). Minimum energy conformation. View in plane of imidazole. Electrostatic potential energy difference contoured at +/- 2 Kcal/mol.

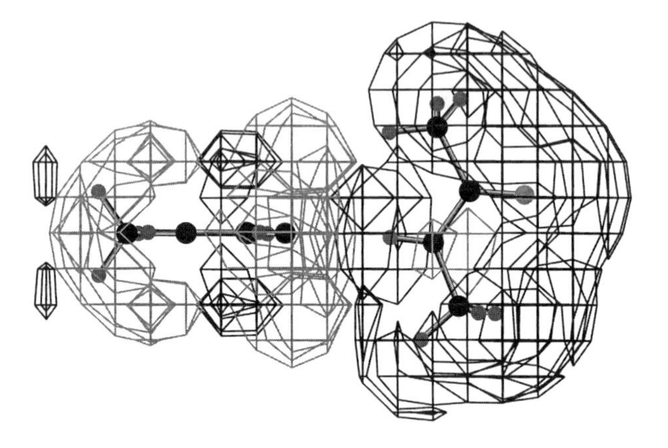

Figure 7a. N-methyl acetamide (NH donor) / methyl imidazole (ring -N= acceptor). View in plane of acetamide. Electrostatic potential energy difference contoured at +/- 2 Kcal/mol.

Figure 7b. N-methyl acetamide (NH donor) / methyl imidazole (ring -N= acceptor). View in plane of imidazole. Electrostatic potential energy difference contoured at +/- 2 Kcal/mol.

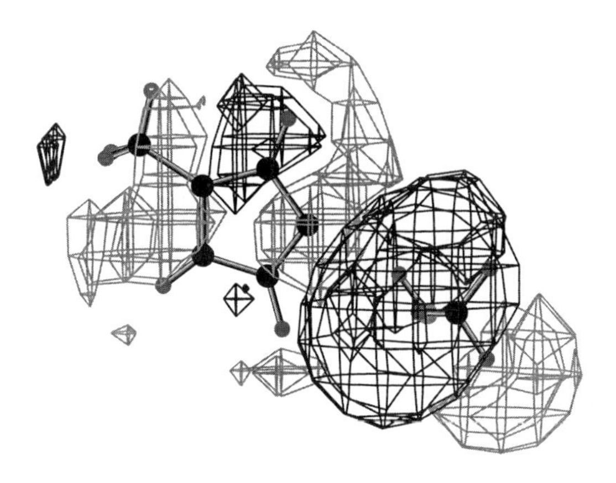

Figure 8. Methanol (OH donor) / methyl imidazole (-N= acceptor). Electrostatic potential energy difference contoured at +/- 2 Kcal/mol.

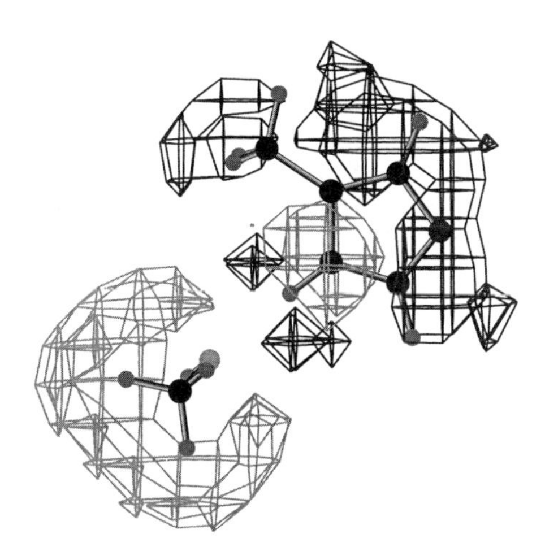

Figure 9. Methyl imidazole (NH donor) / methanol (O acceptor). View in plane of imidazole. Electrostatic potential energy difference contoured at +/- 2 Kcal/mol.

acetamide complex (Figure 6b) but the difference is not as great as when it is an acceptor. In all cases the pattern of shift in electrostatic potential energy is similar.

The impact on the methanol group is larger than in the N-methyl acetamide-methanol complexes when methanol is donor, but smaller than when methanol is acceptor. In the cases where the perturbation is larger (Figure 5 and Figure 8), the effect on the methyl group is not uniform with one proton showing contours opposite to the rest.

Ionic complexes. All of the above complexes involve neutral monomers. The perturbation resulting from an ionic monomer is significantly larger than from a neutral one. Therefore, difference potentials for N-methyl acetamide / methyl ammonium (Figure 12) and acetate / N-methyl acetamide (Figure 10) were contoured at +/-5 Kcal/mol. Acetate / N-methyl acetamide shows a decrease in potential energy across the length of N-methyl acetamide on the face distal to the other monomer. This is nearly independent of the orientation of the acetate group (Figs 10 vs. 11). There is an increase in electrostatic potential energy over the acetate group which extends over the amide nitrogen and proton.

The N-methyl acetamide / methyl ammonium complex shows a similar polarization, i.e., a region of decrease in potential energy about the donor that extends over the carbonyl oxygen and a region of increase about the acceptor.

When both hydrogen bond monomers are ionic (a salt-bridge), perturbation is greater and difference potentials were contoured at +/-8 Kcal/mol. Acetate and methyl guanidinium ions can form double hydrogen bonded complexes. In this geometry (Figure 13) the effect of complexation increases potential energy around the donor and decreases energy around the acceptor, with the region of decrease extending to the hydrogens of the donor.

When one of the two possible hydrogen bonds is broken (Figure 14) the perturbation is significantly smaller and is shown contoured at +/-5Kcal/mol. There is now a small decrease in the contact region, a decrease in the 'far' side of the acceptor and an increase around the donor.

Acetate / methyl ammonium also forms a nearly symmetric doubly hydrogen bonded complex (Figure 15), with two NH protons each hydrogen bonded to an acetate oxygen. The perturbation is similar to that of the guanidinium-acetate ion pair with a broad decrease in electrostatic potential around the acceptor and an increase around the donor.

When acetic acid / methyl amine difference potentials are contoured (Figure 16) the major perturbation is now in the contact region with smaller contour volumes around the other parts of the complex.

Gradients. The contours describe a limiting value of EP energy difference. Outside the contour the absolute values are smaller, inside they are greater. Therefore, in the contact region between the monomers of a complex where there is no contour, the EP energy difference is lower in absolute value. In some complexes, at the same levels of energy contoured, there is very little space between the positive and negative contours. Thus there is a larger gradient in EP energy difference (note that this is not an electrostatic field gradient). In others there is a large distance between positive and negative contours resulting in a smaller gradient. The largest gradients are found between ion pairs and neutral-ion complexes. In the neutral complexes those with imidazole show a larger contour volume and there is a larger effect on N-methyl acetamide when it is a hydrogen bond donor. The largest gradient between neutrals is with acetic acid / methyl amine neutral complex, suggesting that the larger pKa difference for this pair is a factor in the perturbation of the molecular electrostatic potential.

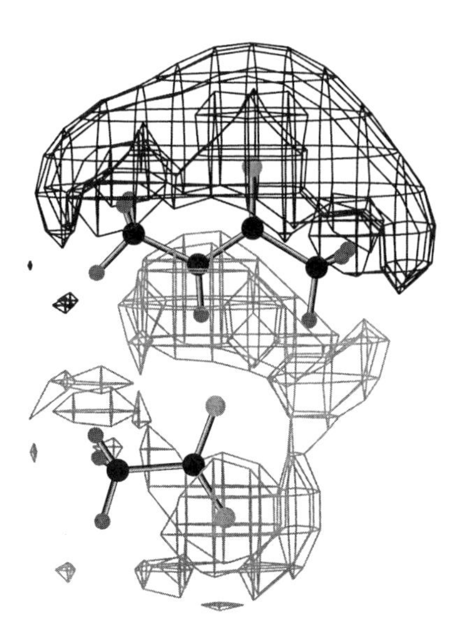

Figure 10. N-methyl acetamide (NH donor) / acetate ion. Non-minimum energy geometry. Electrostatic potential energy difference contoured at +/- 5 Kcal/mol.

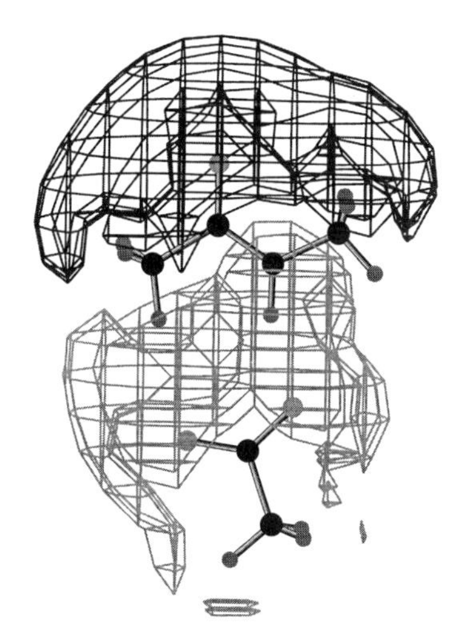

Figure 11. N-methyl acetamide (NH donor) / acetate ion. Minimum energy geometry. Electrostatic potential energy difference contoured at +/- 5 Kcal/mol.

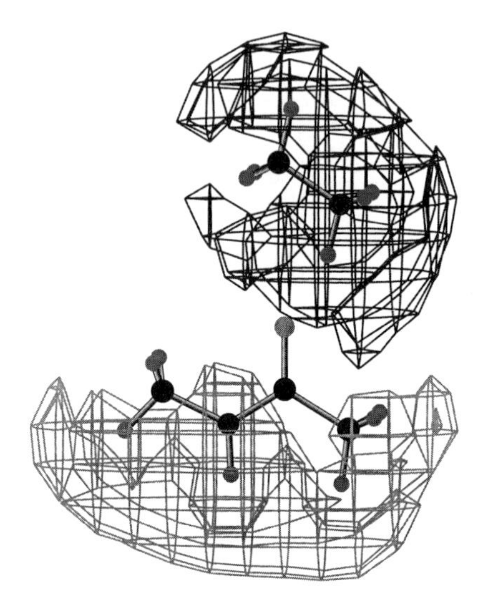

Figure 12. Methyl ammonium ion (NH donor) / N-methyl acetamide (carbonyl oxygen acceptor). Electrostatic potential energy difference contoured at +/- 5 Kcal/mol.

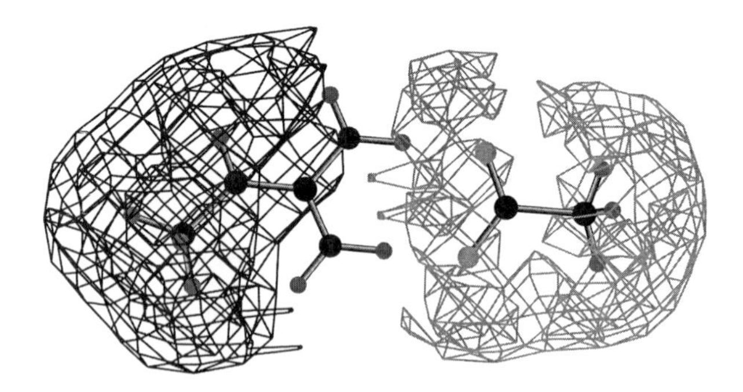

Figure 13. Methyl guanidinium ion / acetate ion. Both acetate oxygens forming a hydrogen bond in plane of guanidinium group. Electrostatic potential energy difference contoured at +/- 8 Kcal/mol.

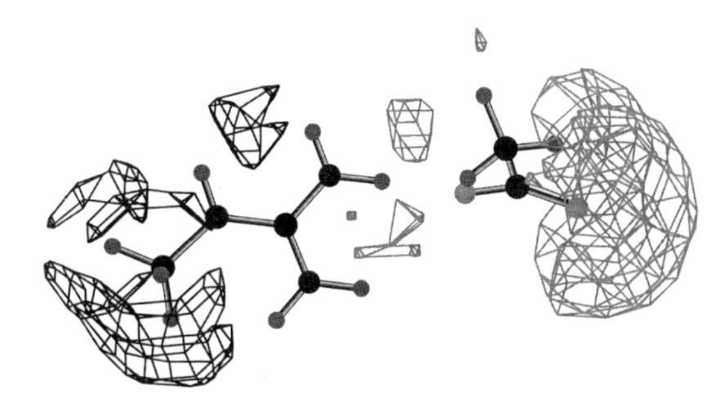

Figure 14. Methyl guanidinium ion / acetate ion. Acetate ion twisted out of plane of guanidinium group by 90°. Electrostatic potential energy difference contoured at +/- 5Kcal/mol.

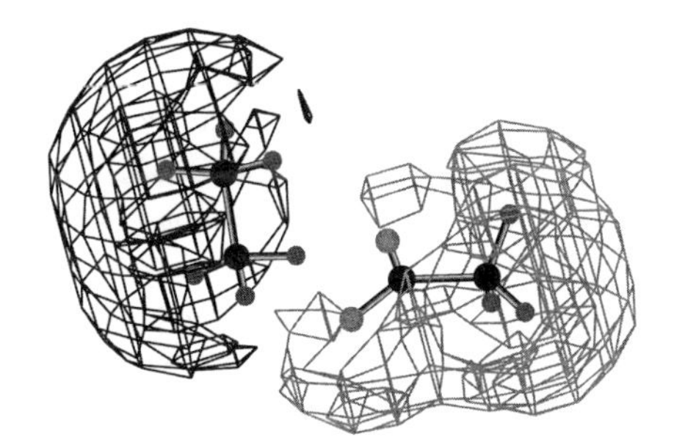

Figure 15. Methyl ammonium ion / acetate ion. Ammonium hydrogens forming two hydrogen bonds to acetate in a nearly symmetric conformation. Electrostatic potential energy difference contoured at +/- 10 Kcal/mol.

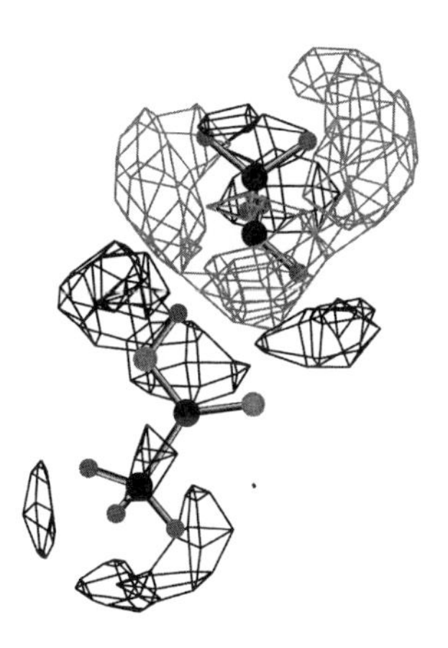

Figure 16. Acetic acid (OH donor) / methyl amine neutral pair. Electrostatic potential energy difference contoured at +/- 2 Kcal/mol

Although a large gradient might be expected to be manifested in changes of the charges on the atoms of the hydrogen bond this is not so, as may be seen in Table I. While there is only a very small change in the charge on the atoms of the hydrogen bond in the first six rows, only row six of these displays a large gradient graphically. Similarly, there is a large gradient for N-methyl acetamide / acetate but this is not reflected in the charges.

The size of the contour volumes is related to the magnitude of the complexation energy when it is large: contours are much larger for ion pairs or ion-neutral pairs. However within neutral pairs there is no clear relation. The complexation energy ΔEscf of V is 2.16 Kcal/mol lower than for I but the contours are much larger. Similarly, ΔEscf of VII is 3.29 Kcal/mol less than for I but the volumes of VII are bigger still. The most stable neutral pair, VI, is more stable than I by 0.42 Kcal/mol but also has a larger contour.

Table I. Changes in dipole moments, monomer total charge and atomic charges of hydrogen bonded atoms

#	Complex	$\Delta\mu$	$\Delta\mu_a$	μ_{REF}-μ	$\Delta(\Sigma q)_B$	$\Delta q(AH)$	$\Delta q(B)$
1	meoh-nma	0.66	5.5	0.80	0.045	0.015	-0.022
2	"	0.43	4.2	0.54	0.027	0.007	-0.030
3	"	0.78	5.4	0.96	0.054	0.045	-0.040
4	"	0.70	6.9	0.89	0.041	0.005	-0.016
5	nma-meoh	0.48	4.0	0.67	0.053	0.06	-0.07
6	im-nma	1.24	3.4	1.33	0.06	0.05	-0.10
7	nma-im	1.31	1.6	1.34	0.13	0.43	-0.23
8	meoh-im	0.58	6.0	0.84	0.11	0.34	-0.1
9	im-meoh	0.89	3.0	0.95	0.06	0.07	-0.13
10	nma-ac	*-3.52*	*17.1*	*3.89*	0.05	-0.037	0.08
11	nma-ac	*3.72*	*83.4*	*3.7*	0.07	-0.087	0.063
12	nma-amh	*-2.22*	*0.8*	*2.22*	0.14	0.001	-0.13
13	guan-ac-pl	-4.41	1.10	4.4	0.05	-0.054	0.16
14	guan-ac-tw	-5.51	0.6	5.51	0.25	0.067	0.16
15	ac-amh	-3.88	0.6	3.88	0.22	0.14	-0.05
16	ach-am	0.75	31.7	1.03	0.14	0.50	-0.15

$\Delta\mu$ = Change in dipole moment; $\Delta q(B)$ = Change in atomic charge on Acceptor atom. The first molecule of a pair is the hydrogen bond donor; nma = N-methyl acetamide; meoh = Methanol; im = Methyl imidazole; ac = Acetate ion; amh = Methyl ammonium ion; guan = Methyl guanidinium ion; ach = Acetic acid; amh = Methyl amine. Groups (1-4), (10,11) and (13,14) are alternate geometries

Charge Shift. A simple method of accounting for the perturbation resulting from complex formation might be to allow for a small degree of charge shift between the components of a complex. However the data would not suggest this as a viable method. No assignment of partial atomic charge is unique since charge is not an observable quantity. Despite this the charges from fitting to electrostatic potentials are sensible from a chemical reactivity perspective. Charges are fitted to a rigorously computable one electron property, the electrostatic potential, with the constraint that total molecular charge is conserved; therefore free monomers maintain their total charge in the uncomplexed state. Upon complex formation the fitting process only requires that the total charge of the complex be maintained. The monomers within a given complex are not constrained to preserve their original charge and when the monomer charges within a complex are summed (Table I) the total deviates from the original free monomer total charge. From the point of view of the molecular electrostatic potential this may seem to be a shifting of charge, particularly as it is in the direction of the donor becoming more positive. This shift can be large in the case of ionic complexes. For neutral pairs the deviations are much smaller except where imidazole is a donor and in an ion-neutral pair where N-methyl acetamide is an acceptor. It has been estimated from other methods (*28*) that the amount of charge transferred ranges from 0.01 to 0.03 electrons with the strong hydrogen bonds of a salt bridge exceeding 0.1 electrons. The current analysis, which reflects an alternative partitioning, generally agrees with this; however, the larger shift of charge in a neutral pair does not correlate with the strength of the hydrogen bond as determined by the quantum mechanical complexation energy. Complexes 7 and 8 and 1 show shifts of 0.13, 0.11 and 0.045 electrons respectively. Although complex 8 shows a larger shift than 1 it is less stable than complex 1 by 3.34 Kcal/mol. It is interesting that perturbation effects involving N-methyl acetamide are otherwise generally larger when it is a donor. Qualitatively the magnitude of the charge shift correlates with the magnitude of the perturbations based on the size of the contours.

Summary of Qualitative Results. The most important effects are the decrease in electrostatic potential about the acceptor and the increase about the donor with the impact on the donor often larger than the acceptor. The contours reveal that interactions with a complex will be perturbed by up to several Kcal/mol in the contact region between the complex and another molecule. Although the most important effects are overall changes in electrostatic potential, these effects are not uniform. When an imidazole is involved an alternant pattern of perturbation to electrostatic potential is observed, suggesting a polarization of one component by the other. The π system of imidazole would be expected be more polarizable. Morokuma (*29*) analyzed hydrogen bonding in terms of electron density differences. The total electron density difference reflects an overall similar view of hydrogen bonding with increases in electron density around the donor and decreases around the acceptor. When electron density was partitioned into exchange, polarization and charge transfer the latter two were seen as dominant. Electron density effects fall off much more rapidly than electrostatic potential effects and the impact of hydrogen bonding on force field calculations in their current implementations will be dominated by Coulomb terms that decrease at from r^{-1} to r^{-2} depending on the effective dielectric constant used making visualization of electrostatic potential differences more important in the region beyond the VDW zone.

Energetic consequences.

When charges used in a force field are derived from fitting to the molecular electrostatic potential representative fragments of molecules are frequently used or, when amino acids are studied, whole capped amino acids are used (*30, 31*). However, as seen above, there is substantial polarization of the electrostatic potential of the constituents of a hydrogen bonded pair. Therefore a pair's interaction energy with a third group should be altered by this polarization. The extent of this may be estimated by calculating the interaction energy of the bimolecular complex with another molecule first using the charge set of the complex, then using the monomer charge sets. The difference in EP interaction energy reflects the impact of hydrogen bond formation.

Most of the stabilization energy of a hydrogen bond pair can be described by the electrostatic component (*32*). This may be seen in Figures 17 and 18 where the quantum mechanical complexation energy is plotted against the classical Coulomb interaction energy of the complex, calculated with the monomer charges in a constant dielectric of 1 and a distance dependent dielectric. Although a distance dependent dielectric was used for the contouring, in keeping with the typical usage for protein simulations, this results in a poorer correlation between ΔEscf and ΔEc. The interaction energies of the bimolecular complexes with N-methyl acetamide were therefore calculated with a constant dielectric of 1.

The Coulomb interaction energy between the bimolecular complex and N-methyl acetamide was calculated with two charge sets, the reference set determined for the free monomers and the complex set determined for the complex as a whole. The reference interaction energy, ΔEqref, and the change upon substituting the charge set of the complex, ΔΔE, are shown in Table II. The magnitude of this perturbation is expressed as a percent of the reference interaction energy and is shaded light gray for ion-neutral pairs and darker gray for ion-ion pairs.

Table II. Effect of complexation of dimer on energy of trimer

#	Complexes	ΔEscf	ΔEqref	ΔΔE	%ΔE	Δμ	rrms(%)
1	meoh-nma	-7.59	-4.311	-0.249	5.8	0.66	8.6
2	"	-5.74	-4.304	-0.18	4.2	0.43	8.9
3	"	-5.43	-4.344	-0.212	4.9	0.78	7.9
4	"	-5.77	-4.296	-0.273	6.4	0.70	8.4
5	nma-meoh	-5.52	-5.339	-0.365	6.8	0.48	6.6
6	im-nma	-8.01	-5.496	-0.332	6.0	1.24	8.5
7	nma-im	-4.35	-6.084	-0.355	5.8	1.31	5.3
8	meoh-im	-6.96	-6.348	-0.317	5.0	0.58	8.3
9	im-meoh	-6.14	-5.465	-0.165	3.0	0.89	9.7
10	nma-ac	-20.61	-6.348	-1.351	21.3	*-3.52*	10.9
11	nma-ac	-22.91	-6.174	-1.532	24.8	*-3.1*	4.0
12	nma-amh	-29.74	-6.627	-1.389	21.0	*-2.22*	1.3
13	guan-ac-pl	-103.35	-9.908	2.016	-20.3	-5.51	1.5
14	guan-ac-tw	-76.85	-10.146	1.272	-12.6	-4.41	1.2
15	ac-amh	-123.34	-7.751	1.36	-17.5	*-3.88*	4.6
16	acH-am	-10.75	-4.658	-0.231	5.0	0.75	2.5

Abbreviations are the same as used in Table I. ΔE = Change in energy; rrms = relative rms error in fitted charges.

N-methyl acetamide was attached to each complex at 2.1Å from a free donor or acceptor site of the complex with linear hydrogen bond angles with no other close contacts. In complexes 1-4 the initial nma is a donor (NH) to the additional nma carbonyl oxygen. In complex 5 the second nma is a donor (NH) to the carbonyl oxygen of the initial nma. In complexes 6, 7, 8 and 9 methyl imidazole is hydrogen bonded, as donor in 7 and 8 and as an acceptor in 6 and 9, to the additional nma. In complexes 10 and 11 the second nma is a donor to the acetate oxygen. In complexes

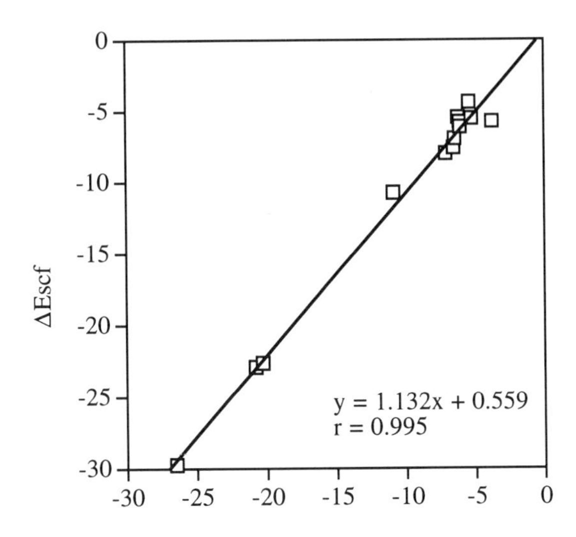

Figure 17. Plot of ab initio vs. classical Coulomb complexation energies (Kcal/mol). Classical energies evaluated from monomer based charges using a constant dielectric of 1.

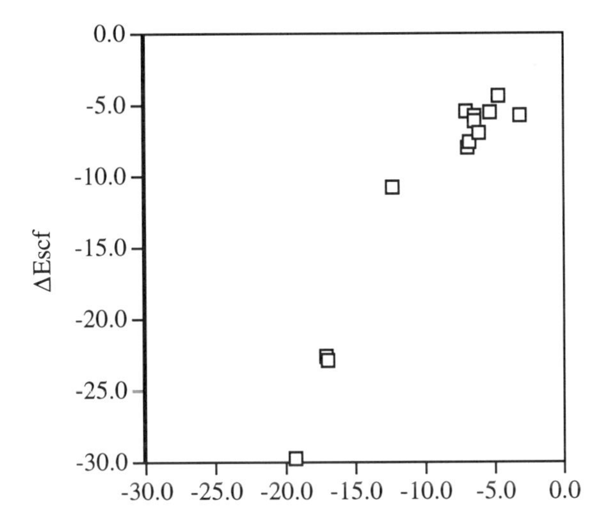

Figure 18. Plot of ab initio vs. classical Coulomb complexation energies (Kcal/mol). Classical energies evaluated from monomer based charges using a distance-dependent dielectric.

12 the second nma is an acceptor to NH as donor. Thus in complexes 1 through 12 the central molecule is a donor to one partner and an acceptor to the other. In complexes 13 and 14 guanidinium ion is a donor to both acetate and the added nma. In complexes 15 the acetate carbonyl group is an acceptor to both the added nma and to the ammonium ion. In complex 16 it is an acceptor to the added nma and a donor via the acidic proton to methyl amine.

The neutral pairs only generate a 3% to 7% perturbation on the third molecule, but the complexes involving ion-pairs or ion-neutral pairs show a much large perturbation of 12 to 25%. The effect of complexation upon the interaction energy of the complex with N-methyl acetamide is to increase the interaction energy, making the complex more stable, with most pairs, including the neutral-ion pairs. However, the ion-pair N-methyl acetamide trimer complexes are destabilized by substituting the charge set of the dimer complex.

This is in keeping with the observed changes in magnitude and direction of dipole moments of the bimolecular complexes upon electronic relaxation. In all but the ion-pair complexes the dipole moment increases due to the monomers inducing a dipole in each other. The larger complex dipole leads to a larger Coulomb interaction energy with a third group. In ion-pairs the dipole moment decreases with essentially no rotation, indicating a partial neutralization, leading to a decrease in the magnitude of the interaction with a third molecule.

These results may be compared with early observations on the non-additivity of complexation energies for water trimers (*33*). It was observed that when a central molecule of a trimer is both a donor and an acceptor in a sequential arrangement of hydrogen bonds the complex is stabilized relative to the components sums, but when the central molecule is either a double donor or a double acceptor the complex is destabilized. Complexes 13, 14 are double donors and complexes 10, 11 and 15 are variations of double acceptors where two atoms within the same molecule are accepting a hydrogen bond rather than the original classification of the same atom in a molecule accepting two hydrogen bonds. The current work deals only with the Coulomb interactions in a force field context and not SCF energies of trimers, but suggests that the non-additivities are unfavorable when ion pair components are involved such as in complex 13 or 14 which are strongly destabilized but double donors yet the double acceptors are stabilized, the neutal ion pairs much more so.

Conclusions

Quantitation of the difference contours by integration of the volumes is in progress, but qualitative conclusions are possible. The molecular electrostatic potential of an amide backbone, as modeled by N-methyl acetamide, is perturbed significantly by hydrogen bonding. The perturbation is not uniform and is strongly dependent on the side of the amide that is involved. It is much larger when the amide group is a donor than an acceptor. Transmission of this effect through the amide π-system also appears to be less important than inductive-type effects when the amide is an acceptor. The large polarizable π-system of methyl imidazole results in larger difference volumes than with N-methyl acetamide in both the other molecule of the pair and in methyl imidazole itself.

The destabilizing of ion-pair-X interactions would likely be attenuated on the surface of a protein where charged groups interact predominantly with water resulting in fewer ion-ion pairs and more neutral (i.e. water)-ion pairs. However, in the interior of a protein there are many more neutral pairs and neutral-ion pairs. These may stabilize the protein more than current models describe. Formation of an ion-pair is energetically very favorable in vacuo relative to separated ions, but not relative to the neutral pairs (*34*.) in a low dielectric. Complex 15 is -123.3 Kcal/mol more stable than the ions, but +8.25 Kcal/mol less stable than the corresponding neutral pair #16 in vacuo. Ion pair formation in the protein hydrophobic core, a low dielectric medium, where other interactions are possible, may decrease binding energy of the pair with other groups such that formation of two ion-neutral pairs by components of an ion-ion pair might be more favorable resulting in increased stabilization with other groups. Related work supports this conclusion (*35*). Thermodynamic integration experiments may help to determine the balance between factors resulting in stabilization and factors resulting in destabilization.

The magnitude of these effects, as described by the methodology currently used to generate fitted charges for a force-field, is limited by the quality of the original fitted charges. The rrms error in fitting to the quantum mechanical potentials at points around the molecule or complex is 5-10% (Table III). This reflects the limitation of a set of distributed monopoles in reproducing completely the molecular electrostatic potential. However, this magnitude of error has been seen to be satisfactory (*36*) for force-fields. The fitting error is due in part to incomplete sampling of the quantum mechanical potentials in the algorithm used to distribute points around the molecule, but even when much denser point distributions are used the error is reduced somewhat without altering significantly the values of atomic charge by simply adding more points to the fit. Another major source of error is the inability of atom-centered monopoles to completely describe the molecular electrostatic potential. Distributed multipoles on atom centers and non atom-centered charges are able to give better descriptions of the molecular electrostatic potential (*37, 38,39*) but these are not currently used in commonly available force-fields and are more expensive to use in evaluating Coulomb interactions. The ability of the charge set to reproduce the quantum mechanical potentials will be assessed by calculating difference potentials from the quantum mechanical potential energy grid directly and contouring these (in progress).

The main conclusions, that perturbation of ion-pair interactions with a third molecule is large, and that perturbation of other interactions are much smaller, is not altered by consideration of the rrms errors. They are very small for the ion-pairs. For the ion-neutral and neutral-neutral pairs the perturbation of the electrostatic interaction energy is on the order of the rms errors and a more detailed interpretation requires difference potentials calculated directly from the quantum mechanical potentials.

The polarization effects seen herein suggest that current force fields using atom centered charges with no accounting for polarization do not adequately describe the electrostatic environment. The impact of this on energy calculations is difficult to estimate due to competing effects. Hydrogen bonding involving ion-pairs results in a decreased stabilization of interaction with neighboring groups in the model studied here, whereas all other interactions result in a stabilization of interactions with other groups. Charged residues are found more often on the surface of a protein where they can be readily solvated resulting in ion-neutral pairs. Therefore the results suggest that interaction should be more attractive than currently implemented resulting in a protein structure that is not as compact as is should be. Mitigating this is the fact that charges are commonly derived from a 6-31G* basis set that overestimates dipole moments by 12% to 15%. This is larger than the magnitude of the perturbation for ion-neutral and neutral-neutral pairs which is 5% - 10% and may then be excessive. Average energetic consequences may not be large but need to be studied in greater detail with thermodynamic averaging techniques.

The polarizations resulting from hydrogen bond formation are anisotropic and while average energies may be adequately represented due to cancellation of errors, static descriptions of the electrostatic environment will clearly be perturbed by complexation relative to the model based on no polarization. Such a perturbation will have an effect on ligand binding to a receptor site; in particular the preferred binding orientation may depend on this perturbation and is not compensated for by averaging.

The simple two body model presented here suggests qualitative the effects to be considered, but multibody effects must need to be investigated.

Table III. Errors in fitted charges for monomers

Monomer	rrms(%)
CH3CO2H	6.7
CH3CO2-	1.1
CH3NH2	21.5
CH3NH3+	1.1
Methyl guanidinium	0.9
CH3OH	11.9
nma	7.3
Methyl imidazole	10.3

Literature Cited

1 Guo, H.; Karplus, M. *J Phys Chem* **1992**, *96*, 7273-7287.
2 Scheiner, S. In *Reviews in Computational Chemistry*; K. B. Lipkowitz and D. B. Boyd, Ed.; VCH Publishers, Inc.: New York, NY, 1991; Vol. 2; pp 165-218.
3 Reynolds, C. A.; Ferenczy, G. G.; Richards, W. G. *Theochem-J Mol Struct* **1992**, *88*, 249-269.
4 Mitchell, J. B. O.; Thornton, J. M.; Singh, J.; Price, S. L. *J. Mol. Biol.* **1992**, *226*, 251-262.
5 Mitchell, J. B. O.; Price, S. L. *J. Comp. Chem.* **1990**, *11*, 1217-1233.
6 Weiner, P. K.; Kollman, P. A.; Nguyen, D. T.; Case, D. A. *J. Comp. Chem.* **1986**, *7*, 230.
7 Brooks, B. R.; Bruccoleri, R. E.; Olafson, B. D.; States, D. J.; Swaminathan, S.; Karplus, M. *J. Comp. Chem.* **1983**, *4*, 187.
8 Dauber-Osguthorpe, P.; Roberts, V. A.; Osguthorpe, D. J.; Wolff, J.; Genest, M.; Hagler, A. T. *Proteins: Struc. Funct. & Genet.* **1988**, *4*, 31.
9 Clementi, E.; Corongiu, G.; Lie, G. C.; Niesar, U.; Procacci, P Motecc, E. Clementi Ed. **1989**, 363
10 Jorgensen, W. L., Chandrasekhar, J.; Madura, J. D.; Impey, R. W.; Klein, M. L. *J. Chem. Phys.* **1983**, *79*, 927
11 Colonna,F.; Angyan, J. G.; Tapia, O. *Chem. Phys. Lett.* **1990**, *172*, 55-61
12 Merz, K. M. *J Comput Chem* **1992**, *13*, 749-767.
13 Williams, D. E. In *Reviews in Computational Chemistry*; K. B. Lipkowitz and D. B. Boyd, Ed.; VCH Publishers, Inc: New York, NY, 1991; Vol. 2; pp 219-271.
14 Alkorta, I.; Villar, H. O.; Arteca, G. A. *J Comput Chem* **1993**, *14*, 530-540.
15 Westbrook, J. D.; Levy, R. M.; Kroghjespersen, K. *J Comput Chem* **1992**, *13*, 979-989.
16 Stouch, T. R.; Williams, D. E. *J Comput Chem* **1992**, *13*, 622-632.
17 Stouch, T. R.; Williams, D. E. *J Comput Chem* **1993**, *14*, 858-866.
18 Reynolds, C. A.; Essex, J. W.; Richards, W. G. *J. Am. Chem. Soc.* **1992**, *114*, 9075-9079.
19 Gaussian 92, Revision D.2, Frisch, M. J.; Trucks, G. W.; Head-Gordon, M.; Gill, P. M. W.; Wong, M. W.; Foresman, J. B.; Johnson, B. G.; Schlegel, H. B.; Robb, M. A.; Replogle, E. S.; Gomperts, R.; Andres, J. L.; Raghavachari, K.; Binkley, J. S.; Gonzalez, C.; Martin, R. L.; Fox, D. J.; Defrees, D. J.; Baker, J.; Stewart, J. J. P.; Pople, J. A. Gaussian, Inc.: Pittsburgh, PA, 1992.
20 Hariharan, P. C.; Pople, J. A. *Chem. Phys. Lett.* **1972**, *66*, 217.

21 Singh, U. C.; Kollman, P. A. *J. Comp. Chem.* **1984**, *5*, 230.
22 Montagnani, R.; Tomasi, J. *Theochem-J Mol Struct* **1993**, *98*, 131-138.
23 Sybyl 6.0, Tripos Associates: 1699 S. Hanley Road, Suite 303, St. Louis, MO 63144, 1993.
24 Solmajer, T.; Mehler, E. L. *Int J Quantum Chem* **1992**, *44*, 291-299.
25 Geissner-Prettre, C.; Pullman, A. *Theoret. Chim. Acta (Berl.)* **1972**, *25*, 83-88
26 Krijn, M. P. C. M.; Feil, D. *J. Chem. Phys.* **1988**, *89*, 4199-4208.
27 Duffy, E. M.; Severance, D. L.; Jorgensen, W. L. *J Am Chem Soc* **1992**, *114*, 7535-7542.
28 Scheiner, S. In *Aggregation Processes in Solution.*; E. Wyn-Jones and J. Gormally, Ed.; Elsevier: Amsterdam, 1983; pp 462-508.
29 Yamabe, S.; Morokuma, K *J Am Chem Soc* **1975**, *97*, 4458-4465.
30 Chipot, C.; Maigret, B.; Rivail, J. L.; Scheraga, H. A. *J Phys Chem* **1993**, *97*, 3452.
31 Williams, D. E. *Biopolymers* **1990**, *29*, 1367-1389.
32 White, J. C.; Davidson, E. R. *Theochem-J Mol Struct* **1993**, *101*, 19-31.
33 Hankins, D.; Moskowitz, J.W. *J. Chem. Phys.* **1970**,*53*, 4544-4554.
34 Scheiner, S.; Xiofeng, D. *Biophys. J.* **1991**, *60*, 874-883
35 Lockhart, D. J.; Kim, P. S. *Science* **1993**, *260*, 198-202.
36 Cieplak, P.; Kollman, P. *J Comput Chem* **1991**, *12*, 1232-1236.
37 Rauhut, G.; Clark, T. *J Comput Chem* **1993**, *14*, 503-509.
38 Sokalski, W. A.; Shibata, M.; Ornstein, R. L.; Rein, R. *Theor Chim Acta* **1993**, *85*, 209-216.
39 Sokalski, W. A.; Keller, D. A.; Ornstein, R. L.; Rein, R. *J Comput Chem* **1993**, *14*, 970-976.

RECEIVED July 29, 1994

Chapter 5

Competing Intra- and Intermolecular Hydrogen Bonds for Organic Solutes in Aqueous Solution

Peter I. Nagy[1], Graham J. Durant[1,3], and Douglas A. Smith[1,2,4]

[1]Center for Drug Design and Development, Department of Medicinal and Biological Chemistry, and [2]Department of Chemistry, University of Toledo, Toledo, OH 43606
[3]Cambridge Neuroscience, Cambridge, MA 02139

For molecules stabilized by intramolecular hydrogen bond(s) in the gas phase, there is a competition in aqueous solution between the intramolecular hydrogen bonds and those which may form to the water solvent. Whether the intramolecular hydrogen bond is maintained or disrupted for the most stable solute structure in solution depends on the chemical nature of the molecule. 1,2-Ethanediol, 2-hydroxybenzoic acids and the neutral and protonated conformers and tautomers of 2-(4(5)-imidazolyl)-ethylamine (histamine) are discussed from this respect in this paper. The structures of monohydrated polar molecules obtained in gas-phase optimization are compared with the solution structure around the solute in water for the 4-hydroxybenzoic acid and imidazole. Geometries have been optimized at the ab initio HF and MP2 levels using the 6-31G* basis set. The corresponding energies were obtained for several systems in single point calculations using the 6-311++G** basis set. Solution simulations and relative hydration free energies were obtained using the Monte Carlo method and the statistical perturbation theory utilized in the BOSS program.

A large number of molecules have polar groups in positions favorable for formation of intramolecular hydrogen bonds. Well known examples include vicinal diols and their monomethylethers, sugars, ortho disubstituted benzene derivatives, 1,2-disubstituted ethane derivatives, some hydroxy and dicarboxylic acids, etc. For these systems in aqueous solution there is a competition between forces which lead to intramolecular hydrogen bonding and intermolecular forces which lead to hydrogen bonds with the water solvent. Because chemical equilibria are typically much different in solution compared to the gas phase proper estimation of solvent effects is a key problem in the theoretical calculation of free energy changes upon solvation for the determination of the equilibrium mixture of the conformers in aqueous solution.

[4]Current address: Concurrent Technologies Corporation, 1450 Scalp Avenue, Johnstown, PA 15904

0097–6156/94/0569–0060$08.00/0

Despite the large number of solution phase calculations in the literature, the methods used can be classified fundamentally into one of two types. In the first, the solvent is represented by explicit molecules and the solution is considered as a large ensemble of solvent molecules around the solute(s), typically for Monte Carlo (*1,2*) or molecular dynamics calculations (*3,4*). The other main approach is to consider the solvent as a polarizable continuous medium and the solute is placed in a cavity created in the continuum solvent. There are several implementations for this latter approximation (*5-13*). Recent results suggest that combining continuum solvent calculations with ab initio calculations of the solute internal energy is successful when estimating the energy changes of physico-chemical processes in aqueous solution (*14,15*). The benefit of the continuum solvent calculations is the considerable reduction in computer time. No information can be obtained, however, in these cases about the structure of the solvent around the solute. In the present study the explicit consideration of the solvent molecules was chosen. It has been postulated that in explaining the possible disruption of the internal hydrogen bonds consideration of the solvent structure, at least in the first hydration shell of the solute, is extremely important.

Relative free energies of three molecules in the gas phase and in aqueous solution are presented here, based on original reports, to estimate their equilibrium mixtures in the two phases. 1,2-Ethanediol (*16,17*), 2-hydroxybenzoic acid (*18*) and histamine (2-(4(5)-imidazolyl)-ethylamine) (*74*) (Figure 1) can form intramolecular hydrogen bonds in their lowest energy gas phase conformations but not in certain other geometric arrangements. The question whether the disruption of these bonds and the concomitant conformational/tautomeric change takes place in aqueous solution is one subject of the present study. The first two molecules have been chosen because they are simple models for biologically important compounds (sugars, drugs containing salicylic acid as building block (*19-21*)) while histamine itself is of biological significance (22).

Two other questions are also addressed in this paper: *a*) how informative are the ab initio optimized monohydrate structures when studying solvation in bulk water, and *b*) if intramolecular hydrogen bonds are disrupted upon solvation, what is the driving force. Question *a* deserves some attention since monohydrates are used generally to obtain parameters for the intermolecular potential function used for simulations in water. An answer to question *b* would highlight whether a direct interaction between solute and solvent molecules is responsible for the change in the solute conformation leading to the disruption of the intramolecular hydrogen bond or if a restructured solvent around the solute can better accommodate the preferred conformation. In the latter case the conformational change may be considered driven by solvent-solvent rather than solute- solvent interactions. Radial and energy pair distribution functions have been analyzed, coordination numbers and numbers of hydrogen bonds were calculated and single snapshots and averages over several thousand snapshots were obtained to study the solution structure and to locate favorable hydration sites. These results, combined with the calculation of the relative free energies of the conformers and/or tautomers, give insight into the process when internally bound structures undergo conformational changes upon hydration.

Methods and Calculations

Methods applied for studying the three molecules are described in detail in the original reports (*16-18*). A brief summary of them may be appropriate, however, even here.

Ab initio calculations in the gas phase were carried out using the MONSTERGAUSS (*23*) and Gaussian 86-90 programs (*24*). Geometries for the isolated molecules were optimized at the HF/6-31G* level and single point energies were calculated for the different conformers/tautomers at the MP2/6-31G*//HF/6-31G* and MP2/6-311++G**//HF/6-31G* levels (*25-28*). Geometries of the monohydrates of imidazole (*29*), acetic and benzoic acids (*30*) were obtained at the MP2/6-31G* level and single point energies and basis set superposition error (*31,32*) corrected energy values were obtained at the MP2/6-311++G**//MP2/6-31G* and MP2/6-31G*//MP2/6-31G* levels, respectively.

Normal frequency analysis and thermal corrections to obtain free energies at 298 K and 1 atm were carried out using the rigid rotor-harmonic oscillator approximation (*33*). Normal frequencies obtained with the HF/6-31G* basis set were scaled by a factor of 0.9 (*25*) in calculating zero point energy (ZPE) corrections. The thermal corrections give the free energy contribution, G(T,p), between 0 and 298 K as a difference of the enthalpy, H(T,p) and entropy TS(T,p) terms: G(T,p) = H(T,p) - TS(T,p). In calculating the total relative free energies of the conformers/tautomers in the gas phase at 298 K and 1 atm the formula $\Delta G_{gas} = \Delta E(0) + 0.9\ \Delta ZPE + \Delta H(298, 1atm) - 298\ \Delta S(298, 1atm)$ was used.

Monte Carlo solution simulations were carried out for the isobaric-isothermal (NPT) ensembles (*34-36*) using the BOSS program of Jorgensen, versions 2.0-3.1 (*37*). 255-262 water molecules and one solute molecule were considered to simulate dilute aqueous solutions. The TIP4P model for the water molecules (*38,39*) and optimized solute geometries obtained from ab initio calculations considering, however, united CH_x (x=1-3) atoms, were used. Periodic boundary conditions and preferential sampling (*40,41*) were used considering $4.5\text{-}7\times10^6$ configurations for characterizing the solution structures. The 12-6-1 intermolecular potential was used with either the OPLS parameters (*42-44*) or those developed in the original reports (*16-18*). Cut-off radii were taken as 9.75 Å and 8.50 Å for the solute-solvent and solvent-solvent interactions, respectively.

Relative free energies and their components, relative enthalpy and entropy, were calculated in solution using the free energy perturbation method (*45*). Forward and backward simulations were carried out with a linear coupling parameter allowing changes in the geometric and potential parameters of 5-20% between their starting and final values. Free energy increments were obtained considering 3×10^6 configurations.

Structures and conformers studied here are outlined in Figure 1. Table I gives the relevant torsional angles and the symbols for the conformers. Letters t and g or g' refer to trans and gauche arrangements of the H-O-C-C atoms in 1,2-ethanediol. T and G have the corresponding meanings for the O-C-C-O torsional angle. Histamine g and t symbols refer to gauche and trans conformations about the C5-Cα-Cβ-N8 bond (τ2). 1H and 3H refer to the location of the tautomeric hydrogen atom on the N1 and N3 atoms, respectively. Structures protonated at the amine group are referred to by a + sign.

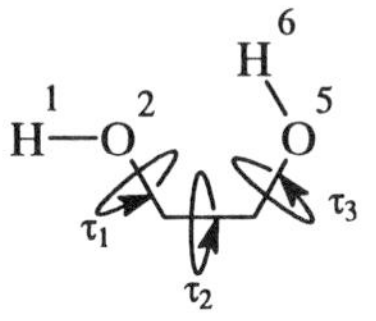

1,2-ethanediol

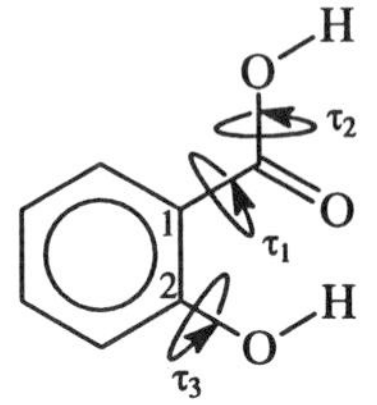

2-hydroxy benzoic acid

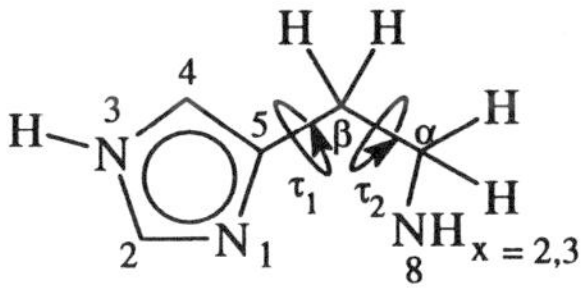

gauche 3H-histamine

Figure 1. 1,2 ethanediol. Torsional angles: H1-O2-C-C (τ1), O-C-C-O (τ2), C-C-O5-H6 (τ3). The conformer in the figure corresponds to τ1 = 180°, τ2 = 0°, τ3 =0°. τ1 and τ3 are designated by t (about 180°), g (about 60°) and g' (about -60°). τ2 is designated by T (about 180°) and G (about 60°). 2-hydroxybenzoic acid O=C-C1-C2 (τ1), H-O-C=O (τ2), H-O-C2-C1 (τ3). Structure **1** (in the figure): τ1=0°, τ2=0°, τ 3=0°, Histamine N1-C5-Cβ-Cα (τ1), C5-Cβ-Cα-N8 (τ2). Conformers are designated by g and t when τ2 is about 60° (gauche) and 180° (trans), respectively. NH_x group with x=2 and 3 refer to neutral and protonated histamine. Names of the protonated structures bear an additional + sign in Table VI.

Results and Discussion

Conformational/Tautomeric Equilibrium.

1,2-ethanediol. Six conformers are ordered in Table II by increasing ΔG_{gas} values. There are two low energy conformers, tGg' and gGg', with intramolecular hydrogen bonds. Structures lacking stabilization by intramolecular hydrogen bonding are higher in energy than the lowest energy form by 3.4-4.7 kcal/mol in energy and 2.8-4.0 kcal/mol in free energy. Consideration of the relative zero point energies decreases the energy separation by up to 0.5 kcal/mol. Thermal corrections of up to 0.3 kcal/mol in absolute value are generally of negative sign. The only larger value is 0.7 kcal/mol for the tTt conformer (a corrected value as compared to that published in ref. 16) where hindered rotations instead of vibrations were considered for the two H-O-C-C torsions. The sum of the relative ZPE and thermal correction is indicated by Δ G(T) in Table II. The calculated ΔG_{gas} results are in good agreement with those available from electron diffraction (*46*), microwave (*47*) and infrared (*48,49*) experiments suggesting a mixture of the tGg' and the gGg' forms in the gas phase.

Table I. HF/6-31G* optimized torsional angles.[a]

		τ1	τ2	τ3
1,2-ethanediol[b]				
	tGg'	-170.9	60.6	-53.9
	gGg'	76.0	59.7	-46.7
	gGg	44.0	49.0	44.0
	tTt	180.0	180.0	180.0
	tGg	-178.3	64.0	59.1
	tGt	-166.2	72.6	-166.2
2-hydroxybenzoic acid[c]				
	1	0.0	0.0	0.0
	2	180.0	0.0	0.0
	7	0.0	0.0	0.0
4-hydroxybenzoic acid[c,d]				
	3	0.0	0.0	0.0
histamine[e]				
	g1H	-46.6	67.7	
	g3H	-63.8	68.1	
	t3H	-66.9	177.5	
	t1H	-72.6	177.7	
	g3H+	-41.0	62.8	
	t3H+	-50.5	170.5	

[a]For definition of the τ1, τ2 and τ3 torsional angles, see Figure 1 and the footnote in it. [b]Refs. 16, 17. [c]Ref. 18. [d]τ3 = H-O-C4-C3. [e]Ref. 74.

In aqueous solution all conformers are favored by solvent effects as compared to tGg', the structure most stable in the gas phase. Though some conformers are without intramolecular hydrogen bonding and thus can be better hydrated, the disruption of this bond in aqueous solution is not a unique structural change governing the relative hydration free energy. A small stabilization is observed going from tGg' to gGg without internal hydrogen bond, while gGg', another structure with intramolecular hydrogen bonding is stabilized by almost 1 kcal/mol. A similar value of 1.22 kcal/mol was obtained upon the disruption of the internal bond for the tTt form that can be reached, however, by a basic G to T rotation about the C-C bond. Relatively small geometric changes lead to stabilization by 3.4-5.2 kcal/mol for tGt and tGg. The number of the solute-water hydrogen bonds, H_b, increases slightly with the disruption of the internal hydrogen bond and the more negative ΔG_{sol} values. This index may be important despite its small changes in the series, since energy pair distribution functions for 1,2-ethanediol (*16,17*) show interaction energies of -8 to -3 kcal/mol for diol-water pairs with intermolecular hydrogen bonding. Formation of a new solute-water intermolecular hydrogen bond should decrease the total solute-solvent interaction energy of about 5 kcal/mol on average, thus an increase of H_b by about 1 unit for tGg is in line with the ΔG_{sol} value of -5.2 kcal/mol and a stabilization free energy of 0.95 kcal/mol for gGg' is close to the value proportional with the increase of 0.3 H_b. On the other hand, small ΔG_{sol} values as compared to their H_b values for gGg and tTt indicate the importance of other factors, too.

Table II. Relative energy terms for 1,2-ethanediol in the gas phase and in solution.[a,b]

	ΔE	$\Delta G(T)$	ΔG_{gas}	ΔG_{sol}	ΔG_{tot}	H_b
tGg'	0.00	0.00	0.00	0.00	0.00	3.5
gGg'	0.24	0.05	0.29	-0.95	-0.66	3.8
gGg	3.55	-0.75	2.80	-0.24	2.56	4.0
tTt	3.36	0.18	3.54	-1.22	2.32	4.0
tGg	4.65	-0.64	4.01	-5.23	-1.22	4.4
tGt	4.45	-0.43	4.02	-3.40	0.62	4.2

[a]Energies from ref. 16,17 with corrected $\Delta G(T)$ value for tTt. $\Delta E = \Delta$ (MP2/6-31G*//HF/6-31G*), $\Delta G(T) = \Delta$ (0.9*ZPE + H(T) - TS(T)) (T=298 K), $\Delta G_{gas} = \Delta E + \Delta G(T)$, $\Delta G_{tot} = \Delta G_{gas} + \Delta G_{sol}$. Values in kcal/mol. H_b is the number of the solute-solvent intermolecular hydrogen bonds. [b]Refs. 16, 17.

The total relative free energy, ΔG_{tot}, is most negative for tGg and indicates a prevailing O-C-C-O heavy atom linkage without intramolecular hydrogen bonding in aqueous solution. The calculations (*16,17*) predict 64% tGg, 25% gGg' and 8% tGg' in the equilibrium mixture. Solution NMR experiments by Chidichiomo, et al. (*50*) found gauche O-C-C-O skeleton for the solute in lyotropic liquid crystalline solution while Pachler and Wessels found 12% trans conformer in heavy water (*51*). Raman spectroscopic results by Melaknia, et al. (*52*) were interpreted as revealing "virtually complete absence of any intramolecular hydrogen bonds" in aqueous solution. The calculated equilibrium mixture is in only partial agreement with the experimental results. The theoretical results do not indicate a considerable trans conformational fraction and do describe 33% of the conformers with intramolecular hydrogen bond. Thus the most important result of the combined ab initio + Monte Carlo method for 1,2-ethanediol is that stable conformers exist in the gas phase only with intramolecular hydrogen bonds while the most stable and prevailing structure in solution still maintains the gauche O-C-C-O linkage even though the intramolecular hydrogen bond is disrupted.

2-hydroxybenzoic acid. No experimental results are available with which to compare the calculated values in Table III. The theoretical results are in line with those for 1,2-ethanediol and histamine (see next section): the most stable conformer of the isolated molecule is stabilized by an intramolecular hydrogen bond. For 2-hydroxybenzoic acid the most stable form is structure **1** with the syn carboxylic group and the phenolic OH group making a hydrogen bond to the carbonyl oxygen (*18*). Comparison of this structure with the most stable form of the isomeric 4-hydroxybenzoic acid, **3**, which can not form an intramolecular hydrogen bond, is informative. The MP2/6-31G*//HF/6-31G* relative energies and total free energies in the gas phase are 5.2 and 4.4 kcal/mol, respectively. Thus, the inability of **3** to form a hydrogen bond leads to a considerable increase in the total internal energy. The C=O...H-O(phenolic) hydrogen bond is also disrupted in structure **7**, and the relative energy increases to 11.1 kcal/mol. The significantly larger relative value as compared to **3** is explained by the unfavorable orientation of the oxygen lone pairs in the phenolic group after rotating the hydrogen by 180° when going from **1** to **7**. No such problem with the oxygen lone pairs exists in **3**. Structure **2** is the second lowest energy planar conformer of 2-hydroxybenzoic acid. Though it also has an internal hydrogen bond, its relative free energy is 2.9 kcal/mol. This considerably increased value in the gas phase emphasizes the importance of the type of the hydrogen bond created within the molecule. The $\Delta G(T)$ values are negative for structures in Table III and decrease the ΔE separations by 12-16% to form the relative ΔG_{gas} values.

Table III. Relative energy terms for 2- and 4-hydroxybenzoic acids in the gas phase and in solution.[a,b]

	ΔE	$\Delta G(T)$	ΔG_{gas}	ΔG_{sol}	ΔG_{tot}	H_b
1	0.00	0.00	0.00	0.00	0.00	2.0
2	3.28	-0.38	2.90	-1.46	1.44	2.7
7	11.10	-1.47	9.63	-5.48	4.15	4.2
3	5.21	-0.86	4.35			4.3

[a]For definitions of the energy terms, see Table II. [b]Ref. 18.

Del Bene has pointed out (*53,54*) the importance of diffuse functions in the basis set used for describing hydrogen bonds between simple polar molecules. Diffuse functions in the basis set reduced the binding energies of the complexes by about 1 kcal/mol depending on their chemical character. Application of the 6-311++G** basis set instead of the 6-31G* at both the HF and MP2 levels for the histamine conformers (see next section) led to a decrease in the relative energies of 0.7-1.0 kcal/mol for conformers *without* intramolecular hydrogen bond, but the change in the stabilization among different conformers *with* intramolecular bond was no more than

0.3 kcal/mol. This finding suggests that use of diffuse functions is important for proper estimation of the energy separation of conformers with *versus* without intramolecular hydrogen bonds, while calculations based on the 6-31G* basis set are acceptable for conformers with different types of internal hydrogen bonds. Since the relative energies of conformers without these hydrogen bonds are considerable for both the 1,2-ethandiol and hydroxybenzoic acid systems, an overestimation of their relative energies by about 1 kcal/mol does not affect the qualitative conclusions concerning their participation either in the equilibrium gas phase or solution mixture.

Solvation relative free energies for **1**, **2** and **7** show that hydration favors more the structures which are less stable in the gas phase. Thus **7**, without an intramolecular hydrogen bond, is stabilized by 5.5 kcal/mol and **2**, with a probably less favorable intramolecular hydrogen bond than in **1**, is stabilized by 1.5 kcal/mol. In contrast to the case of the 1,2-ethandiol conformers, the stabilization due to hydration is not enough to stabilize any structure without intramolecular hydrogen bonding so much that it becomes the prevailing form in solution. Instead, **1** is predicted as the most abundant form even in solution. This conclusion appears still valid even allowing for the decrease in ΔG_{gas} by some tenths of a kcal/mol for **2** due to the basis set effect discussed above. The importance of the solute-water hydrogen bond formation is seen by considering the H_b values (Table III) which are in line with the change in the ΔG_{sol} term: the larger number for H_b leads to proportionally more negative ΔG_{sol} values.

Histamine. Four neutral and two protonated conformers/tautomers are compared in Table IV. In the gauche conformations histamine is capable of forming intramolecular hydrogen bonds that are not possible in the trans forms. Experimental results suggest a complicated equilibrium of the different conformers and tautomers which include both protonated and neutral forms in aqueous solution. The molecule has two basic centers: the side chain NH_2 group (pK_a=9.40) and the basic ring nitrogen (pK_a=5.80) (*55*). In aqueous solution at pH = 7.4 almost 97 percent of the solute molecules are in the monocation form, protonated at the amine group (*56*). X-ray diffraction experiments for the neutral (*57*), monocationic (*58*), and dicationic (*59,60*) histamine structures all found the extended, trans conformation in the crystalline phase. On the contrary, the structurally related L-histidine hydrochloride salt (*61*) and the neutral L-histidine zwitterion (*62,63*) take the gauche arrangement with an intramolecular hydrogen bond in the latter. Based on the experimental results consideration both of the neutral and protonated forms seemed to be relevant. There is no possibility of intramolecular hydrogen bond formation in the dicationic form, thus this structure was disregarded.

The neutral forms, the lowest energy gauche and trans conformers for the 1H and 3H tautomers, were selected based on previous 3-21G optimized structures of Vogelsanger et al. (*64*) Protonated structures were chosen based on the more stable neutral forms in aqueous solution.

Gauche structures g1H and g3H are the lowest energy neutral forms in Table IV. There are two different hydrogen bonds in these: N(ring)-H···N(amine) for g1H and N(ring)···H-N(amine) for g3H. As was discussed (*74*), the former hydrogen bond type is stronger and is considered here as the primary reason for the lower energy, 2.5 kcal/mol at the MP2/6-311++G**//HF/6-31G* level, of g1H as compared to g3H. Trans structures, t3H and t1H, without internal hydrogen bonding are higher in energy than even the more weakly hydrogen bound g3H by 0.9 and 1.8 kcal/mol, respectively. The ΔG(T) terms are considerable for these conformers and decrease the ΔG_{gas} values by 25-33% compared to ΔE.

Table IV. Relative energy terms for histamine in the gas phase and in solution.[a,b]

	ΔE	ΔG(T)	ΔG_{gas}	ΔG_{sol}	ΔG_{tot}	H_b
g1H	0.00	0.00	0.00	0.00	0.00	4.7
g3H	2.53	-0.73	1.80	-3.74	-1.94	4.7
t3H	3.40	-1.12	2.28	-5.36	-3.08	5.0
t1H	4.28	-1.08	3.20	-4.48	-1.28	5.2
g3H+	0.00	0.00	0.00	0.00	0.00	3.8
t3H+	14.20	-1.17	13.03	-10.96	2.07	4.6

[a]ΔE = Δ (MP2/6-311++G**//HF/6-31G*). For definitions of the other terms, see Table II. [b]Ref. 74.

Solvation terms are large and display an opposite trend than the internal energies: high free energy trans species, t1H and t3H are stabilized the most upon solvation. As a result t1H, without intramolecular hydrogen bond, becomes favorable with respect to the g1H form. Furthermore, t3H becomes the most stable form of the four in aqueous solution. These are cases where the solvation compensates for the increase in the internal energy upon the change in the conformation and tautomeric form going from g1H to t3H. The g3H structure is also more stable in solution than g1H. In g3H both the 3H proton of the ring and the lone pair region of the amine group are exposed to hydration. Solution structure simulations for imidazole (*29*) and methylamine (*65-67*) proved that these sites are more preferred for hydration than the basic N3 of the ring and the N-H bond of the amine open to hydration in g1H. H_b values are not informative in this case. They are equal for g1H and g3H and is larger by only 0.3 units for t3H even though this latter structure has two further sites formally open for hydration. The discrepancy due to the similar H_b values can be explained by assuming that, although the numbers of the solute-water hydrogen bonds are similar, their strength should be different for the three structures. In fact, distribution functions indicate differences both for the pair energies and for the distribution of the water molecules around the hydration sites of the three structures.

Protonated structures g3H+ and t3H+ differ in the gas phase by 14.2 kcal/mol in energy and 13.0 kcal/mol in free energy. The large stabilization of the g3H+ form is reasonable: the protonated amine group enters into a hydrogen bond to distribute the excess positive charge. In t3H+ the charge must largely be localized in the NH_3^+ cationic head and solvation stabilizes this conformer by 11.0 kcal/mol. In this extended conformation two new hydration sites are available, one of which allows the hydration of a N-H bond in the positively charged group. The sum of the ΔG_{gas} and ΔG_{sol} terms remains positive, 2.1 kcal/mol, indicating the preference of the g3H+ form over the t3H+ form in aqueous solution at a pH about 7. Thus, for the protonated histamine the maintenance of the intramolecular hydrogen bond is favored over formation of more hydrogen bonds to the water in the trans form but with considerably higher internal free energy.

The theoretical results for the protonated forms are in good agreement with those obtained from solution NMR experiments (*56*). Interpretation of the latter suggests that the trans form comprises about the 45% of the equilibrium mixture in 0.4 M D_2O solution at pH=7.9. Calculated equilibrium mixture (*74*) contains 34% t1H+ form (not discussed here), 2% t3H+ and 64% g3H+, meaning a total trans fraction of 36%. vs. the experimental prediction of 45%.

Monohydrate Structures - Solution Structures. In this section monohydrate structures will be contrasted with solution structures in which the solute is surrounded by a large number of water molecules. The changes in the intermolecular hydrogen bonding pattern are the primary consideration of this comparison.

Table V shows considerably different binding energies for different monohydrates. The OPLS intermolecular potential was used to calculate this energy in some selected (non optimized) geometries for 1,2-ethandiol-water dimers with different conformations of the organic solute. These structures may be, however, close to the lowest energy arrangements as concluded from the onset values of the energy pair distribution functions (*16,17*). The onset values, -6.5 kcal/mol for tTt and about -7.3 kcal/mol for tGg' and gGg', indicate slightly different binding energies for the monohydrated 1,2-ethanediol depending on the solute conformation. Furthermore, calculations indicate some differences in the binding energy for dimers with donor or acceptor water. Averages over snapshots in solution simulations find that hydrogen bonds with acceptor water are present in significantly larger percentage than when water molecules donate a proton. (*16,17,67*). These result suggest more stable hydrogen bonds in the former arrangement.

Ab initio binding energies with optimized geometries are available for the acetic acid, benzoic acid and imidazole monohydrates (Table V). Figure 2 shows the HF/6-31G* and MP2/6-31G* optimized benzoic acid monohydrate geometries with syn and anti conformations of the acid group. Rather similar structures were obtained for the corresponding acetic acid hydrates (*30*). The O(ac)-H(ac)···O(w) distance is larger at both optimization levels with the anti acid group and a second, relatively long (2.0-2.1 Å) and highly bent bond can be formed between the water hydrogen and the carbonyl oxygen atoms with the syn acid group that is missing in the anti conformation. This distorted six membered ring structure is considered as the main source of the binding energy difference for the acid - monohydrate conformers. Table V indicates that the binding energies of the syn-acid monohydrates are more negative than those of the anti form by 0.9-1.2 kcal/mol for acetic acid and 1.6-1.7 kcal/mol for benzoic acid considering MP2/6-311++G**//MP2/6-31G* or basis set superposition error corrected MP2/6-31G*//MP2/6-31G* values (30). Partial

Table V. Energies of the monohydrate formation in kcal/mol.

	MP2/6-311++G**// MP2/6-31G*	MP2/6-31G*// MP2/6-31G*	OPLS
1,2-ethanediol[a]			
tTt			-5.9, -6.3
tGg'			-6.4, -6.6
gGg'			-7.0
Acetic acid[b]			
syn	-10.0	-14.3 (-9.3[c])	-8.8[d]
anti	-8.8	-10.8 (-8.4[c])	-8.2[d]
Benzoic acid[b]			
syn	-10.4	-14.5 (-9.6[c])	
anti	-8.7	-11.0 (-8.0[c])	
Imidazole[e]			
N1	-7.6	-8.6 (-6.8[c])	-8.3
N3	-7.4	-9.7 (-5.9[c])	-5.6

[a]Ref. 16. [b]Ref. 30. [c]BSSE corrected value. [d]Ref. 68. O-H···Ow kept linear. [e]Ref. 29.

geometry optimization of the acetic acid - water dimers (when the O-H···Ow bond was kept linear) (*68*) using the OPLS pair potential leads to a binding energy of 0.6 kcal/mol more negative for the syn monohydrate. Even in this case where the Hw···O= interaction should be weak due to the geometry restriction, the more favorable binding of a water molecule in the syn arrangement is demonstrated. However, Monte Carlo simulations for the acetic acid conformers in aqueous solutions using electrostatic potential fitted charges (*68*) find more favorable hydration of the anti form by about 4 kcal/mol. A solution simulation using combined quantum mechanical/molecular mechanics interaction potential (*69*) predicts that the gas phase preference of the syn form by 5.9 kcal/mol drops in relative free energy to about 1.1 kcal/mol upon solvation in bulk water. Considering that the syn monohydrate is more stable than the anti form by more than 7 kcal/mol (*30*) this means that the monohydration slightly increases the difference between the syn and anti conformers while it will be considerably reduced in dilute aqueous solution. These results indicate that calculated conformers energies based on the total geometry optimized structures of monohydrates may be useful in studying gas-phase dimers but suggest hydration effects in contrast to that in bulk water.

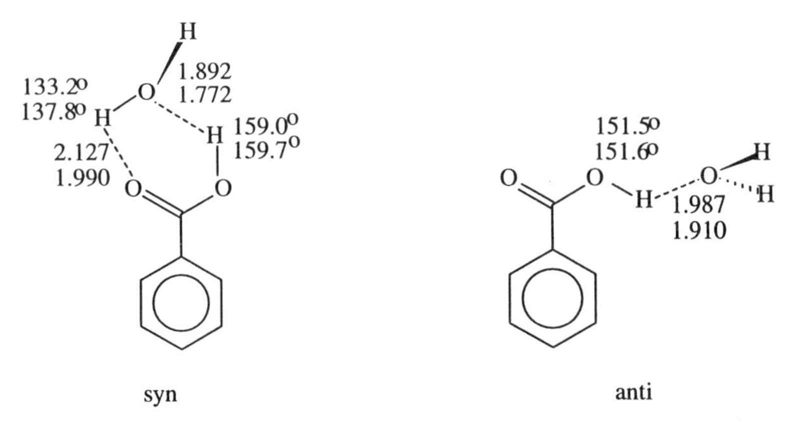

Figure 2. Syn and anti benzoic acid monohydrates. Upper and lower values for the intermolecular geometric parameters were obtained in optimizations at the HF/6-31G* and MP2/6-31G* levels, respectively.

Snapshots of 4-hydroxybenzoic acid in aqueous solution (Figure 3) show separately hydrated carboxylic OH and carbonyl group. Thus, if there are a large number of water molecules for hydration they preferentially hydrate the polar sites with nearly linear hydrogen bonds rather than form bifurcated structures with at least one strongly bent intermolecular hydrogen bond (*18,30,67,70,71*). This conclusion is not based on considering a simple snapshot but on the statistical analysis of 4000 snapshots used to locate the most frequent positions of the water molecules around the hydroxybenzoic acid solute (*18*). (The hydration pattern for the 4-hydroxy group is similar to that found for alcohols (*16,17,66,67*): one acceptor and two more loosely bound donor water molecules.)

Monohydration of imidazole (*29*) at the N1 rather than N3 hydration site seems to be preferred by 0.2-0.9 kcal/mol based on the best results in Table V. The table shows that only BSSE corrected MP2/6-31G* values are in close agreement with the MP2/6-311++G** binding energies. The OPLS calculations with parameters fitted to obtain good experimental measurables (liquid density, heat of vaporization) give results in qualitative agreement with the ab initio ones. HF and MP2 geometry optimizations using the 6-31G* basis set give an almost linear O(w)¨H-N1 and a considerably bent N3¨H(w)-O(w) hydrogen bond for the two monohydrates. Furthermore, the O(w)¨H(N1) distance is shorter by about 0.07 Å at both levels than the N3¨H(w) separation. The shorter and more linear hydrogen bond is assumed to be stronger, providing a rationale for the more negative binding energy.

Monohydrates of imidazole without an evident polar site nearby to N1 or N3 do not demonstrate bifurcated structures (*29*). The constant deviation of the N3¨H(w)-O(w) bond from linearity by 30° at both the HF and MP2 levels leaves, however, some suspicion with respect to a non polar neighbor to N3. More interesting is that in

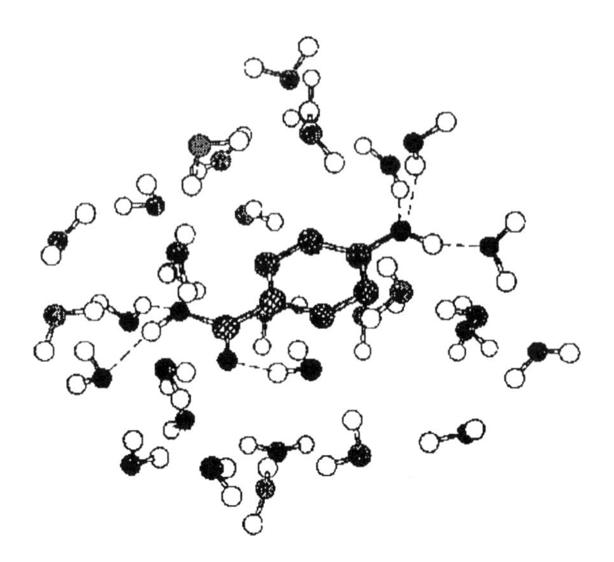

Figure 3. Snapshot for the aqueous solution of 4-hydroxybenzoic acid. Hydrogen bonds are indicated by dashed lines.

(Reproduced with permission from reference 18. Copyright 1993 American Chemical Society.)

the imidazole N3 dihydrate (Figure 4) the N3···H(w)-O(w) bond deviates from the linear by less than 20° and the N3···H(w) distance shortens by 0.1 Å! The C2 atom, in between two nitrogen atoms, bears a large positive charge and attracts the negative (oxygen) site of the water in the *monohydrate* resulting in a stretched and fairly bent N3···H(w)-O(w) bond. In the *dihydrate* a more flexible arrangement is possible: the first water moves closer (by 0.1 Å) to the N3 site and a less bent hydrogen bond becomes possible. The interaction with the C2 site will be the primary role of the second water (*29*).

Thus, if there is a center of large, opposite polarity close to the primary hydration site then a single water interacts with both sites either forming two hydrogen bonds (acetic and benzoic acids) or one hydrogen bond and a strong interaction with the second polar site (imidazole N3). In both cases the geometry moves toward a distorted ring structure. This can be resolved by adding further water molecules to the system. Already the second water has large effect, but the rigid and ordered structure disappears in solution as demonstrated by the snapshot for the imidazole solution (Figure 5) and by the statistical analysis over 4000 solution configuration (*29*).

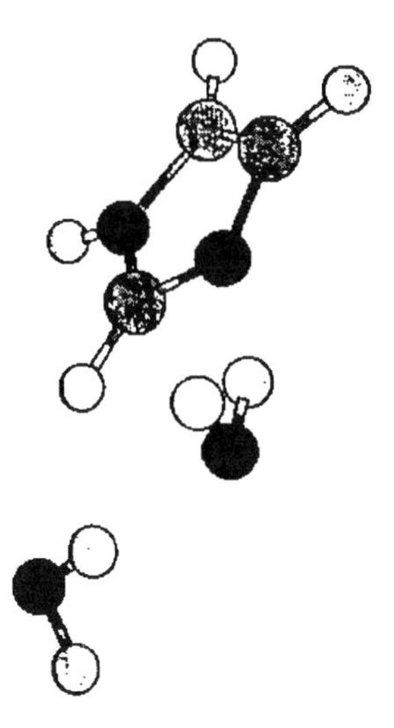

Figure 4. HF/6-31G* optimized structure of the imidazole dihydrate hydrating the N3 site.
(Reproduced with permission from reference 29. Copyright 1993 American Chemical Society.)

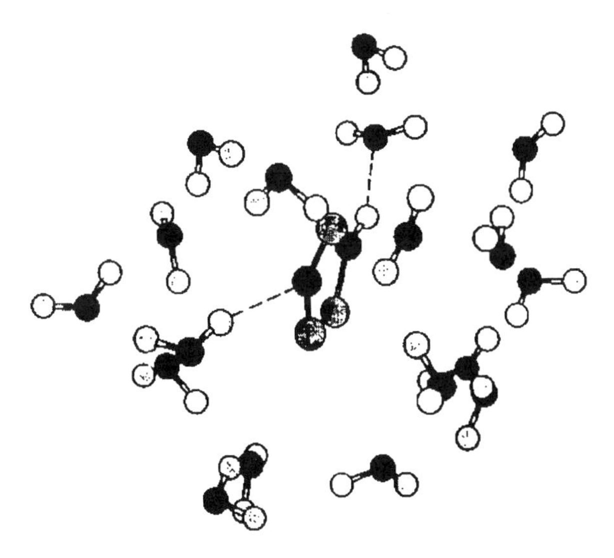

Figure 5. Snapshot for the aqueous solution of imidazole. Hydrogen bonds are indicated by dashed lines.
(Reproduced with permission from reference 29. Copyright 1993 American Chemical Society.)

Solvation Thermodynamics. In this section we analyze the calculated ΔG, ΔH and TΔS terms upon the solvation of different conformers to find the driving force for the geometric changes of some solutes in water.

As was discussed in connection with the disruption of the intramolecular hydrogen bond for the three solutes under study, the structures most favored by hydration have the largest number of the solute-water hydrogen bonds, H_b. While they may not all be of maximal strength, a rough correlation was found between the increasing number of the solute-water hydrogen bonds and the relative hydration free energy. This correlation suggests a driving force to maximize the number and strength of the solute-water hydrogen bonds, thereby achieving the maximum negative solute-solvent interaction energy. This may be opposed, however, by the energetic requirement for reorganization of bulk water leading to the increase of the solvent-solvent interaction energy. Thus, since $\Delta H = \Delta E_{SX} + \Delta E_{SS}$, where the two terms are the changes in the solute-solvent and the solvent-solvent energies, respectively, optimization of ΔH is more likely than maximization of the number of the solute-water hydrogen bonds. The other factor affecting the change in the relative free energy is the TΔS entropy term which has no such clear connection to simple terms and can only be calculated indirectly from the ΔH - ΔG difference. (ΔH is determined by umbrella sampling separately.) TΔS represents the disorder of the molecules in solution and is found to be weakly correlated with the solvent reorganization energy for some systems (*72,73*).

The data in Table VI lead to two general observations. First, the σ values (standard error) in parentheses are some tenths of a kcal/mol for ΔG's and amount generally to no more than 10% of the data while σ is some kcal/mol both for ΔH and TΔS and is commensurate with the data itself in many cases. Second, due the relationship $\Delta G = \Delta H - T\Delta S$, if ΔH and TΔS are of the same sign, as happens for most transformations in Table VI, they are of compensating nature for ΔG. Due to the rather large values, and correspondingly rather large errors for ΔH and TΔS, ΔG is a small difference of two large and uncertain numbers. Luckily, ΔG is obtained directly from the free energy perturbation calculations while its components ΔH and TΔS are estimated with the error indicated in Table VI. Accordingly, these results allow an estimate of the entropy or enthalpy dominance only in some cases.

Errors in ΔH and TΔS are larger than the data themselves for 1,2-ethanediol in the tGg' to tGt and tGg' to tTt transformations. Only for the transformation tGg' to tGg a prediction of ΔH dominancy seems more sound. The small geometric change would not require large reorganization of the solvent structure, in line with the small TΔS term. Also, the nearly unit increase in H_b for this transformation makes the interpretation, concluding the ΔH dominancy, reasonable.

Though the relative errors for many terms in Table VI are large, enthalpy or entropy dominancy can be determined even considering these uncertain values provided the same sign for ΔH and TΔS. This is fulfilled with one exception for the further transformations. Considering the transformation processes with negative ΔG, the process is enthalpy dominated if both the ΔH and TΔS terms are of negative sign and is entropy dominated if both terms are of positive sign. Due to the change of sign of the entropy term in calculating the free energy change, $\Delta G = \Delta H - T\Delta S$, a negative ΔG can only be produced if the absolute value of ΔH is larger than that of TΔS having a common negative sign. The opposite applies to cases when both terms have positive sign. This means that even though the error limits would allow for the reversal of the order of the ΔH and TΔS terms (taking e.g. one at the upper limit and the other at the lower limit defined by the error) only those combinations are possible that produce the given ΔG. This requires, however, the same deviation from the average values for the enthalpy and entropy terms. Thus, although the relevant ΔH and TΔS values may be uncertain according to the standard errors in parentheses, their difference should be maintained. (No such a restriction can be imposed when the two terms have opposite signs, as is seen for 1,2-ethanediol.)

In the **1** to **2** transformation for the 2-hydroxybenzoic acid conformers the forward simulation predicts enthalpy dominancy while the backward simulation is not

Table VI. Thermodynamics of the structural changes in aqueous solution.[a]

	ΔG	ΔH	TΔS
1,2-ethanediol[b]			
tGg' to tGg	-5.2 (0.3)	-4.3 (3.1)	0.9 (3.1)
tGg' to tGt	-3.4 (0.3)	-1.6 (3.3)	1.8 (3.3)
tGg' to tTt	-0.8 (0.5)	0.2 (3)	1.0 (3)
	-1.6 (0.5)		
2-hydroxybenzoic acid[c]			
1 to **2**	-1.1 (0.2)	-3.3 (1.9)	-2.2 (1.9)
	-1.8 (0.2)	-1.1 (2.1)	0.7 (2.0)
1 to **7**	-4.6 (0.4)	14.6 (3.9)	19.2 (4.0)
	-6.4 (0.5)	-23.8 (4.9)	-17.4 (5.0)
histamine[d]			
g1H to g3H	-3.4 (0.3)	3.0 (3.5)	6.5 (3.5)
	-4.1 (0.3)	-8.0 (3.1)	-3.9 (3.1)
g3H to t3H	-0.7 (0.3)	-28.7 (4.5)	-28.0 (4.5)
	-2.5 (0.3)	-11.8 (4.5)	-9.3 (4.6)
g3H+ to t3H+	-10.6 (0.3)	4.7 (4.2)	15.3 (4.2)
	-11.3 (0.3)	2.3 (4.2)	13.6 (4.1)

[a]Energy values in kcal/mol. Std. deviation in parentheses. ΔG = ΔH - TΔS Pairs of the values from forward and backward simulations. [b]Refs. 16,17. [c]Ref. 18. [d]Ref. 74.

conclusive due to opposite signs for the terms. It is possible to choose values within the error limits, e.g. ΔH = -0.1 kcal/mol and TΔS = 1.7 kcal/mol, to still give ΔG = -1.8 kcal/mol, for which the dominancy has changed as compared to that suggested by the average values. Thus cases in which the enthalpy and entropy terms are of opposite sign but have close average values with large relative errors are not conclusive. Another source of the uncertainty: the average values in forward and backward simulations suggest different dominancies as was found for the **1** to **7** transformation.

The g1H to g3H tautomeric and the g3H to t3H conformational changes for histamine provide examples of the problems discussed above. Forward and backward simulations for g1H to g3H suggest different dominancies, while a small enthalpy dominancy is predicted for the g3H to t3H conformational change. Large relative errors prevent, however, the estimation of the real contributions of ΔH and $T\Delta S$ terms to ΔG. The conformational change g3H+ to t3H+ shows an overwhelming entropy dominancy, in contrast to the previous two transformations for histamine with more balanced contributions.

In summary, in cases studied here for the disruption of intramolecular hydrogen bonding upon hydration or stabilization of isomeric structures with different kinds of intramolecular hydrogen bonds a common driving force, i.e. decrease of enthalpy or increase of entropy, has not been identified. In many cases the dominancy could not be determined due to the large relative errors in the ΔH and $T\Delta S$ terms or the different prediction in the forward and backward simulations. Nonetheless, both enthalpy and entropy dominancies were identified in the hydrogen bond disruption processes.

Conclusions

Hydration favors conformers without intramolecular hydrogen bonds or tautomers when the strongly polar sites are more exposed to the solvent. The intramolecular hydrogen bond is disrupted when the relative solvation free energy becomes more negative than the relative internal free energy stabilizing the internally bound structure, e.g. for the tGg' to tGg transformation of 1,2-ethanediol or the g3H to t3H conformational change for histamine. The internal stabilization remains the dominant effect for the **1** structure of 2-hydroxybenzoic acid and for the protonated g3H+ structure of histamine.

Optimized monohydrate structures reflect several important aspects of the hydration. The hydrogen bonds, however, may be stretched and bent for the gas-phase dimers as compared to the in-solution structures if there are neighboring polar sites in the solute. The monohydrate structures indicate a trend to form more solute-water hydrogen bonds if possible.

Conformational/tautomeric changes in solution are often accompanied by relatively large enthalpy and entropy changes as compared to the that for the solvation free energy itself. This is possible because solvation enthalpies and entropies are of considerably compensating character in these cases. Though changes in the number of the solute-water hydrogen bonds is a major factor, not a single energy term or structure characteristic has been found responsible for the changes in the solvation free energy for isomeric solutes.

Acknowledgments

DAS would like to thank the Ohio Supercomputer Center for several generous grants of time in support of this work. We gratefully acknowledge the assistance of Mr. Joseph Bitar in the preparation of this manuscript.

Literature Cited

1. Beveridge, D. L.; Jorgensen, W. L. (Eds.) Computer Simulations of Chemical and Biomolecular Systems. *Ann. N. Y. Acad. Sci.* **1986,** *482.*
2. Beveridge, D. L.; DiCapua, F. M. *Rev. Biophys. Biophys. Chem.* **1989**, *8*, 431.
3. Van Gunsteren, W. F.; Weiner, P. K. (Eds.) *Computer Simulations of Biomolecular Systems,* Escom: Leiden, 1989.
4. McCammon, J. A.; Harvey, S. C. *Dynamics of Proteins and Nucleic Acids*; Cambridge Univerity Press: Cambridge,1987.
5. Tapia, O. In *Quantum Theory of Chemical Reactions*; Daudel, R.; Pullman, A.; Salem, L.; Veillard, A. (Eds); Reidel: Dordrecht, 1980; Vol. II, p.25.
6. Miertus, S.; Scrocco, E.; Tomasi, J. *Chem. Phys.* **1981**, *55*, 117.
7. Sanchez Marcos, E.; Terryn, B.; Rivail, B. *J. Phys. Chem.* **1985**, *89*, 4695.
8. Pascual-Ahuir, J. L.; Silla, E.; Tomasi, J.; Bonaccorsi, R. *J. Comput. Chem.* **1987**, *8*, 778.
9. Rashin, A. A. *J. Phys. Chem.* **1990**, *94*, 1725.
10. Still, W. C.; Tempczyk, A.; Hawley, R. C.; Hendrickson, T. *J. Am. Chem. Soc.* **1990**, *112*, 6127.
11. Wong, M. W.; Frisch, M.; Wiberg, K. B. *J. Am. Chem. Soc.* **1991**, *113*, 4776.
12. Cramer, C. J.; Truhlar, D.G. *J.Comput. Chem.* **1992**, *13*, 1089.
13. Cramer, C. J.; Truhlar, D. G. *J. Am. Chem. Soc.* **1993**, *115*, 2226.
14. Tunon, I.; Silla, E.; Tomasi, J. *J. Phys. Chem.* **1992**, *96*, 9043.
15. Tunon, I.; Silla, E.; Pascual-Ahuir, J. L. *J. Am. Chem. Soc.* **1993**, *115*, 2226.
16. Nagy, P. I., Dunn, W. J. III, Alagona, G., Ghio, C. *J. Am. Chem. Soc.* **1991**, *113*, 6719.
17. Nagy, P. I., Dunn, W. J. III, Alagona, G., Ghio, C. *J. Am. Chem. Soc.* **1992,** *114*, 4752.
18. Nagy, P. I.; Dunn, W. J. III, Alagona, G.; Ghio, C. *J. Phys. Chem.* **1993**, *97*, 4628.
19. Berger, J.; Rachlin, A. I.; Scott, W. E.; Sternbach, W. E.; Goldberg, M. W. *J. Am. Chem. Soc.* **1951**, *73*, 5295.
20. Youssefyeh, R. D.; Campbell, H. F.; Airey, J. E.; Klein, S.; Schnapper, M.; Powers, M.; Woodward, R.; Rodriguez, W.; Golec, S.; Studt, W.; Dodson, S. A.; Fitzpatrick, L. R.; Pendley, C. E.; Martin, G. E. *J. Med. Chem.* **1992**, *35*, 903.
21. Price, N. S. *Biochem. J.* **1979**, *177*, 603.
22. Cooper, D. G.; Young, R. C.; Durant, G. J.; Ganellin, C. R. in *Comprehensive Medicinal Chemistry*, Vol. 3. pp 323-421, Pergamon Press, 1990.
23. *MONSTERGAUSS*, Peterson, M. R.; Poirier, R. A., Dept. of Chemistry, University of Toronto, Toronto, Ontario, Canada.
24. *Gaussion 90*, Revision J: Frisch, M. J.; Head-Gordon, M.; Trucks, G. W.; Foresman, J. B.; Schlegel, H. B.; Raghavachari, K.; Robb, M.; Binkley, J. S.; Gonzalez, C.; Defrees, D. J.; Fox, D. J.; Whiteside, R. A.; Seeger, R.; Melius, C. F.; Baker, J.; Martin, R. L.; Kahn, L. R.; Stewart, J. J. P.; Topiol, S.; Pople, J. A., Gaussian, Inc., Pittsburgh PA, 1990.

25. Hehre, W. J.; Radom, L.; Schleyer, P. v. R.; Pople, J. A. *Ab Initio Molecular Orbital Theory*, John Wiley and Sons: New York, 1986.
26. Hariharan, P. C.; Pople, J. A. *Theor. Chim. Acta* **1973**, *28*, 213
27. Moller, C.; Plesset, M. S. *Phys. Rev.* **1934**, *46*, 618.
28. Pople, J. A.; Binkley, J. S.; Seeger, R. *Int. J. Quant. Chem. Symp.* **1976**, *10*, 1.
29. Nagy, P. I.; Durant, G. J.; Smith, D. A. *J. Am. Chem. Soc.* **1993**, 115, 2912.
30. Nagy, P. I.; Smith, D. A., Alagona, G.; Ghio, C. *J. Phys. Chem.*, **1994**, *98*, 486.
31. Boys, S. F.; Bernardi, F. *Mol. Phys.* **1970**, *19*, 553.
32. Sokalski, W. A.; Roszak, S.; Hariharan, P. C.; Kaufman, J. J. *Int. J. Quant. Chem.* **1983**, *23*, 847.
33. McQuerrie, D. *Statistical Mechanics*; Harper and Row: New York, 1976.
34. Jorgensen W. L.; Madura, J. D. *J. Am. Chem. Soc.* **1983**, *105*, 1407.
35. Jorgensen W. L.; Swenson, C. *J. Am. Chem. Soc.* **1985**, *107*, 1489.
36. Jorgensen W. L.; Gao, J. *J. Phys. Chem.* **1986**, *90*, 2174.
37. Jorgensen, W. L. User's Manual for the BOSS Program Version 3.1.
38. Jorgensen, W.L.; Chandrasekhar, J.; Madura, J. D.; Impey, R. W.; Klein, M. L. *J. Chem. Phys.* **1983**, *79*, 926.
39. Jorgensen, W. L.; Madura, J. D. *Mol. Phys.* **1985**, *56*, 1381.
40. Owicki, J. C.; Scheraga, H. A. *Chem. Phys. Lett.* **1977**, *47*, 600.
41. Jorgensen W. L. *J. Phys. Chem.* **1983**, *87*, 5304.
42. Jorgensen, W. L.; Tirado-Rives, J. *J. Am. Chem. Soc.* **1988**, *110*, 1657.
43. Blake, J. F.; Jorgensen, W. L. *J. Am. Chem. Soc.* **1990**, *112*, 7269.
44. Briggs, J. M.; Nguyen, T. B.; Jorgensen, W. L. *J. Phys. Chem.* **1991**, *95*, 3315.
45. Jorgensen, W. L., Ravimohan, C. *J. Chem. Phys.* **1985**, *83*, 3050.
46. Bastiensen, O. *Acta Chem. Scand.* **1949**, *3*, 415.
47. Caminati, W.; Corbelli, G. *J. Mol. Spectrosc.* **1981**, *90*, 572.
48. Frei, H.; Ha, T.-K.; Meyer, R.; Gunthard, Hs. H. *Chem. Phys.* **1977**, *25*, 271.
49. Takeuchi, H.; Tasumi, M. *Chem. Phys.* **1983**, *77*, 21.
50. Chidichimo, G.; Imbardelli, D.; Longeri, M.; Saupe, A. *Mol. Phys.* **1988**, *65*, 1143.
51. Pachler, K. G. R.; Wessels, P. L. *J. Mol. Struct.* **1970**, *6*, 471.
52. Maleknia, S.; Friedman, B. R.; Abedi, N.; Schwartz, M. *Spectrsosc. Lett.* **1980**, *13*, 777.
53. Del Bene, J. E. *J. Chem. Phys.* **1987**, *86*, 2110.
54. Del Bene, J. E. *J. Comput. Chem.* **1989**, *10*, 603.
55. Paiva, T. B.; Tominaga, M.; Paiva, A. C. M. *J. Med. Chem.* **1970**, *13*, 689.
56. Ganellin, C. R.; Pepper, E. S.; Port, G. N. J.; Richards, W. G. *J. Med. Chem.* **1973**, *16*, 610.
57. Bonnet, J. J.; Ibers, J. A. *J. Am. Chem. Soc.* **1973**, *95*, 4829.
58. Prout, K.; Critchley, S. R.; Ganellin, C. R. *Acta. Cryst.* **1974**, *B30*, 2884.
59. Veidis, M. V.; Palenik, G. J.; Schaffrin, R.; Trotter, J. *J. Chem. Soc.* **1969 A**, 2659.
60. Yamane, T.; Ashida, T.; Kakudo, M. *Acta. Cryst.* **1973**, *B29*, 2884.
61. Donohue, J.; Caron, A. *Acta. Cryst.* **1964**, *17*, 1178.
62. Madden, J. J.; McGandy, E. L.; Seeman, N. C. *Acta. Cryst.* **1972**, *B28*, 2377.

63. Madden, J. J.; McGandy, E. L.; Seeman, N. C.; Harding, M. M.; Hoy, A. *Acta. Cryst.* **1972**, *B28*, 2382.
64. Vogelsanger, B.; Godfrey, P. D.; Brown, R. D. *J. Am. Chem. Soc.* **1991**, *113*, 7864.
65. Dunn, W. J. III, Nagy, P. I. *J. Phys. Chem.* **1990**, *94*, 2099.
66. Dunn, W. J. III, Nagy, P. I. *J. Comput. Chem.* **1992**, *13*, 468.
67. Nagy, P. I. *Acta Chim. Hung.* **1992**, *129*, 429.
68. Pranata, J. *J. Comput. Chem.* **1993**, *14*, 685.
69. Gao, J.;Pavelitis, J. J. *J. Am. Chem. Soc.* **1992**, *114*, 1912.
70. Alagona, G.; Ghio, C.; Kollman, P. A. *J. Am. Chem. Soc.* **1985**, *107*, 2229., **1986**, *108*, 185.
71. Alagona, G.; Ghio, C. *J. Mol. Liquids.* **1990**, *47*, 139.
72. Nagy, P. I.; Dunn, W. J. III; Nicholas, J. B. *J. Chem.Phys.* **1989**, *91*, 3707.
73. Dunn, W. J. III; Nagy, P. I.; Collantes, E. R. *J. Am. Chem. Soc.* **1991**, *113*, 7898.
74. Nagy, P. I., Durant, G. J., Hoss, W. P., Smith, D. A. *J. Am. Chem. Soc.*, **1994**, *116*, 4898.

RECEIVED July 29, 1994

Theoretical Methods and Graphical Analysis

Chapter 6

Energetics and Structure in Model Neutral, Anionic, and Cationic Hydrogen-Bonded Complexes

Combined Ab Initio SCF/MP2 Supermolecular, Density Functional, and Molecular Mechanics Investigation

Nohad Gresh[1,4], Martin Leboeuf[2,3], and Dennis Salahub[2,3]

[1]Laboratoire de Biochimie Théorique, Institut de Biologie Physicochimique, 13, rue Pierre et Marie Curie, Paris 75005, France
[2]CERCA (Centre de Recherche en Calcul Appliqué), 5160 Boulevard Décarie, Montréal, Québec H3X 2H9, Canada
[3]Département de Chimie, Université de Montréal, C.P. 6128, Succ. A, Montréal, Québec H3C 3J7, Canada

A scrutiny of the energetical and structural factors governing the formation of several representative neutral, anionic, and cationic H-bonded complexes is reported. We utilize three distinct theoretical methodologies : a) large-basis set *ab initio* SCF supermolecule computations. The decomposition of ΔESCF into its four separate first- and second-order components is performed with the help of the recently developed Restricted Variational Space Analysis procedure *(1)*. The energy computations are complemented by MP2 computations to evaluate the contribution due to the dispersion energy, and an evaluation of the counterpoise correction is made at both the HF and MP2 levels; b) Density Functional computations which incorporate the nonlocal Perdew corrections for exchange and correlation *(2)*, with an additional estimate of the counterpoise correction; c) molecular mechanics computations using the SIBFA *(3)* procedure, initially formulated on the basis of *ab initio* computations, in which the interaction energy is computed as a sum of five separate contributions.

For each complex investigated, the evolution of each individual energy contribution is monitored as a function of appropriate stepwise radial or angular variations, or as a function of well-defined configurations of binding. This is done for ten distinct bimolecular complexes. The nonadditive behavior of the binding energy and its components is furthermore assessed in the complexes of two anionic and two neutral ligands with three water molecules. The overall agreement between the three procedures relating to these fourteen complexes is extremely encouraging. The optimized binding energies are in overall very good agreement with available experimental gas-phase complexation enthalpies.

[4]Current Address: Laboratoire de Pharmacochimie Moléculaire et Structurale, Faculté de Pharmacie, 4, Avenue de l'Observatoire, 75270 Paris, France

0097–6156/94/0569–0082$09.98/0

The present investigation is devoted to a series of representative neutral, anionic, and cationic complexes encountered in molecular recognition. Each complex is studied with three distinct theoretical methodologies. First we perform large-basis set *ab initio* SCF computations, and decompose the binding energy into its separate Coulomb, exchange, polarization, and charge-transfer contributions. For that purpose we make use of the Restricted Variational Space Analysis procedure (denoted as RVS below) developed by Stevens and Fink *(1)*. The operational definition of these four contributions does not involve the extraneous E_{mix} term found in the Morokuma procedure *(4)*, and furthermore allows for an energy decomposition in complexes involving more than two molecules : this additional feature will enable us to quantify the nonadditive behavior of the second-order term in such complexes. In order to evaluate the dispersion energy contribution, MP2 computations *(5)* are further carried out. The magnitude of the Basis Set Superposition Error (BSSE) *(6)* is evaluated by the standard counterpoise method, at both uncorrelated and correlated levels. A careful choice of Gaussian basis sets is mandatory to ensure an adequate balance between first- and second-order contributions. As in our previous investigation devoted to a series of Zn^{2+} - ligand complexes *(7)*, the heavy atom outer shells, as well as the 1s hydrogen shell, consist of 4 Gaussians, split into 3-1, and the heavy atom outer shells are supplemented with two uncontracted 3d polarization functions (see refs. *7, 8*). Inner-core heavy atom electrons are replaced by the model core potentials derived by Stevens et al. *(9,10)*. Each complex is investigated by confronting the energetics of distinct competitive H-bonding configurations (e.g., mono- versus bidentate) , or by monitoring, in a well-defined monodentate configuration, the evolution of the binding energy and its components as a function of in- or out-of-plane angular variations at a preset donor-acceptor distance. In this manner, we will be able to quantify the corresponding individual trends of each separate energy term, and its bearing on the total configurational preference.

A crucial issue for the prospect of extensive, reliable simulations on large molecular complexes of biochemical and pharmacological interest is the capacity of molecular mechanics procedures to reproduce the detailed features of the *ab initio* SCF supermolecular energy surface. In our opinion, it is advantageous for a consistent and general approach if each of the *separate ab initio* energy components has a distinct formulation of its own in terms of its molecular mechanics counterpart. A major objective of the present work is to evaluate to what extent this could be achieved with the SIBFA (Sum of Interactions Between Fragments A*b initio* computed) procedure *(3)*, in which the interaction energy is formulated in terms of five separate energy terms, as recalled below. Because refinements to the procedure were recently brought about in the course of an extensive study devoted to a series of Zn^{2+} -- ligand complexes *(7)*, we needed to evaluate how such refinements would now translate in the field of H-bonded complexes, using a set of parameters remaining consistent with the one derived in the preceding study. Refined molecular mechanics procedures have also been put forward by others *(11-13 and review in 14)*, but their use has so far been restricted to a limited subset of H-bonded complexes. With the exception of *(12)*, these did not include an explicit charge-transfer term.

Density Functional Theory (DFT) *(2, 15, and references therein)* is emerging as a most promising tool for molecular simulations, and can tackle significantly larger molecular systems than standard *ab initio* procedures. Let us recall that if N denotes the total number of basis functions, DFT scales formally as N^3, as contrasted to N^4 and N^5 or higher at uncorrelated (HF) and correlated levels respectively. Also, and most importantly, the correlation energy is inherently built-in together with the exchange term, within the so-called XC operator. Two neutral intermolecular H-bonded complexes, the linear water dimer, and the formamide-water complex, were recently investigated by DFT computations performed with nonlocal gradient corrections (2a), providing accurate energetic and structural results. In the prospect of large-scale applications of DFT, it is highly instructive to extend its use here to the same neutral and charged complexes as studied by HF/MP2 as well as SIBFA, and assess the mutual consistency between these three procedures. The BSSE correction will also be evaluated.

We wish to acknowledge that *ab initio* supermolecule computations have already been reported on the H-bonded complexes dealt with here *(12-22)*. These were often performed in conjunction with simplified molecular mechanics computations derived *a posteriori* from them for the purpose of Monte-Carlo simulations *(18d-f, 19)*. With some exceptions, however, *(12, 13, 23)*, a breakdown of the total energy into its components was not reported, the studies focusing on one, or a limited set of alternative, binding configurations. Estimates for the BSSE correction were generally not provided, even though as shown below, it can be significant, as in anion-ligand complexes.

Procedure

***Ab initio* SCF Supermolecule Computations.** Energy-decomposition at the supermolecule level is performed with the RVS procedure of Stevens and Fink *(1)*. Recent applications of this procedure are a joint *ab initio* SCF and molecular mechanics investigation of the binding of Zn^{2+} to representative ligands *(7, 24)*, and a quantitative evaluation of the hardness and softness behavior of a series of neutral and anionic ligands towards Mg^{2+}, Ca^{2+}, Zn^{2+}, and Cd^{2+} *(25)*. Within this procedure, for each binding configuration of monomer A to monomer B, we perform the six following computations, providing, respectively : the separate Coulomb energy, E_c; the first-order energy $E1 = E_c + E_e$, obtained upon freezing the occupied orthogonalized MO's of all interacting entities, which includes the exchange contribution E_e; the polarization energy of A, $E_{pol}(A)$, obtained upon relaxing the occupied and virtual MO's of A; the polarization and charge-transfer of A towards B, obtained upon freezing only the occupied MO's of B, whence $E_{ct}(A)$; and, conversely, $E_{pol}(B)$ and $E_{ct}(B)$ by proceeding analogously with monomer B instead of A. Correlation effects were computed using the Moller-Plesset perturbation procedure *(5)*.

The 1s core electrons of C, N, and O, and the 1s, 2s and 2p core electrons of S, are replaced by the compact effective pseudopotentials derived by Stevens et al. *(9)*. The H basis set has 4 Gaussians, contracted to *(3, 1)*. The 2s and 2p valence orbitals of C, N, and O, as well as the 3s and 3p valence orbitals of S, also consist of 4 Gaussians, contracted to *(3, 1)*. Two 3d polarization orbitals are used on the first and second-row heavy atoms. They have exponents of 0.75 and 0.15 for C, 0.77 and 0.15 for N, 0.80 and 0.20 for O, and 0.70 and 0.13 for S. The use of a second diffuse polarization function is necessary to approach the SCF limit of the dipole moment of water. The *ab initio* SCF and intermolecular RVS computations were done using a modified version of the Hondo *(26)* suite of programs (W. J. Stevens, Center for Advanced Research in Biotechnology, Rockville, Maryland).

Density Functional computation details. The calculation were performed with deMon, a Linear Combination of Gaussian Type Orbital (LCGTO) code developed by St-Amant and Salahub *(2)*. All calculations were done at the non-local level, with density gradient type corrections included self-consistently. The functional of Perdew and Wang *(27)* was used for exchange, and that of Perdew *(28)* for correlation. Numerical instabilities arising when the ratio of the density gradient to the density becomes too large are overcome by the introduction of a damping factor *(2)*.

The precise choice (or formulation) of non-local functional remains a matter of some uncertainty. Our own experience for a variety of bonding situations indicates that the damped PW-86 *(27)* exchange functional coupled with Perdew's correlation functional *(28)* is robust and yields results that are often in somewhat better agreement with experiment than those from Becke-Perdew- BP *(29, 28)* or PW-91 calculations *(30)*. Others *(31)* have come to a similar conclusion. For the specific case of the water dimer *(29)* there is about a kcal/mole difference in the binding energy between PW-86 and BP calculations. After correction for the BSSE, PW-86 yields 5.6 kcal/mol while BP yields 4.2 kcal/mol (for reference - the Local Density Approximation is seriously overbound, at 8.7 kcal/mol). The accepted experimental value is 5.4 ±0.7 kcal/mol, very close to the PW-86 value. However, the experimental error bar is large and there have been some indications from high-level correlated ab initio calculations for values in the lower range. Rybak et al. *(32)* used symmetry-adapted perturbation theory to predict 4.7 kcal/mol whereas Feller *(33)* has estimated the complete basis set full-CI limit at 5.1 kcal/mol. Clearly, more work is needed on all fronts to obtain absolute accuracies in the 0.1 kcal/mol range for this, or any other, particular system. More systematic comparisons of the merits of various functionals will be forthcoming on representative neutral and ionic complexes; however, in the present work we are mainly interested in the consistency of our three procedures for a variety of interactions and we shall see that PW-86 is very satisfactory in this respect.

The same basis sets were used as in our previous studies of hydrogen bonds *(2a)*. Orbital basis sets comparable to the standard 6-31G** basis were used. These have the (5211/411/1) pattern for C, O, and N and the (41/1) pattern for H. Auxiliary basis sets of the type (k_1, k_2; l_1, l_2), where $k_1(l_1)$ is the number of s-type Gaussians in the CD(XC) basis and $k_2(l_2)$, the number of s- p- d-type Gaussians constrained to have the same exponent in the CD(XC) basis, are used for the fitting of the charge density (CD) and exchange-correlation potential (XC). These have the pattern (5,2; 5,2) for C, O, and N and (5,1; 5,1) for H. While the charge density is fitted analytically, the exchange-correlation terms are fitted numerically on a grid comprising 32 radial shells, each having 26 angular points, for each atom. At the end of the scf procedure, the XC contribution to the energy and gradient is calculated on an augmented grid, consisting of the same 32 radial shells, but now having a total of 2968 points per atom.

Molecular Mechanics Computations. The molecular mechanics computations were performed with the SIBFA (Sum of Interactions Between Fragments *Ab initio* computed) procedure *(3)*, in which the binding energy is computed as five separate terms:

$$\Delta E = E_{mtp} + E_{rep} + E_{pol} + E_{ct} + E_{disp}$$

We will briefly outline here recent refinements to the expression of these terms, brought about in the course of our above-mentioned study of Zn^{2+} - ligand interactions *(7, 24)*:

E_{mtp}, the electrostatic (multipolar) term, is computed as a sum of multipole-multipole interactions. The multipoles, up to quadrupoles, are derived from the molecular orbitals of each individual monomer, and distributed on the atoms and barycenters of the chemical bonds, using a procedure due to Vigne-Maeder and Claverie *(34)*. The monomer MO's result from *ab initio* SCF computations using the same large Gaussian basis set as the RVS ones. The multipoles thus derived supersede those of our standard library of multipoles *(3)* which were derived from more restricted Gaussian basis sets.

E_{rep}, the short-range repulsion energy, is computed as a sum of bond-bond, bond-lone pair, and lone-pair-lone-pair interactions, conforming to the developments carried out in Ref. *(3)*. We have further elaborated on this expression in Ref. *(7)*, by now taking into account the effect of hybridization on the chemical bonds as well as on the lone-pairs, whereas it was initially limited *(3)* solely to the lone-pairs, and we have used an S^2/R rather than an S^2 formulation. Thus, the repulsion between hybrids A and B is expressed as :

$$rep(A,B) = N_{occ}(A)\, N_{occ}(B)\, S^2(A,B)/R_{AB}$$

$N_{occ}(A)$ and $N_{occ}(B)$ denoting the occupation numbers of hybrids A and B : 2 in the general case, 1 for each lobe of a π lone pair, S(A, B) denoting a function of the overlap and R_{AB} being the distance between the centroids of hybrids A and B *(see Ref. 7 for discussion.)*. Thus, if A denotes a chemical bond between atoms P and Q, its centroid coincides with that of the Boys orbital localized on this bond, obtained as an output of the Hondo computer program. If it denotes a lone pair hybrid, it coincides with the 'tip' of the lone pair.

$E_{pol,}$ the polarization energy contribution, is computed by using anisotropic polarizabilities, themselves derived from the monomer MO's, and distributed on the barycenters of the monomer localized orbitals, according to a procedure due to Garmer and Stevens *(35)*, rather than with the help of experimental isotropic polarizabilities as was done before *(3, 36)*. The necessity of resorting to anisotropic and distributed polarizabilities was put forward by Stone *(37)*. A Gaussian screening of the field is used, as in our previous work *(7)* devoted to Zn^{2+} - ligand complexes. We use here a simplified expression of the polarizing field, limited to monopoles, rather than the more complete expansion of *(3, 36, 37)*. Consistent with this approximation, the effect of the higher-order induced dipoles *(37a)* is not taken into account. A more complete expansion of E_{pol}, also including a coupling of E_{pol} to E_{ct} through a reduction of the effective charges of the polarizing entities at higher order is being sought presently. However, the satisfactory results presented below justify the present simplifications.

In the expression of $E_{ct,}$ the charge-transfer contribution, an explicit coupling to E_{pol} is introduced. This is done essentially by means of an expansion of the effective radius of the electron donor atom, proportional to the polarizing field and along its direction *(7)*.

$E_{disp,}$ the dispersion energy contribution, is now computed as a C_6/Z^6, C_8/Z^8, and C_{10}/Z^{10} expansion, as formulated in refs. *(38a,b)*. Here $Z = R_{IJ}/\sqrt{(W'_I W'_J)}$, where R_{IJ} is the distance between I and J, W'_I and W'_J the effective radii used for E_{disp}, and C_6, C_8, and C_{10} are empirical coefficients. We will, furthermore, following a recent formulation due to Creuzet and Langlet (S. Creuzet, and J. Langlet, manuscript in preparation), reduce each of these terms by an exponential damping factor having the form :

$$E_{damp}(n) = (1/R^n)\, R_{KI}\, R_{KJ} \exp(-a_{damp}(n)\, T(I,J))$$

R_{KI} and R_{KJ} the atom-specific parameters used for the dispersion energy *(3b)*, n has the values 6, 8, or 10, and T(I,J) is a function of the interatomic distance R_{IJ} and of the effective radii W'_I and W'_J :

$$T(I,J) = ((\, W'_I + W'_J)\, b_{damp}/R_{IJ}) - 1$$

$a_{damp}(n)$ and b_{damp} are empirical parameters.

Furthermore, an explicit exchange-repulsion term *(38 b)* is added to E_{disp}:

$$E_{exch\text{-}disp}=R_{KI}R_{KJ}\ (1\text{-}Q_I/N_{val}(I)\ (1\text{-}Q_J/N_{val}(J))\ C_{exch}\ \exp(-\beta Z)$$

Q_I and Q_J denoting the net charge of atoms I and J, $N_{val}(I)$ and $N_{val}(J)$ the number of their valence electrons. Finally, directionality effects are accounted for by introducing additional interactions involving fictitious atoms at the barycenters of the heteroatom lone pairs (S. Creuzet, and J. Langlet, manuscript in preparation), affected by a multiplicative constant, C_{lp}. The values of the reduced effective radii W' of the lone-pair bearing atoms, and those of the fictitious atoms, are the same as in the original derivation of the SIBFA procedure *(3a)*, where they were used for the computation of the intramolecular repulsion energy.

Calibration of the SIBFA Procedure

We will use below the abbreviated notation SMM to denote the components of the SIBFA molecular mechanics procedure.

Repulsion Energy. The heavy-atom effective radii were determined in the course of our study of Zn^{2+} -- ligand complexes. We now need to adjust E_{rep} for the case of H-bonded complexes, in which at least one H atom is directly involved in the intermolecular complex. Because of the presence of penetration terms, and conforming to our earlier work *(3, 36)*, we calibrate E_{rep} in such a way that it matches as closely as possible the difference between E1(SCF) and E_{mtp}. E_{rep} depends on three parameters : the multiplicative factor C1; the exponent in the exponential, α; and the effective radius, W_H, of the hydrogen atom. We have resorted to three different water-water complexes to calibrate these constants : 1) the linear water dimer, the θ angle being set at 140° *(3)*, variations of the O(w1) -- H(w2) distance are performed; 2) a complex in which one H atom from one molecule faces directly another H from the other water, in such a way that the four atoms Ow1-Hw1-Hw2-Ow2 are collinear, for various H-H distances; 3) a complex in which two water molecules are disposed at right angles to one another. This is the configuration found in octahedral complexes of a Zn^{2+} ion, which we have investigated previously *(7)*. The oxygens point inward, in such a way that their external bisectors intersect and are perpendicular to the plane of the other molecule. We compare in Table I the values of E1(SCF) to those of E1(SMM) for these three kinds of complexes, indicating a close match. The K_{PM} multiplicative factors involving the H and O atoms on the one hand, and the S atom on the other hand, are taken as products of the individual atomic parameters $K_z(O)$, $K_z(H)$, and $K_z(S)$ (see 3 and references therein). The value of $K_z(S)$ was calibrated so that E1(SMM) matches E1(SCF) upon performing variations of the S--Hw distance in the CH_3SH -- H_2O complex, water being coplanar with the C-S-H plane, the Hw--S bond being along the external bisector of the C-S-H angle. (The effective radii for S(neutral) and S(anionic) remain the same as calibrated upon studying the complexes of methane thiol and methane thiolate with Zn^{2+} *(24)*).

The value of W_H is 1.7 Å, actually the same as in Ref. *(7)*. The value of the exponential factor α is 10.88. The values of the multiplicative constants, C1, of E_{rep}, are 244.16, 487.80, and 975.61, for bond-bond, bond-lone-pair, and lone-pair-lone-pair repulsions, respectively. The ratios of 1, 2, and 4 of these values were adopted to account for the presence of a $1/\sqrt{2}$ normalization factor in the expansion of a bond orbital on its two atomic centers. The value of $K_z(S)$ is 2.40.

Charge-Transfer. The value of the multiplicative factor S_M involving nonmetallic electron-acceptors is calibrated so that E_{ct}(SMM) matches E_{ct}(SCF) in the linear water dimer, θ = 140°, by performing variations of the Ow1--Hw2 distance. A value of

Table I. Comparison of E1(SCF) and E1(SMM) values in the water-water complexes used to calibrate Erep. Energies in kcal/mole. See text for definition of the approaching modes of Wb towards Wa.

a) Linear								
d(O-H)	1.6	1.7	1.8	1.9	2.0	2.1	2.2	2.3
E1(SCF)	8.0	3.6	0.8	-1.0	-2.0	-2.5	-2.8	-2.9
E1(SMM)	8.7	3.9	0.9	-0.9	-2.0	-2.5	-2.8	-2.9
b) H <--> H								
d(H-H)	1.7	1.8	1.9	2.0	2.1	2.2	2.3	
E1(SCF)	6.3	5.2	4.4	3.8	3.3	2.8	2.5	
E1(SMM)	6.8	5.4	4.5	3.8	3.2	2.8	2.4	
c) O <--> O								
d(O-O)	2.64	2.70	2.77	2.83	2.90	2.97		
E1(SCF)	8.5	7.5	6.6	5.8	5.1	4.6		
E1(SMM)	8.1	7.1	6.2	5.5	4.9	4.4		

0.66 was adopted. The exponent η retains the value of 9.50 that was determined previously *(7)*.

Dispersion. The novel formulation of E_{disp} by Creuzet and Langlet was originally calibrated so that it matches the value of the dispersion energy computed by the Symmetry Adapted Perturbation Theory (SAPT) *(39)* on a series of water dimer complexes. The parameters given below show only minor differences from the ones originally derived by these authors.

Values of C_6, C_8, and C_{10} : 0.15, 0.31, and 0.18, respectively. Values of C_{exch} and β 240. and 9.40. Values of $a_{damp}(6)$, $a_{damp}(8)$, and $a_{damp}(10)$: 1.23, 1.8, and 1.36. Value of b_{damp} 1.37. Value of C_{lp}, 0.16.

RESULTS AND DISCUSSION

The results of our computations are reported in Tables II-XII. These Tables report :

a) the values of the supermolecule binding energy at the Hartree-Fock level, ΔEHF(0) and those of its first (E1) and second- (E2) order components, the magnitude of the BSSE correction, BSSE, the BSSE-corrected value of ΔEHF(0), ΔEHF, the value of the interaction energy at the MP2 correlated level, ΔEMP2(0), that of the BSSE correction at the MP2 level, BSSE1, and the values of ΔEMP2(0) after correlated and uncorrelated BSSE corrections, ΔEMP2(1) and ΔEMP2, respectively;
b) the value of the DFT interaction energy, ΔEDFT(0), that of the DFT BSSE correction, and that of the BSSE-corrected interaction energy, ΔEDFT;
c) the value of the SMM interaction energy in the absence of the dispersion energy term, ΔESMM(0) and of its first- and second-order components E1(SMM) and E2(SMM), the separate value of the dispersion energy, and the total value of the interaction energy, ΔESMM.

The BSSE correction at the MP2 correlated level will be seen to attain very large values, in some instances equalling the energy gain due to correlation. This is not the case, on the other hand, in the DFT computations. For this reason, we will resort preferentially to the values of ΔEMP2 rather than ΔEMP2(1) in order to compare our HF/MP2 results to the DFT and SMM ones. The large MP2, BSSE values have already been discussed by others *(32, 33, 40, 41)*. It is likely that the true correction lies in between the uncorrelated and correlated BSSE. However, our pragmatic choice to use the uncorrelated BSSE yields results consistent with those of the other two approaches so we have not pursued the question of the exact MP2 (or higher-level) BSSE further.

Water-water. A wealth of theoretical studies have been devoted to the water dimer as a paradigm for H-bonded complexes (see e.g.*2, 12, 13, 32, 33, 35* and references therein). We compare here the binding energetics obtained by the three computational procedures in the linear, bifurcated, and cyclic dimers. For consistency with our previous study *(3b)*, in the linear water dimer, the angle θ, between the external bisector of the H-bond acceptor, and the O--O line, is set at 140°, and the two O's and the 'donated' proton are collinear.

The results are reported in Table II. A very good agreement of the three procedures can be seen. The sole discrepancy relates to the relative ordering of bifurcated versus cyclic dimers, which in the MP2 procedure favors the cyclic one by 0.5 kcal/mole, whereas both dimers are virtually indistinguishable in the DFT and SMM computations. The binding energy of -5.4 kcal/mole is consistent with the experimental water dimerization energy of -5.4 kcal/mole, as discused above, with a large uncertainty of ±0.7 kcal/mole *(42)*. It is highly instructive to observe that the preference for the linear over the cyclic and bifurcated dimers is *not* due to the first-

Table II. Binding energies in the linear, bifurcated, and cyclic water dimers Energies in kcal/mole. See text for definitions.

	Linear	*Bifurcated*	*Cyclic*
ΔEHF(0)	-3.9	-2.5	-2.9
Ec	-6.8	-4.3	-3.9
Ee	4.8	2.6	1.7
E1	-2.0	-1.7	-2.2
POL	-0.9	-0.3	-0.2
CT	-1.0	-0.5	-0.4
E2	-1.9	-0.8	-0.6
BSSE	-0.5	-0.4	-0.4
ΔEHF	-3.4	-2.1	-2.5
ΔEMP2(0)	-5.9	-3.7	-4.2
BSSE1	-2.0	-1.3	-0.8
ΔEMP2(1)	-3.9	-2.4	-3.0
ΔEMP2	-5.4	-3.3	-3.8
ΔEDFT(0)	-5.7	-3.5	-3.6
BSSE	-0.4	-0.4	-0.1
ΔEDFT	-5.3	-3.1	-3.5
ΔESMM(0)	-3.7	-2.4	-2.3
Emtp	-5.9	-3.6	-3.5
Erep	3.9	2.0	1.8
E1	-2.0	-1.7	-1.6
Epol	-0.9	-0.4	-0.4
Ect	-0.8	-0.3	-0.2
E2	-1.7	-0.7	-0.6
Edisp	-1.6	-1.3	-1.2
ΔESMM	-5.3	-3.7	-3.5

order term, E1 but, rather, to the second-order term E2. This is due to the fact that the much more favorable values of the electrostatic term in the linear dimer are counterbalanced by the short-range repulsion term, leaving out E1 with a weight of less than fifty percent in this dimer. This is accounted for in the SMM computations. These observations can also be made upon examining the Anisotropic Site Potential results of Millot and Stone *(13)*, but there is, to our knowledge, no other precedent to them.

Formamide-water. The formamide-water complex has lent itself to numerous theoretical computations, using both *ab initio* SCF supermolecule *(16)* and molecular mechanics techniques *(16b, and 14 for a review),* owing, *inter alia,* to its importance as a model for the solvation of peptides and proteins. The most extensive *ab initio* SCF study to date, due to Jasien and Stevens in 1985, performed gradient-energy minimization and Configuration Interaction *(16b).* The recent paper by Sim et al. *(2a)* investigated its monohydration energies using for the first time the Density Functional Theory with nonlocal gradient corrections. Since the first *ab initio* SCF supermolecule paper by Alagona et al *(16a),* four well-defined monohydration sites could be defined and ranked in terms of their relative energies. The first (I), is the bidentate mode, with water acting simultaneously as an H-bond donor to the carbonyl O, and as an H-bond acceptor from the nitrogen-bound H, Hc, that is *cis* to C=O. The second, (II), is a monodentate mode in which water is H-bonded through one H to the C=O group, at a C-O-Hw angle close to 120°. In modes III and IV, water acts as an H-bond acceptor from the hydrogens, Ht and Hc, that are *trans* and *cis,* respectively, to C=O. As concerns the last two positions, water approaches the NH bond along its external bisector, and we did not perform an optimization of angular variables, since, as stressed above, we are essentially interested here in probing the mutual consistency of the three procedures in well-defined binding positions, rather than exploring in-depth the potential surface as was done by the preceeding investigators. The SCF/MP2 and DFT computations were performed at one single position in configuration I, resulting from an in-plane energy-minimization with the SMM procedure. This restriction is justified on account of the comparative nature of our study, and the fact that, furthermore, the energy surface for formamide-water was reported to be very flat around the energy-minimum (16). The search for the optimal θ value in mode II was done by keeping the C=O oxygen, the bound water H, and the water O collinear.

Our results are reported in Table III. The three procedures provide the same ordering I > II> III >IV, with a pronounced preference in favor of I over II, whereas that of III over IV is weak (a reverse IV > III ordering could possibly be obtained by a more complete optimization, but this is not relevant here). The binding energies in the four positions, have closely similar numerical values. They also compare well with, on the one hand, the SCF/CI computations of Jasien and Stevens, and, on the other hand, the DFT computations of Sim et al. It should be noted that the values of Jasien and Stevens are closer to the Perdew calculations of Sim et al. than to the Becke-Perdew calculations of these authors. Numerical energy differences can be due to the use of different basis sets (thus, e. g., our SCF computations use the same basis set as Jasien and Stevens, but with an additional split d polarization function on heavy atoms) and internal geometries, and the fact that, as discussed above, we did not perform here an extensive search of the energy minima on the hypersurface. The magnitude of the BSSE correction of about 0.6 kcal/mole, is similar to that found by Jasien and Stevens. On the other hand, the MP2 contribution in the range 1.8-2.6 kcal/mole is larger than the CI gain in energy computed by them, which is in the range 0.8-1.6 kcal/mole *(16b).* This may be due to the very diffuse additional 3d polarization function introduced here on the heavy atoms, and/or the more approximate nature of the MP2 procedure with respect to the CI one to compute the gain in energy due to correlation. The weight of E1 within ΔEMP2 or ΔESMM is

Table III. Binding energies in the formamide-water complexes. Energies in kcal/mole. See text for definitions.

	I. Bidentate	*II.*	*III.*	*IV.*
ΔEHF(0)	-8.0	-5.9	-4.4	-3.5
Ec	-11.4	-8.6	-7.5	-6.4
Ee	6.5	5.2	5.3	5.2
E1	-4.9	-3.4	-2.2	-1.2
POL	-1.1	-0.8	-0.7	-0.7
CT	-1.9	-1.7	-1.3	-1.3
E2	-2.9	-2.5	-2.0	-2.0
BSSE	-0.7	-0.7	-0.6	-0.5
ΔEHF	-7.3	-5.2	-3.8	-3.0
ΔEMP2(0)	-10.6	-8.2	-6.6	-5.8
BSSE1	-2.4	-2.2	-2.1	-2.0
ΔEMP2(1)	-8.2	-6.0	-4.5	-3.8
ΔEMP2	-9.9	-7.5	-6.0	-5.3
ΔE(HF/CI) (Ref. 16 b)	-9.5	-6.7	-6.0	-6.2
ΔEDFT(0)	-10.5	-7.1	-5.9	-5.2
BSSE	-0.5	-0.5	-0.5	-0.5
ΔEDFT	-10.0	-6.6	-5.4	-4.7
ΔESMM(0)	-6.8	-5.5	-4.3	-3.6
Emtp	-10.1	-7.8	-6.4	-5.5
Erep	5.7	4.0	4.0	3.9
E1	-4.5	-3.7	-2.4	-1.6
Epol	-1.4	-1.0	-1.2	-1.2
Ect	-0.9	-0.8	-0.7	-0.8
E2	-2.3	1.8	-1.9	-2.0
Edisp	-3.3	-2.5	-1.9	-2.0
ΔESMM	-10.1	-8.0	-6.2	-5.6

modest, amounting at best to a little less than fifty percent in configuration I, and being less than one-third in IV. The preference for the bidentate mode is supported by gas-phase microwave studies of the formamide-water complex *(43)*.

Methane thiol-water. In the CH_3SH-H_2O complex, either monomer can behave as the proton donor or as the proton acceptor. We will focus here only on the mode with sulfur as the proton acceptor. We report in Table IV the evolution of the binding energies as a function of out-of-plane variations of the position of water, as defined by the ϕ = Hw-S-C-H angle, the S, Hw, and Ow atoms being collinear, and the interatomic distance being set at 2.9 Å. The three procedures provide binding energies that are identical to within 0.2 kcal/mole over the whole range of angular values, and all locate the energy minimum at $\phi = 105°$, i.e., close to the perpendicular to the C-S-H plane passing through S. Note that the same optimal value of ϕ was found previously upon investigating the binding of Zn^{2+} to methane thiol *(24)*. The best binding energy, amounting to -3.6 kcal/mole, has the same value as the MP2 energy published by Zheng and Merz using a 6-31G* basis set *(17)*. The preference for ϕ = 105° is due to E1, and within E1, to the electrostatic terms, E_c and E_{mtp}. E_{pol} and E_{ct} can be seen to have, within E2, virtually equal weights in both the SCF and SMM procedures.

Formate-water. The formate ion is ubiquitous as the terminal part of the aspartate and glutamate side-chains, in anionic saccharides, as well as in the structure of drugs and enzyme inhibitors, etc. Together with that of its cation-binding properties *(18b, 44)*, an understanding of its binding energetics to the neutral prototypic ligand, water, is warranted. The first *ab initio* SCF supermolecule computations devoted to formate-water, due to Port and Pullman in 1974 *(18a)*, compared for the first time its stabilization energies in well-defined bi- and monodentate configurations. Computations using extended basis sets were published by Pullman and Berthod in 1981 *(18b)*. Comparisons between large-basis set *ab initio* and SIBFA computations were provided later (3b), but on a more limited set of points than in the present study, and without decomposing E1 and E2 into their individual components. *Ab initio* SCF computations, supplemented in several cases by molecular mechanics Monte-Carlo computations were also reported by several other authors *(17, 18)*, the most recent study being due to Zheng and Merz in 1992 *(17)*.

In Table V we compare the binding energies of formate-water in the bidentate (bridge, 'B') position and in six monodentate positions in which the θ angle C-O-Hw is varied by five15° steps starting from $\theta = 105°$. The d-OHw distance is preset at 1.9 Å , water remaining coplanar with formate, and the O-Hw-Ow angle remains set at 180°. As in the preceeding studies *(3b, 18a)*, the Hw-Ow bond is external ('E') to that between the carbon atom and the unbound anionic oxygen. The numerical agreement between the three procedures is very satisfactory. The largest discrepancy, occurring in the bridge position, is that between ΔEMP2, on the one hand, and ΔEDFT and ΔESMM, on the other hand, and amounts to 2.9 kcal/mol. That between the DFT and SMM computations is only 1.3 kcal/mol. The values computed in these two configurations are extremely close to those published by others.

In the monodentate 'E' position, all three procedures indicate $\theta = 120°$ as being the best angular value, but an extremely shallow well can be seen in the whole 105° 180° range by the three of them. Whatever the binding mode, and despite the anionic nature of the complex, the numerical values of E1(SCF) as well as of E1(SMM) never exceed sixty per cent of the total binding energy. In the HF computations, the shallow minimum of E1 in the monodentate configuration is due to a compensation between the electrostatic term, favoring the two smallest θ values, and the short-range repulsion, which decreases monotonically as θ increases. E_{ct} has its minimum at θ = 105°, close to the location of one anionic oxygen lone-pair, whereas E_{pol} as well as

Table IV. Methane thiol-water. Evolution of the binding energies as a function of variations of the ϕ = H-S-C-H(w) angle. Energies in kcal/mole, angles in degrees. See text for definitions.

ϕ	*90°*	*105°*	*120°*	*135°*	*150°*	*165°*	*180°*
ΔEHF	-2.2	-2.4	-2.2	-1.9	-1.7	-1.5	-1.5
Ec	-2.9	-3.1	-2.7	-2.3	-1.9	-1.5	-1.6
Ee	1.5	1.4	1.3	1.1	1.0	0.7	0.9
E1	-1.4	-1.6	-1.4	-1.2	-0.9	-0.8	-0.8
POL	-0.3	-0.3	-0.3	-0.3	-0.3	-0.3	-0.3
CT	-0.4	-0.4	-0.4	-0.4	-0.4	-0.4	-0.4
E2	0.8	-0.8	-0.8	-0.7	-0.7	-0.7	-0.7
BSSE	-0.3	-0.3	-0.3	-0.3	-0.3	-0.3	-0.3
ΔEHF	-1.9	-2.1	-1.9	-1.6	-1.4	-1.2	-1.2
ΔEMP2(0)	-3.8	-3.9	-3.7	-3.3	-3.1	-2.9	-2.9
BSSE1	-1.2	-1.2	-1.2	-1.2	-1.2	-1.2	-1.2
ΔEMP2 (1)	-2.6	-2.7	-2.5	-2.1	-1.9	-1.7	-1.6
ΔEMP2	-3.5	-3.6	-3.4	-3.0	-2.8	-2.6	-2.6
ΔEDFT(0)	-3.5	-3.7	-3.5	-3.1	-2.8	-2.5	-2.4
BSSE		-0.2	-0.1	-0.1	-0.1	-0.1	
ΔEDFT		-3.5	-3.4	-3.0	-2.7	-2.4	
ΔESMM(0)	-2.2	-2.4	-2.0	-1.7	-1.4	-1.2	-1.1
Emtp	-2.6	-2.7	-2.3	-1.9	-1.5	-1.3	-1.2
Erep	1.1	1.0	0.9	0.8	0.6	0.6	0.6
E1	-1.5	-1.7	-1.4	-1.2	-0.9	-0.7	-0.7
Epol	-0.5	-0.3	-0.3	-0.3	-0.3	-0.3	-0.3
Ect	-0.3	-0.4	-0.4	-0.3	-0.2	-0.2	-0.1
E2	-0.8	-0.7	-0.7	-0.6	-0.5	-0.5	-0.3
Edisp	-1.4	-1.4	-1.3	-1.3	-1.3	-1.3	-1.4
ΔESMM	-3.6	-3.8	-3.4	-3.0	-2.7	-2.5	-2.5

Table V. Formate-water. Binding energies in the bidentate 'B' configuration and in the monodentate 'E' configuration as a function of the θ = C-O-Hw angle. Energies in kcal/mole, angles in degrees. See text for definitions.

	'B'	*'E'*					
		θ 105°	*120°*	*135°*	*150°*	*165°*	*180°*
ΔEHF(0)	-16.5	-13.7	-14.3	-14.4	-14.0	-13.7	-13.8
Ec	-26.2	-17.0	-17.1	-16.6	-15.9	-15.4	-15.2
Ee	17.1	9.5	8.8	8.1	7.2	6.7	6.4
E1	-9.2	-7.5	-8.3	-8.5	-8.7	-8.7	-8.8
POL	-4.6	-4.0	-3.9	-3.8	-3.8	-3.7	-3.4
CT	-2.5	-2.0	-1.8	-1.7	-1.4	-1.2	-1.1
E2	-7.1	-6.0	-5.7	-5.5	-5.2	-4.9	-4.5
BSSE	-1.5	-0.8	-0. 9	-0.8	-0.8	-0.7	-0.7
ΔEHF	-15.0	-12.9	-13.4	-13.6	-13.2	-13.0	-13.1
ΔEMP2(0)	-19.6	-16.3	-16.6	-16.3	-15.9	-15.6	-15.6
BSSE1	-3.8	-2.6	-2.5	-2.2	-2.1	-2.1	-2.1
ΔEMP2(1)	-15.8	-13.7	-14.1	-14.1	-13.8	-13.5	-13.5
ΔEMP2	-18.1	-15.5	-15.7	-15.5	-15.1	-14.9	-14.9
ΔEDFT(0)	-21.8	-16.0	-16.3	-16.1	-15.7	-15.3	-15.2
BSSE	-0.7	-0.7	-0.6	-0.6	-0.7	-0.7	-0.7
ΔEDFT	-21.0	-15.3	-15.8	-15.5	-15.0	-14.6	-14.5
ΔESMM(0)	-14.8	-12.8	-14.1	-14.3	-13.9	-13.6	-13.6
Emtp	-21.5	-15.4	-15.4	-14.9	-14.2	-13.6	-13.5
Erep	13.3	7.5	6.1	5.3	4.7	4.2	4.1
E1	-8.2	-7.9	-9.3	-9.6	-9.5	-9.4	-9.4
Epol	-4.3	-2.6	-2.7	-2.7	-2.6	-2.6	-2.7
Ect	-3.2	-2.3	-2.2	-2.0	-1.7	-1.6	-1.5
E2	-7.5	-4.9	-4.9	-4.7	-3.9	-4.2	-4.2
Edisp	-5.2	-3.2	-2.9	-2.7	-2.6	-2.5	-2.5
ΔESMM	-20.9	-16.0	-17.0	-17.0	-16.5	-16.5	-16.1

the MP2 correction are virtually flat in the range 105°-165°. The angular trends of each individual *ab initio* component are correctly mirrored by their SMM counterparts. E1(SMM) is slightly larger than E1(SCF), but this is compensated for by a correspondingly smaller value of E2(SMM) than E2(SCF).

The energy difference between the B and E configurations amounts to 2.4 -4.2 kcal/mol, a range of values in agreement with that from the other *ab initio* calculations *(17, 18)*.

Methoxy-water. Anionic sp^3 oxygens connected to a tetravalent carbon are encountered, *inter alia,* upon formation of tetrahedral acyl intermediates in metallo-enzyme catalyzed cleavage of the peptide bond, and these can be modelled by the CH_3O^- anion. Such intermediates are stabilized, besides their binding to a divalent cation, by H-bonding interactions with neighboring proton donors, including one water molecule (see, e.g., *45*, and Refs. therein). This has prompted us to investigate the binding properties of CH_3O^- with one water molecule, so as to complement our previous study of its binding to Zn^{2+} *(24)*. Furthermore, values of its hydration enthalpies were determined from both experimental gas-phase measurements *(46, 47)* and recent large basis-set *ab initio* computations from others *(17, 18a, 19),* against which it will be instructive to compare the results of our combined methodological approach. Similar to formate-water in the 'E' configuration, and to methylthiolate-water reported below, we monitor the evolution of the binding energies as a function of the C-O-Hw angle in the range 105°-165°. The O-Hw distance is set at 1.80 A, O, Hw, and Ow are collinear, and water is coplanar with its ligand. The results are reported in Table VI .

The angular preferences are more accentuated than in formate-water. The magnitude of the BSSE correction, amounting to 1.5 kcal/mole in the SCF computations, is greater than in formate-water (0.9 kcal/mole), but remains similar as concerns the DFT computations. The energy minimum is found between 105° and 120° for ΔEDFT, and at 120° for both ΔEMP2 and ΔESMM. The numerical agreement between the three procedures, concerning the binding energies, is particularly striking. The value of -21.9 kcal/mole is intermediate between those of the two published experimental gas-phase hydration enthalpies of -19.9 determined from ion-cyclotron resonance by Caldwell et al. in 1984 *(46)*, and of -24 kcal/mole determined from thermochemistry by Meot-Ner (Mautner) in 1986 *(47)*. Somewhat larger monohydration enthalpies than ours, namely -25.8 kcal/mole and -24.3 kcal/mole were obtained by Gao et al. and Zheng and Merz, resp., *(18e, 17)*, but these were uncorrected for BSSE.

In both *ab initio* SCF and SMM computations, the weight of E1 within the total energy is further reduced, at the best value of θ, with respect to the formate-water case, amounting to less than half of the total ΔE, despite the more localized character of the anionic charge of the ligand. Such a greater local character, by increasing the value of the electrostatic term, results in an approximately 0.1 Å closer proximity of Hw than in formate-water, whence a faster increase of E2 and E_{disp} than that of E1, and an accordingly larger relative weight. E_c manifests a more pronounced preference for the smaller values of θ than was the case with formate-water, counteracted by a monotonous decrease of E_e upon increasing θ, a balance of trends resulting in a virtually flat evolution of E1 over the whole 120°-180° range. Similar to formate-water, E_{pol} is very shallow and E_{ct} has a distinct preference for the smaller two values of θ. All these trends are paralleled by the SMM computations.

Methane thiolate-water. Methane thiolate (CH_3S^-) is the terminal group of the cysteinate side-chain, which results from deprotonation of cysteine residues, as observed *inter alia* in the cysteine protease class of enzymes *(48)*. It is also a constitutive group of the class of mercaptan inhibitors, of which thiorphan and

Table VI. Binding energies in the methoxy-water complex, as a function of the θ = C-O-H(w) angle. The O-H(w) distance is set at 1.80 Å. Energies in kcal/mole. See text for definitions.

θ	*105°*	*120°*	*135°*	*150°*	*165°*
ΔEHF(0)	-19.8	-20.3	-20.0	-19.5	-19.1
Ec	-23.6	-23.4	-22.5	-21.5	-20.7
Ee	13.5	12.7	11.8	10.6	9.8
E1	-10.1	-10.6	-10.8	-10.9	-10.9
POL	-6.2	-6.0	-6.1	-5.8	-5.7
CT	-3.4	-3.2	-3.0	-2.6	-2.2
E2	-9.5	-9.1	-9.0	-8.4	-7.9
BSSE	-1.3	-1.4	-1.5	-1.3	-1.2
ΔEHF	-18.5	-18.9	-18.5	-18.2	-17.9
ΔEMP2(0)	-23.2	-23.2	-22.7	-22.0	-21.4
BSSE1	-3.6	-3.6	-3.5	-3.2	-2.9
ΔEMP2(1)	-19.6	-19.6	-19.2	-18.8	-18.5
ΔEMP2	-21.9	-21.8	-21.2	-20.7	-20.2
ΔH (exp)	-19.9 (Ref. 46) -24.0 (Ref. 47)				
ΔEDFT(0)	-21.5	-21.8	-21.6	-21.0	-20.3
BSSE	-0.6	-0.6	-0.6	-0.6	-0.4
ΔEDFT	-20.9	-21.2	-21.0	-20.4	-19.9
ΔESMM(0)	-16.3	-17.0	-16.9	-16.6	-16.5
Emtp	-21.1	-21.1	-20.6	-19.9	-19.3
Erep	12.6	11.6	10.9	10.2	9.6
E1	-8.5	-9.5	-9.7	-9.7	-9.7
Epol	-4.5	-4.1	-4.0	-4.0	-4.0
Ect	3.4	-3.4	-3.2	-3.0	-2.8
E2	-7.9	-7.5	-7.2	-7.0	-6.8
Edisp	-4.8	-4.6	-4.5	-4.4	-4.3
ΔESMM	-21.1	-21.8	-21.6	-21.0	-20.3

analogues are lead compounds *(49)*. We have previously investigated the binding of Zn^{2+} to CH_3S^- and wish to now complement this investigation with that of its binding to water as a model for incoming neutral H-bond donors in, e.g., metalloenzyme recognition sites *(50)*. Table VII reports the evolution of the binding energies as a function of the θ (C-S-Hw) angle, the S-Hw distance being set at 2.4 Å. The BSSE correction in the SCF computations, namely -0.9 kcal/mole, is smaller than in methoxy-water. The greater spatial expansion of the S^- orbitals than of the O^- ones is offset by the increased interatomic separation to the H-donor ligand. On the other hand, the BSSE is, in the DFT computations, larger in methane thiolate-water than in methoxy-water. The energy minimum which occurs at θ = 105° for ΔEMP2 is displaced towards 15° larger values in both DFT and SMM computations. The latter two procedures yield a value (-15.2 kcal/mole) that is 1 kcal/mole higher than ΔEMP2. This numerical value, while close to that from a large MP2 basis set computation by Gao et al. (-15.4 kcal/mole , *Ref. 18e*), is smaller than that reported by Zheng and Merz (-17.4 kcal/mole, *Ref. 17)*, but these two computations were reported in the absence of the BSSE correction. Our computations provide a monohydration enthalpy that is close to the -14 kcal/mole experimental gas-phase monohydration enthalpy of SH^- *(51)*, an anion that was computed *(18a)* to have a closely similar monohydration enthalpy to that of CH_3S^-.

Within the total ΔE, the weight of E1 is smaller than in methoxy-water. E1 has its minimum at 120°, in contrast to the flat behavior it manifested, in methoxy-water, over the whole 120°-180° range. This translates the much more accentuated directionality of the Coulomb term in favor of θ = 105° than was the case in the latter complex. Within E2, E_{ct}, and, to a smaller extent, E_{pol}, display a slightly more accentuated angularity than in methoxy-water. These behaviors are accounted for in the SMM computations. The more accentuated directionality of CH_3S^- -- H_2O than CH_3O^- -- H_2O, as translated by ΔEMP2, ΔEDFT, as well as ΔESMM, is certainly noteworthy, in view of the significantly larger (0.6 Å) increase of the water-ligand separation occurring in CH_3S^- -- H_2O.

Hydroxy-water. Because of the importance of OH^- in chemically- as well as enzymatically-driven catalysis *(52, and refs. therein)*, and as a complement to our prior studies on the Zn^{2+} - OH^- complex *(24)*, we investigate here the binding energetics of this pivotal anion to one water molecule, followed in a succeeding section by its trihydration energetics. We limit ourselves to a model approach of the OH^- ion to one water H, in such a way that the four atoms H-O(-)--Hw-Ow are collinear, and monitor the evolution of the binding energetics only as a function of the O(-)--Hw distance. This is justified since : a) the energy dependence on the θ = H-O(-) --Hw angle was shown to be weak (18e); and (b) because very small anionic or cationic ligands are prone, due to the shortened interatomic distances, to amplify whatever shortcomings may remain in the various theoretical procedures. It was a more crucial issue, prior to a more detailed study, to ascertain first whether these procedures can account correctly for the radial dependencies of the energy and its components. The results are reported in Table VIII.

A substantial value for the BSSE correction, amounting to 3 kcal/mole, is computed at the MP2 minimum. After its subtraction, ΔEMP2 amounts to -27.2 kcal/mole, a value that is very close to the -27.5 kcal/mole value of Gao et al. *(18e)*, and close to the experimental gas-phase monohydration enthalpy of OH^- of -25 kcal/mole *(53)*. Remarkably, the value of E1 within the total energy is the *smallest* within the whole series of anionic ligand monohydrates investigated here. Within E2, E_{ct} has a value that is more than twice that of E_{pol}. On the other hand, the DFT calculations yield a binding energy that is more than 4 kcal/mole weaker than ΔEMP2. This is the most serious discrepancy encountered between the DFT and SCF/MP2 procedures. It may reflect the need, for extremely small anions dealt with in DFT

Table VII. Binding energies in the methane thiolate-water complex, as a function of the θ = C-S-H(w) angle. The S-H(w) distance is set at 2.40 Å. Energies in kcal/mole. See text for definitions.

θ	*105°*	*120°*	*135°*	*150°*	*165°*	*180°*
ΔEHF(0)	-12.0	-11.8	-11.0	-10.0	-9.1	-8.3
Ec	-15.5	-14.9	-13.3	-11.3	-9.5	-8.8
Ee	9.1	8.2	7.1	5.4	4.1	3.5
E1	-6.4	-6.4	-6.2	-5.8	-5.5	-5.2
POL	-3.0	-2.8	-2.6	-2.3	-2.1	-2.0
CT	-2.5	-2.4	-2.1	-1.7	-1.3	-1.2
E2	-5.4	-5.3	-4.7	-4.0	-3.5	-3.2
BSSE	-0.8	-0.8	-0.8	-0.8	-0.8	-0.8
ΔEHF	-11.2	-11.0	-10.2	-9.2	-8.3	-7.5
ΔEMP2(1)	-15.0	-14.6	-13.8	-12.8	-12.0	-11.6
BSSE1	-2.3	-2.3	-2.2	-2.2	-2.2	-2.1
ΔEMP2(1)	-12.7	-12.3	-11.6	-10.6	-9.8	-9.5
ΔEMP2	-14.2	-13.8	-13.0	-12.0	-11.2	-10.8
ΔEDFT(0)	-15.8	-16.4	-16.1	-15.0	-13.5	-12.1
BSSE	-1.2	-1.2	-1.1	-1.1	-1.1	-1.1
ΔEDFT	-14.6	-15.2	-15.0	-13.9	-12.4	-11.0
ΔESMM(0)	-11.7	-11.2	-10.4	-9.4	-8.5	-8.1
Emtp	-14.0	-13.3	-12.0	-10.4	-9.2	-8.7
Erep	8.3	7.7	6.8	5.8	5.1	4.8
E1	-5.7	-5.6	-5.2	-4.6	-4.1	-3.8
Epol	-2.7	-2.6	-2.5	-2.5	-2.4	-2.4
Ect	-3.1	-3.0	-2.7	-2.3	-2.0	-1.9
E2	-5.8	-5.6	-5.2	-4.8	-4.4	-4.3
Edisp	-4.3	-4.2	-4.0	-3.9	-3.7	-3.6
ΔESMM	-15.2	-15.3	-15.2	-14.2	-13.1	-12.1

Table VIII. Binding energies in the hydroxy-water complex as a function of the O(-) -- H(w) distance. Energies in kcal/mole, distances in Å. See text for definitions.

d O(⁻) -- H(w)	*1.5*	*1.6*	*1.7*	*1.8*
ΔEHF(0)	-24.6	-25.7	-25.8	-25.3
Ec	-39.7	-33.9	-29.7	-25.9
Ee	37.5	26.0	18.5	12.9
E1	-2.2	-7.9	-11.2	-13.0
POL	-5.6	-4.7	-4.0	-3.6
CT	-17.0	-13.2	-10.5	-8.6
E2	-22.5	-17.9	-14.5	-12.1
BSSE	-3.8	-3.7	-3.1	-3.0
ΔEHF	-20.8	-22.0	-22.7	-22.3
ΔEMP2(0)	-29.4	-30.3	-30.3	-29.6
BSSE1	-8.1	-7.7	-7.1	-6.8
ΔEMP2(1)	-21.3	-22.6	-23.2	-22.1
ΔEMP2	-25.6	-26.6	-27.2	-26.6
ΔH (exp)	-27.5 (Ref. 53)			
ΔEDFT(0)	-22.1	-23.1	-23.2	-22.7
BSSE	-0.7	-0.7	-0.6	-0.5
ΔEDFT	-21.4	-22.4	-22.6	-22.2
ΔESMM(0)	-18.9	-21.3	-22.2	-22.0
Emtp	-33.1	-29.3	-26.1	-23.3
Erep	28.9	20.3	14.3	10.1
E1	-4.3	-9.0	-11.7	-13.2
Epol	-9.6	-8.1	-6.9	-5.9
Ect	-5.0	-4.2	-3.6	-3.0
E2	-14.6	-12.3	-10.5	-8.9
Edisp	-10.4	-7.6	-5.6	-4.3
ΔESMM	-29.3	-28.9	-27.8	-26.3

computations, for an improved representation through the use of an augmented basis set, and this will be reappraised in a forthcoming study. At shorter distances our SMM binding energies are overestimated with respect to ΔEMP2. The energy minimum occurs at 1.5 Å, as contrasted to 1.7 in MP2. E1(SMM) matches well E1(SCF), but E2(SMM) is smaller than E2(SCF). Subtracting from the latter the entirety of the BSSE correction reduces substantially the gap between the two E2 values. In the absence of E_{disp}, ΔESMM is smaller than the uncorrelated ΔESCF, and the reverse situation in the presence of dispersion appears then due to an exaggeration of E_{disp} in the SMM procedure at the shortest <1.8 Å distances. Within E2(SMM), E_{ct} has a smaller value than E2(SCF). This is in contrast to the SCF computations. We will evaluate below, upon investigating the trihydrates of OH^-, how these discrepancies evolve in more complex, and chemically relevant, systems.

Methylammonium-water. The methylammonium moiety (MMA) is encountered as the terminal group of the lysine side-chain in peptides and proteins, in the structure of pharmacological drugs, where it serves as a privileged ligand for the binding of anionic and polar amino-acids, etc. In line with previous studies from one of our Laboratories (3a), we compare here the binding energetics of MMA to water in three distinct configurations : linear, L, bisector, B, and external, E, at an optimized N-O distance of 2.8 Å. The results are reported in Table IX.

All three methodologies are seen to provide the ordering L > B=E. The binding energies in the B configuration, as computed by the three procedures, are identical, namely -18.3 kcal/mole, a value very close to the experimental gas-phase monohydration enthalpy of MMA of -18.8 kcal/mole determined by Lau and Kebarle *(53)*. A value of -19.3 kcal/mole was published by Ikuta, Gao and Jorgensen, and Zheng and Merz at the 6-31G* level (17 18f, 21), which increased to -22.3 kcal/mole at the MP2 level (17). The values were, however, uncorrected for the BSSE, but our present basis set provides an estimate for the BSSE of less than 0.6 kcal/mole. It is instructive to observe that, reminiscent of the water dimer case (see above), even though E_c (and its E_{mtp} counterpart) attain their best values in the L configuration, this is counteracted by the short-range repulsion term, which is itself largest in it. As a result, E1 has its *smallest* value in the L configuration, amounting to only about half of the total binding energy, and it is the second-order term, E2, that ensures the proper discrimination in favor of L rather than B or E. These features are accounted for in the SMM computations. However, even though the magnitude of E2(SCF) is well reproduced by E2(SMM), E_{ct}(SMM) is definitely underestimated with respect to E_{ct}(SCF), whereas E_{pol}(SMM) is overestimated. This may reflect the limitations of the present formulation of E_{pol} and E_{ct}(SMM). We will leave this issue unresolved for the moment. Its impact for larger molecular complexes will be evaluated below, upon dealing with the trihydrates of charged molecules.

Hydronium-water. The importance of the hydronium ion in chemical problems warrants an investigation of its hydration energetics, for which experimental *(54, 55)* and theoretical *(22)* results are available, against which we wish to compare those of our own combined theoretical approach. The binding energy of the H_3O^+ - H_2O complex was optimized by performing 0.1 Å step-wise variations of the intermolecular Ow-H distance, the hydronium OH bond being collinear with the external bisector of the water molecule. As was the case with the hydroxy-water complex, no angular variations were made. The results are reported in Table X.

All three procedures locate the energy minimum at 1.5-1.6 Å. In the SMM procedure, however, a proper match of E1 to E1(SCF) could only be obtained if the effective radii of the hydronium hydrogens were reduced from 1.7 to 1.2 Å. This can be justified on the basis of the more contracted nature of the 1s orbitals of hydronium hydrogens as compared to standard ones (including those linked to the ammonium

Table IX. Binding energies in the methylammonium-water complex in the linear, bisector, and axial configurations. The O-N distance is 2.8 Å. Energies in kcal/mole. See text for definitions.

	Linear	*Bisector*	*Axial*
ΔEHF(0)	-16.1	-13.4	-13.5
Ec	-19.3	-13.6	-14.4
Ee	10.1	3.0	4.0
E1	-9.2	-10.6	-10.4
POL	-2.6	-1.4	-1.4
CT	-4.0	-1.3	-1.5
E2	-6.5	-2.7	-2.9
BSSE	-0.6	-0.5	-0.6
ΔEHF	-15.5	-12.9	-12.9
ΔEMP2(0)	-19.0	-15.8	-15.7
BSSE1	-2.5	-2.1	-2.2
ΔEMP2(1)	-16.5	-13.7	-13.5
ΔEMP2	-18.4	-15.3	-15.1
ΔH (exp)	-18.8 (Ref. 53)		
ΔEDFT(0)	-18.8	-14.3	-13.9
BSSE	-0.5	-0.4	-0.3
ΔEDFT	-18.3	-13.9	-13.6
ΔESMM(0)	-15.1	-12.1	-12.3
Emtp	-18.0	-13.2	-12.6
Erep	8.7	3.9	2.8
E1	-9.3	-9.3	-9.8
Epol	-4.9	-2.5	-2.4
Ect	-0.8	-0.2	-0.1
E2	-5.7	-2.7	-2.5
Edisp	-3.2	-2.0	-1.6
ΔESMM	-18.3	-14.3	-13.9

Table X. Binding energies in the hydronium-water complex as a function of the H -- O(w) distance. Energies in kcal/mole, distances in Å. See text for definitions.

d H -- O(w)	*1.5*	*1.6*	*1.7*	*1.8*
ΔEHF(0)	-27.1	-26.9	-26.1	-24.9
Ec	-30.8	-27.6	-24.9	-22.6
Ee	21.5	14.5	9.7	6.5
E1	-9.3	-13.1	-15.2	-16.1
POL	-6.8	-4.7	-4.0	-3.3
CT	-11.5	-8.6	-6.6	-5.2
E2	-18.2	-13.4	-10.6	-8.5
BSSE		-0.6	-0.6	-0.6
ΔEHF		-26.3	-25.5	-24.4
ΔEMP2(0)	-30.0	-29.5	-28.6	-27.0
BSSE1	-2.6	-2.4	-2.2	-2.0
ΔEMP2(1)	-27.4	-27.1	-26.4	-25.0
ΔEMP2		-28.9	-28.0	-26.5
ΔH (exp)		-31.5 (Ref. 55)		
ΔEDFT(0)	-31.4	-30.1	-29.2	-27.5
BSSE	-0.7	-0.5	-0.5	-0.5
ΔEDFT	-30.7	-29.6	-28.7	-27.0
ΔESMM(0)	-22.0	-23.4	-23.8	-23.2
Emtp	-29.2	-26.4	-23.9	-21.8
Erep	19.9	13.2	8.8	5.9
E1	-9.3	-13.1	-15.1	-15.9
Epol	-11.6	-9.6	-8.1	-6.8
Ec	-1.0	-0.8	-0.6	-0.5
E2	-12.6	-10.4	-8.7	-7.3
Edisp	-6.5	-4.8	-3.6	-2.8
ΔESMM	-28.5	-28.4	-27.4	-26.0

nitrogen). The three procedures are then shown to provide closely similar minimum energy values, namely -28.9, -30.7, and -28.5 kcal/mole, at the correlated BSSE-corrected MP2 and DFT levels, and with the SMM procedure respectively. These values are close to the experimental determinations of the monohydration enthalpies of H_3O^+ due to Grimsrud and Kebarle, amounting to -31.5 kcal/mole *(55)*. They are also very close to the -29.3 kcal/mole value computed by Kochanski *(22)*, but the latter was computed in the absence of correlation, for which the present MP2 estimate amounts to 3 kcal/mole at 1.5 Å. Similar to hydroxy-water versus methoxy-water, the more localized nature of the charge on hydronium versus methylammonium results in E1 having a weight at equilibrium (<50 percent), smaller than in methylammonium-water. As discussed above, this is due to the shortened equilibrium distance, favoring a steeper increase of E2 than E1. As was the case in methylammonium-water, E_{ct}(SMM) is considerably underestimated with respect to E_{ct}(SCF), compensated for by a larger value of E_{pol}. E_{disp}(SMM) is overestimated with respect to the MP2 correction. Despite the shortened intermolecular distances, the BSSE correction remains small (<0.6 kcal/mole).

Formate-methylammonium. The complex between formate and methylammonium is a prototype of intra- and inter-molecular interactions taking place between aspartate and glutamate side-chains, and lysine side-chains in peptides and proteins, between pharmacological drugs and the recognition site of macromolecular receptors, etc. In a 1980 investigation, Gresh and Pullman *(56)* reported the binding energetics of this complex using a (7s,3p/3s) basis set, but restricted at that time to the bidentate complex. In their 1992 paper, Zheng and Merz reported on their results using a 6-31G* basis set, again restricted to this bidentate complex *(17)*. Their binding energy of -125 kcal/mole was the same as in our 1980 paper. We compare here the binding energetics of formate-methylammonium in both bi- and monodentate modes, in the same manner as was done for formate-water. In the bidentate binding mode, the distance between each bound ammonium proton and the carboxylate oxygen is 1.9 Å. In the monodentate mode the O--H (N) distance is 1.7 Å. The results are reported in Table XI.

For both kinds of configurations, and over the whole range of angular values explored, a most satisfactory agreement between the three procedures is obtained. The overall match is to within 2.5 kcal/mole out of 100, except for $\theta = 105°$, where ΔESMM is overestimated by 4.3 kcal/mole with respect to ΔEMP2. It is thus reassuring to observe that the approximately 2 kcal/mole energy difference out of 20 which was observed between the MP2 and SMM results in formate-water is by no means amplified in formate-methylammonium, even though the energies are more than five times greater.

ΔE(HF) in the bridging position amounts to -125 kcal/mole, the same value as in the two above-mentioned investigations. The MP2 correction adds a further 4 kcal/mole to this value. The BSSE correction of 1.8 kcal/mole is only 0.3 kcal/mole larger than in formate-water. In the monodentate position, and for all three procedures, the binding energy remains virtually flat over the whole range of θ values investigated, being affected by <1 kcal/mole out of 110. The sole exception is $\theta = 105°$ in SMM, as mentioned above. Such a behavior is strikingly similar to the one observed in formate-water, whereas the Zn^{2+} - formate complex was shown to manifest a much more accentuated angular character *(7)*. E1(SCF) increases progressively upon increasing θ, whereas E_c has its best values at 105° and 120°. This is correctly accounted for by the corresponding E_{mtp} and E1(SMM) terms. The angular preference of E2(SCF) for $\theta = 105°$ is not reflected in the SMM computations, E2(SMM) remaining flat in the whole angular range. Its underestimated values stem for a large part from the persistent underestimation of E_{ct}(SMM) whenever a nonmetallic cation is involved. Nevertheless, because this is compensated for here by

Table XI. Formate-methylammonium complex. Binding energies in the bidentate 'B' configuration and in the monodentate 'E' configuration as a function of the θ = C-O-Hw angle. Energies in kcal/mole, angles in degrees. See text for definitions.

	'B'			*'E'*			
	θ	*105°*	*120°*	*135°*	*150°*	*165°*	*180°*
ΔEHF(0)	-124.8	-105.1	-106.3	-106.7	-106.8	-107.0	-107.8
Ec	-132.6	-107.4	-107.5	-106.8	-105.9	-105.5	-105.8
Ee	26.3	21.2	19.8	18.1	16.4	15.1	14.6
E1	-106.3	-86.2	-87.7	-88.7	-89.5	-90.3	-91.3
POL	-12.0	-11.7	-11.6	-11.5	-11.3	-11.2	-11.2
CT	-5.2	-5.8	-5.7	-5.2	-4.6	-4.1	-4.0
E2	-17.2	-17.6	-17.3	-16.6	-15.9	-15.3	-15.2
BSSE	-1.7	-1.1	-1.1	-1.1	-1.0	-1.0	-1.0
ΔEHF	-123.1	-104.0	-105.2	-105.6	-105.8	-106.0	-106.8
ΔEMP2(0)	-129.9	-111.9	-111.7	-111.2	-110.8	-110.8	-111.3
BSSE1	-5.3	-4.0	-3.6	-3.4	-3.3	-3.2	-3.2
ΔEMP2(1)	-124.6	-107.9	-108.1	-107.8	-107.5	-107.6	-108.0
ΔEMP2	-128.2	-110.8	-110.6	-110.1	-109.8	-109.8	-110.1
ΔEDFT (0)	-129.9	-110.2	-110.1	-109.8	-109.5	-109.6	-110.2
BSSE	-1.0	-0.8	-0.8	-0.9	-0.9	-1.1	-1.1
ΔEDFT	-128.9	-109.4	-109.3	-108.9	-108.6	-108.5	-109.1
ΔESMM(0)	-121.9	-100.8	-103.9	-104.8	-104.9	-105.2	-105.3
Emtp	-128.0	-105.9	-105.9	-105.0	-103.8	-103.0	-103.3
Erep	21.9	17.1	14.2	12.3	10.8	9.7	9.7
E1	-106.1	-88.8	-91.7	-92.6	-93.0	-93.3	-93.3
Epol	-13.6	-10.0	-10.2	-10.4	-10.4	-10.5	-10.7
Ect	-2.2	-1.9	-1.8	-1.6	-1.5	-1.3	-1.2
E2	-15.8	-11.9	-12.0	-12.0	-11.9	-11.8	-11.9
Edisp	-7.2	-5.7	-5.2	-4.9	-4.7	-4.5	-4.4
ΔESMM	-129.3	-106.5	-109.1	-109.7	-109.6	-109.7	-109.7

a slight overestimation of E1(SMM) with respect to E1(SCF), a close numerical agreement between the total binding energies can be achieved.

Binding of the Methoxy and Hydroxy Anions and of the Methylammonium and Hydronium Cations to Three Water Molecules. In this section, we will solvate the methoxy and hydroxy anions, and the methylammonium and hydronium cations, with three water molecules, a number corresponding to the first hydration shells *(19, 22)*. This will enable us to assess the extent to which the nonadditive behavior of the total *ab initio* interaction energy can be accounted for. The results are gathered in Table XII. Only one distance will be given for each trihydrate, the distance minima specific for each procedure being at most 0.1 Å different from it, the actual energy being <1 kcal/mole greater.

a) CH_3O^- -- $3H_2O$. Each water molecule is placed at a θ angle of 120°, and the respective dihedral angles H-C-O--Hw have the values 0°, 120°, and 240°. The binding energies computed by the three procedures agree to within 1.1 kcal/mole out of 55. Some improper balance of the weights of E1 and E2(SCF) can be observed in the SMM computations. The increases of E1(SMM) and E2(SMM) are smaller than the corresponding SCF ones, when compared to their values in the monohydrates. Edisp(SMM) is on the other hand, larger than the MP2 correction by 6 kcal/mole. This is larger than the value of 4.8 that would have been anticipated on the basis of the corresponding 1.6 kcal/mole energy difference found in the monohydrate. The binding energies of -55.5, -54.6, and -54.4 kcal/mole computed by the MP, DFT, and SMM procedures respectively, compare well with the experimental trihydration enthalpy of methoxy amounting to -57.9 kcal/mole, as determined by Meot-Ner (Mautner) *(47)*.

b) OH^- -- $3H_2O$. The discrepancy between the MP2 and DFT values simply reflects and amplifies the one preexisting at the level of the monohydrate. The SMM values are somewhat (4.4 kcal/mole out of 65) exaggerated with respect to the MP2 value, also a reflection of the situation occurring with the monohydrate. E2(SMM) is smaller than E2(SCF) by 11.3 kcal/mole, but the difference can be substantially reduced if part of the BSSE correction, amounting to 9.4 kcal/mole, were removed from the latter. The MP2 and SMM values of -64 and -68 kcal/mole are exaggerated with respect to the experimental OH^- gas-phase trihydration enthalpy of -60.6 kcal/mole *(47)*.

c) $CH_3NH_3^+$ -- $3H_2O$. An extremely satisfactory agreement between the three procedures is found. Concerning the SCF and SMM procedures, it occurs already at the levels of the E1 and E2 components, so that the numerical match of total energies is not due to a partial compensation of errors as was the case in the anion-trihydrate complexes. The imbalance of E_{ct} and E_{pol} in the SMM procedure does not impair the nonadditivity of E2(SMM) as compared to E2(SCF).

The first *ab initio* SCF computations devoted to the trihydration of the parent compound, NH_4^+, were due to Pullman and Armbruster in 1975 *(57)*. The trends in reduction of the hydration enthalpy upon progressive occupancy of the first hydration shell were clearly shown by these authors, even though as acknowledged by them, the numerical values of the binding energies were exaggerated, due to the necessary use at that time of a minimal basis set. The total binding energies amount to -49 kcal/mole in the MP2 and SMM procedures, and -47 kcal/mole in the DFT procedures, slightly larger than the experimental value of -45.8 kcal/mole *(54)*.

Table XII. Binding energies in the trihydrate complexes of methoxy, hydroxy, methylammonium, and hydronium ions. Energies in kcal/mole, distances in Ångstroms. See text for definitions.

	CH_3O^- -- $3H_2O$	OH^- -- $3H_2O$	$CH_3NH_3^+$ -- $3H_2O$	H_3O^+ --$3H_2O$
	d O--Hw 1.8	*d O--Hw 1.7*	*d N--Ow 2.9*	*d H--Ow 1.6*
ΔEHF(0)	-50.5	-60.8	-42.5	-68.2
E1	-27.2	-28.2	-29.2	-35.7
POL	-7.4	-9.7	-4.8	-11.1
CT	-15.3	-23.7	-7.9	-21.0
E2	-22.7	-33.4	-12.8	-32.2
BSSE	-4.3	-9.4	-1.5	-1.8
ΔEHF	-46.2	-51.4	-41.0	-66.4
ΔEMP2(0)	-59.8	-73.1	-50.5	-76.1
BSSE1	-11.0	-20.5	-6.9	-7.0
ΔEMP2(1)	-48.8	-52.6	-43.6	-69.1
ΔEMP2	-55.5	-63.7	-49.0	-74.3
ΔH (exp)	-57.9	-60.6	-45.8	-68.5
	(Ref. 47)	(Ref. 47)	(Ref. 54)	(Ref. 58)
ΔEDFT(0)	-56.5	-59.1	-49.1	-78.8
BSSE	-1.9	-1.7	-1.8	-1.9
ΔEDFT	-54.6	-57.5	47.3	-76.9
ΔESMM(0)	-39.3	-51.2	-39.5	-61.0
Emtp	-57.8	-72.4	-47.3	-75.9
Erep	35.4	43.3	17.6	39.7
E1	-22.4	-29.1	-29.7	-36.2
Epol	-8.9	-13.9	-9.8	-22.5
Ect	-8.0	-8.2	-1.7	-2.2
E2	-16.9	-22.1	-11.5	-24.7
Edisp	-15.1	-16.9	-7.8	-14.6
ΔESMM	-54.4	-68.1	-49.1	-75.6

d) H_3O^+ -- $3H_2O$. The three procedures provide trihydration enthalpies for H_3O^+ that agree to within 2.6 kcal/mole out of 75. As was the case in the monohydrate of H_3O^+, whereas E1(SMM) is very close to E1(SCF), E2(SMM) is underestimated with respect to E2(SCF), but this is compensated for by a larger value of E_{disp} with respect to the MP2 correction. Once again, such a relative imbalance does not appear to affect the numerical match of the values of ΔESMM with respect to ΔEMP2 or ΔEDFT. The binding energies of -74.3, -76.9, and -75.6 kcal/mole computed with the MP2, DFT, and SMM procedures are larger than the 1977 experimental gas-phase trihydration enthalpy of H_3O^+ of -68.5 kcal/mole *(58)*. A large drop in the gas-phase hydration enthalpy, amounting to 11.5 kcal/mole, was observed upon passing from one to two water molecules, surprisingly limited to only 3 kcal/mole upon passing then from two to three water molecules *(58)*. Theoretical computations could not account for such a severe decrease : thus, this 11.5 kcal/mole energy drop can be contrasted to the corresponding theoretical drop of only 5.7 kcal/mole which was reported by Kochanski *(22)*.

CONCLUSIONS

Through the scrutiny of the energetic and structural properties of a series of fourteen representative neutral and ionic H-bonded complexes, a very satisfactory agreement was found between the results of three distinct theoretical methodologies, namely *ab initio* SCF/MP2, density functional (DFT), and the SIBFA molecular mechanics procedure (SMM). For each complex, we have investigated, at the *ab initio* SCF level, the dependence of each individual component as a function of configuration, or of a key radial or angular variable. Such dependencies were in the general case well accounted for by that of the corresponding SMM term. Our computed binding energies range from -3.8 kcal/mole for the most weakly bound complex, namely CH_3SH-H_2O, to -130 for the most strongly bound one, formate-methylammonium. Whatever lingering numerical mismatches between *ab initio* SCF/MP2 or SMM results could be observed with a weaker complex, were by no means amplified upon passing to a stronger one, as exemplified by formate-water versus formate-methylammonium. Prior to the MP2 correlation computation, the amplitude of the BSSE correction to the SCF computations was found to be modest, except in the case of the hydrates of the anions OH^-, CH_3O^-, where it reaches appoximately 3.5 kcal/mole. In contrast, the correlated BSSE correction was found to be unreasonably large, and we were led to adopt as a final value for the total SCF/MP2 energy that computed after subtraction of the *uncorrelated* BSSE correction. The BSSE correction in the DFT computations remained persistently small, smaller than 1.1 kcal/mole in the bimolecular complexes. The optimized monohydration energies of the anions OH^-, CH_3O^-, CH_3S^-, and cations H_3O^+, $CH_3NH_3^+$, are in close agreement with the available experimental gas-phase enthalpies, as well as with previous *ab initio* SCF computations using large gaussian basis sets. Possibly the sole serious discrepancy between our three procedures relates to the DFT monohydration enthalpy of OH^-, which is about 5 kcal/mole out of 25 smaller than the MP2 one. The use of a larger basis set on the anion could alleviate this situation, and will be evaluated in a forthcoming study.

A closer comparison between the SCF and SMM components of E2 still reveals some caveats. Thus, E_{ct}(SMM) is underestimated with respect to E_{ct}(SCF) in the complexes involving a cationic molecule, $CH_3NH_3^+$ or H_3O^+, with a neutral one, water, or an anionic one, formate. To a large extent, this is compensated for by a E_{pol}(SMM) term that is correspondingly larger than E_{pol}(SCF), so that E2(SMM) still compares well with E2(SCF). With the exception of the hydroxy-water complex, in the anion-neutral ligand complexes, and *a fortiori* in those involving two neutral ligands, E_{ct}(SMM) does satisfactorily match E_{ct}(SCF). Even hydroxy-water could provide a good match of the Ect values if the BSSE correction were to be entirely subtracted from E_{ct}(SCF). Upon passing from the mono- to the trihydrated complexes, all above-mentioned deficiencies did not impair the agreement of numerical values between *ab initio* and molecular mechanics procedures. Not surprisingly, the least satisfactory results obtained with the SMM procedure are those relating to the hydrates of OH^- and H_3O^+, on account of the very short (<1.7 Å) intermolecular equilibrium distances.

We believe that a key feature for the successful reproduction of the *ab initio* SCF supermolecular interaction energies by molecular mechanics procedures resides in the existence of a *separate* formulation for each of the five distinct *ab initio* contributions, and it can be crucial for that purpose to be able to unravel them by means of an energy-decomposition procedure. This was stressed by ourselves *(3, 7)* and others *(10-13)*. It should ensure the transferability of the parameters to virtually any other complex than the ones at hand. In the present study, we needed to resort to only about fifteen computations on the water dimer to calibrate the coefficients and exponents of E_{rep} and E_{ct}, and the effective radius of hydrogen. The calibration of E_{disp} was done by performing a limited number of distance variations on the linear water dimer at preset θ, by comparison to the results of the SAPT procedure *(39 , and J. Langlet, private communication)*. The effective radius of the H_3O^+ hydrogens was calibrated on the basis of the hydronium ion-water complex. The K_z(S) coefficient (3a) for sulfur used in the expression of E_{rep} and E_{disp} involving this atom with H,C,N, and O, was calibrated on the basis of the CH_3SH-H_2O complex, water approaching along the external bisector of S, and remaining in the C-S-H plane. *All* other parameters had been previously derived from our study devoted to the complexes of Zn^{2+} with a series of molecules bearing the same electron-donating functionalities as investigated here.

Simplified force-fields, in which the energy is expressed as a sum of Lennard-Jones and point-charge Coulomb terms, are widely used in Monte-Carlo simulations devoted to, e.g., the solution behavior of neutral or ionic molecules *(14, 18d-f, and references therein)*. The solute parameters are fitted by comparison to the results of *ab initio* SCF HF or HF/MP2 computations on a series of its monohydrate complexes, so that the total molecular mechanics energy reproduces the total *ab initio* one, and no partitioning of ΔE(SCF) into its separate components is attempted. The water molecules, which, similar to the solute, are devoid of atomic polarizabilities and higher-order electrostatic multipoles, are generally endowed with a deliberately exaggerated imbalance of net atomic charges, as is the case with the OPLS calibration *(16b, and references therein)*. This endows each water molecule with an overestimated dipole moment corresponding to its value in bulk water rather than in isolation. The success of this procedure for bulk solvation properties may not necessarily be extended to, *inter alia,* 'discrete' water molecules entrapped within highly polar cavities, as in metalloenzyme sites. Furthermore, as a consequence of the fitting procedure, the imbalance of charges on the solvent may entail a corresponding imbalance on the solute(s) charges and affect the mutual interactions involving non-water molecules in molecular recognition processes.

In the SMM computations, the distributed multipoles and polarizabilities which we used for each monomer were those determined from the *uncorrelated* HF wave function. This was necessary for consistency purposes, since the RVS intermolecular computations, against which we wish to check the SMM ones, could only be carried out *prior* to the MP2 procedure. It is realized that the energy gain due to MP2 can encompass, in addition to the intermolecular dispersion energy itself, an improvement of the self-energy of each individual monomer due to intramolecular correlation. However, the satisfactory reproduction of the MP2 increment by E_{disp}(SMM), and the close agreement between ΔESMM and ΔEDFT indicates that the present procedure is justified from an operational standpoint. In a future work, we will examine in more detail the effects of the MP2 correlation on the magnitude of the distributed multipoles and polarizabilities and its impact on E_{mtp} and E_{pol}(SMM).

Our results, in line with those of ref. *(2a)*, also demonstrate the accuracy and reliability of the DFT procedure in the computations of H-bonded interactions. We were able to extend the scope of ref. *(2a)* to a series of complexes involving at least one ionic molecule. As a continuation of our combined approach, we intend to study the intramolecular interactions governing the conformational behavior of average-sized molecules. Insight into the individual energy components should be afforded by the energy partitioning of the SMM procedure.

In conjunction with our results on a series of Zn^{2+} -- ligand complexes, this study also indicates the robustness of the SMM procedure in reproducing the results of high-quality *ab initio* SCF supermolecule interactions. Applications to metalloenzyme - inhibitor interactions have been inititated. An updated version of the SIBFA program (present QCPE release under number QCPE 614) should be released in the near future.

ACKNOWLEDGMENTS

N. Gresh thanks CERCA for support during his three-month stay as a visiting scientist. The RVS computations were carried out using an Alliant computer at the Center for Advanced Research in Biotechnology in Rockville, Maryland and were initiated during a postdoctoral stay of Nohad Gresh at CARB. It is a pleasure to thank Drs. Walt Stevens and Morris Krauss for the generous access to these facilities and the use of the RVS code of Walt Stevens. Martin Leboeuf thanks NSERC for a postgraduate scholarship.

LITERATURE CITED

1. Stevens, W. J.; Fink, W. *Chem. Phys. Letts.* **1987,** *139*, 15.
2. a) Sim, F.; St-Amant, A.; Papai, I.; Salahub, D. R. *J. Am. Chem. Soc.* **1992,** *114* , 4392.
   b) St-Amant, A.; Salahub, D. R. *Chem. Phys. Letts.* **1990,** *169*, 387.
   c) St-Amant, A. thesis, Université de Montréal, **1992.**
   d) Salahub, D. R.; Fournier, R.; Mlynarski, P.; Papai, I.; St-Amant, A.; Ushio, J. in *Theory and Applications of Density Functional Approaches to Chemistry*; Labanowski, J. and Andzelm, J., Eds.; Springer Verlag, Berlin, 1991.
3. a) Gresh, N.; Claverie, P.; Pullman, A. *Theoret. Chim. Acta* **1984,** *20*, 1.
   b) Gresh, N.; Claverie, P.; Pullman, A. *Int. J. Quantum Chem.* **1986,** *29*, 101.
4. Kitaura, K.; Morokuma, K. *Int. J. Quantum Chem.* **1985,** *10*, 325.
5. Pople, J. A.; Binkley, J. S.; Seeger, R. *Int. J. Quantum Chem.,* Symp. **1976,** *10*, 1.
6. Boys, S. F.; Bernardi, F. *Molec. Phys.* **1970,** *19*, 553.
7. Gresh, N.; Stevens, W.J.; Krauss, M. *J. Comp. Chem.*, to be submitted.
8. Basch, H.; Garmer, D. R.; Jasien, P. G.; Krauss, M.; Stevens, W. J. *Chem. Phys. Letts.* **163,** 514 (1989).
9. Stevens, W. J.; Basch, H.; M. Krauss *J. Chem. Phys.* **1984,** *81*, 6026.
10. Krauss, M.; Stevens, W. J. *J. Am. Chem. Soc.* **1990,** *112*, 1466.
11. Stone, A. J.; Price, S. L. *J. Phys. Chem.* **1988,** *92*, 3325.
12. Mitchell, J.; Price, S. L. *J. Comput. Chem.* **1990,** *11*, 1217.
13. Millot, C.; Stone, A. *J. Molecular Physics* **1992,***77*, 439.
14. Rullman, J. A. C.; van Duijnen, P. Th. *Reports in Molecular Theory* **1990,** 1.

15. a) Parr, R. G.; Yang, W. *Density Functional Theory of Atoms and Molecules;* Oxford University Press, New-York, 1989
b) Sambe, H.; Felton, R. H. *J. Chem. Phys.* **1975**, *62*, 1122.
c) Dunlap, B. I.; Connolly, J. W. D.; Sabin, J. R. *J. Chem. Phys.* **1979,** *71*, 3396.
16. a) Alagona, G.; Pullman, A.; Scrocco, E.; Tomasi, J. *Int. J. Peptide Protein Res.* **1973,** *5*, 251.
b) Jorgensen, W. L.; Swenson, C. J. *J. Am. Chem. Soc.* **1985,** *107*, 1489.
c) Jasien, P. G.; Stevens, W. J. *J. Chem. Phys.* **1986,** *84*, 3271.
17. Zheng, Y.-J.; Merz, K.M., *J. Comp. Chem.* **1992,** *13*, 1151.
18. a) Port, G.N.J.; Pullman, A. *Int. J. Quantum Chem.*, Quantum Biol. Symp. **1974,** *32* , 1.
b) Berthod, H.; Pullman, A. *J. Comp. Chem.* **1981,** *2*, 87.
c) Lukovits, I.; Karpfen, A.; Lischka, H.; Schuster, P. *Chem. Phys. Letts.* **1979,** *63*, 151.
d) Alagona, G.; Ghio, C.; Kollman, P. A. *J. Am. Chem. Soc.* **1986,** *108*, 185.
e) Gao, J.; Garner, D. S.; Jorgensen, W. L. *J. Am. Chem. Soc.* **1986,** *108*, 4784.
f) Jorgensen, W. L.; Gao, J. *J. Phys. Chem.* **1986**, *90*, 2174.
g) Cybulski, S. M.; Scheiner, S. *J. Am. Chem. Soc.* **1989,** *111*, 23.
19. Jorgensen, W. L.; Briggs, J. M. *J. Am. Chem. Soc.* **1989,** *111*, 4190.
20. Rohlfing, C. M.; Allen, L. A.; Cook, C. M.; Schlegel, H. B. *J. Chem. Phys.* **1983,** *78*, 2498 and Refs. therein.
21. Ikuta, S. *Chem. Phys. Letts.* **1983,** *95*, 604.
22. Kochanski, E. *Chem. Phys. Letts.* **1987,** *133*, 143.
23. Umeyama, H.; Morokuma, K. *J. Am. Chem. Soc.* **1977,** *99*, 1316.
24. Gresh, N.; Stevens, W. J.; Krauss, M. to be submitted
25. Garmer, D. R.; Gresh, N. *J. Am. Chem. Soc.*, submitted
26. a) Dupuis, M.; King, H. F. *Int. J. Quantum Chem.* **1977,** *11*, 613.
b) Dupuis, M; King, H. F. *J. Chem. Phys.* **1978,** *68*, 3998.
27. Perdew, J. P.; Wang, Y., *Phys. Rev. B* **1986,** *33*, 8800.
28. Perdew, J. P. *Phys. Rev. B*, **1988,** *33*, 8822; erratum in *Phys. Rev. B* **1986,** *38*, 7406.
29. Becke, A.D. *J. Chem. Phys.* **1988,** *88*, 1053.
30. a) Wang, Y.; Perdew, J.P. *Phys. Rev. B,* **1991**, *44*, 13298.
b) Perdew, J.P.; Wang, Y. *Phys. Rev. B,* **1992**, *45*, 13244.
c) Perdew, J.P.; Chevary, J.A.; Vosko, S.H.; Jackson, K.A.; Pederson, M.R.; Singh, D.J.; Fiolhas, C. *Phys. Rev. B,* **1992**, *46*, 6671.
31. Mijoule, C.; Latajka, Z.; Borgis, D. *Chem. Phys. Letts*, **1993**, *208*, 364.
32. Rybak, S.; Jeziorski, B.; Szalewicz, K. *J. Chem. Phys.* **1991**, *95*, 6576.
33. Feller, D. *J. Chem. Phys.* **1992,** *96*, 6104.
34. Vigne-Maeder, F.; Claverie, P. *J. Chem. Phys.* **1988,** *88*, 4988.
35. Garmer, D. R.; Stevens, W. J. *J. Phys. Chem.* **1989,** *93*, 8263.
36. Gresh, N.; Claverie, P.; Pullman, A. *Int. J. Quantum Chem. Symp.* **1979,** *13*, 243.
37. a) Stone, A. J. *Mol. Phys.* **1985,** *5*, 1065.
b) Stone, A. J. *Chem. Phys. Letts.* **1989,** *155*, 111.
38. a) Creuzet, S.; Langlet, J.; Gresh, N. *J. Chim. Phys.* **1991,** *88*, 2399.
b) Hess,O.; Caffarel, M.; Langlet, J.; Caillet, J.; Huiszoon, C. and Claverie, P. in *Modelling of Molecular Structures and Properties,* J.-L. Rivail, Ed. **1990,** Vol. 71, 323.
39. Hess, O.; Caffarel, M.; Huiszoon, C; Claverie, P. *J. Chem. Phys.* **1990,** *92*, 6049.

40. Saebo, S.; Fong, W.; Pulay, P. *J. Chem. Phys.* **1993**, *98,* 2170.
41. Cook, D.B.; Sordo, J.A.; Sordo, T.L. *Int. J. Quantum Chem.* **1993,** *48*, 375.
42. a) Curtiss, L. A.; Frurip, D. J.; Blander, M. *J. Chem. Phys.* **1979,** *71*, 2703.
b) Dyke, T. R.; Muenter, J. S. *J. Chem. Phys.* **1974,** *60*, 2929.
43. Lovas, F. J.; Suenram, R. D.; Fraser, G.T.; Gilles, C. W.; Zozom, J. *J. Chem. Phys.* **1988,** *88,* 722.
44. (a) Gresh, N.; Pullman, A.; Claverie, P. *Int. J. Quantum Chem.* **1985,** *28,* 785.
(b) Cabaniss, S.; Pugh, K.; Charifson, P.; Pedersen, L.; Hiskey, R. *Int. J. Peptide Protein Res.* **1990,** *36,* 79.
45. a) Kim, H.; Lipscomb, W. N. *Biochemistry* **1990,** *29*, 5546.
46. Caldwell, G.; Rozeboom, M.D.; Kiplinger, J. P.; Bartness, J. E. *J. Am. Chem. Soc.* **1984,** *106,* 4660.
47. Meot-Ner, M. (Mautner), *J. Am. Chem. Soc.* **1986,** *108,* 6189.
48. Fersht, A. *Enzyme Structure and Mechanism*; W.H. Freeman, New York, **1985**.
49. Roques, B. P.; Fournie-Zaluski, M.-C.; Soroca, E.; Lecomte, J. M.; Malfroy, B.; Llorens, C.; Schwartz, J. C. *Nature,* **1980,** *288,* 286.
50. Roderick, S. L.; Fournie-Zaluski, M.-C.; Roques, B. P.; Matthews, B. W. *Biochemistry,* **1989,** *28*, 1493.
51. Crampton, M. R. in *Chemistry of the Thiol Group*, S. Patai, (Ed.), Wiley, New-York, 1974; Vol. 1.
52. Lau, Y.; Ikuta, S.; Kebarle, P. *J. Am. Chem. Soc.* **1982,** *104*, 1462
53. Lau, Y.K.; Kebarle, P. *Can. J. Chem.* **1981,***59*, 151.
54. Meot-Ner, M. (Mautner), *J. Am. Chem. Soc.* **1984,** *106*, 1257, 1265.
55. Grimsrud, E. P.; Kebarle, P. *J. Am. Chem. Soc.* **1973,** *95*, 7939.
56. Gresh, N.; Pullman, B. *Biochim. Biophys. Acta* **1980,** *625*, 345.
57. Pullman, A.; Armbruster, A.-M. *Chem. Phys. Letts.* **1975,** *36,* 558.
58. Kebarle, P. *Ann. Rev. Phys. Chem.* **1977,** *28*, 455.

RECEIVED June 2, 1994

Chapter 7

SAM1: General Description and Performance Evaluation for Hydrogen Bonds

Andrew J. Holder[1] and Earl M. Evleth[2]

[1]University of Missouri—Kansas City, Department of Chemistry, Kansas City, MO 64110
[2]Dynamique des Interactions Moléculaires, Université Pierre et Marie Curie, 4 Place Jussieu, Tour 22, Paris 75240, France

The underlying model of the new semiempirical method SAM1 recently introduced by M.J.S. Dewar *et. al.* is described in further detail. The performance of SAM1 is evaluated in the context of a number of hydrogen bonding interactions. These interactions include the water dimer; hydration studies on the hydronium, ammonium, acetate, formate, nitrate, and hydroxide ions; ammonia / ammonium interaction; and the methanol dimer. This data shows SAM1 to be superior to the previous PM3 and AM1 parameterizations in predicting strong chemical effects (ΔH_f, ionization potentials, dipole moments) as well as weaker interactions such as hydrogen bonding.

Computational chemistry grows in utility and accuracy as the models implementing the underlying theories grow in efficiency and reliability. Theoretical chemistry utilizes many approaches and models in a continuing effort to predict answers to chemical questions without recourse to the laboratory. This has two important positive outcomes. First, it allows computer simulations of chemical processes that are more rapid, less expensive, and less dangerous than actual laboratory experiments. Second, failures in current theoretical models often lead to deeper understanding of the underlying causes of chemical phenomena, resulting in further refinements and improvements of the models. Theories vary greatly in complexity and accuracy. Some models are essentially conceptual, requiring no computations and are best suited to paper and pencil. Others only become usable when supercomputers are applied to their solution. At present, the best and most flexible chemical models are derived from quantum mechanics and are based on the Schrödinger Equation. The Schrödinger Equation relates the properties of electrons to be described as waves and permits a mathematical description of atomic, and by extension, molecular characteristics. The standard form of the Schrödinger Equation can only be solved exactly for the simplest case, the hydrogen atom. Significant approximations are required to apply the Schrödinger wave function approach to

0097–6156/94/0569–0113$08.00/0

problems of interest. A commonly used and essentially standard set of approximations and assumptions is known as Hartree-Fock (HF) Theory and this is the basis for the majority of work done in quantum chemistry at present.

Semiempirical quantum mechanical methods are based on the Hartree-Fock Theory, but ignore some of the less significant aspects of HF theory that full *ab initio* treatments explicitly compute, so that fewer actual calculations are performed. The semiempirical approach uses empirically determined parameters and parameterized functions to replace some of the most time and disk intensive portions of the more complete HF treatments. The semiempirical approximations result in great savings in computational effort. The speed differential between *ab initio* and semiempirical HF methods is an important consideration, as an *ab initio* geometry optimization on a moderately sized system[1] would require in the range of 1,000 times as much CPU time as the equivalent semiempirical calculation.

While there is no general dispute as to the overall accuracy advantage of *ab initio* calculations over semiempirical calculations when they are performed using reasonable basis sets and accounting for correlation, the time and computational resources required are prohibitive for many researchers. Further, while semiempirical methods are clearly inferior to *ab initio* approaches in reproducing absolute results, they have been shown by experience to be quite reliable at reproducing the trends in experimental data that are so important in the application of computational models to real situations. The increasing power of computational resources will probably allow *ab initio* methods to routinely supplant semiempirical approaches on problems of small to medium size (less that 30 atoms) in a few years. This increase of power will at the same time allow semiempirical examination of very large systems, now almost exclusively the domain of classical molecular mechanics. Molecular mechanics is severely limited in that specific parameters for each *type* of interaction must be derived (either from experimental data or from *ab initio* calculations) before a computation can be undertaken. The semiempirical models are element–based and are thus much more general. Also, semiempirical methods are built around the activities of electrons rather than arbitrary potential functions and are able to predict properties that are not possible for molecular mechanics. It is quite possible to envision semiempirical modeling of the molecules of life (enzymes, proteins, membranes) or determining the bulk properties of polymers by treating physically meaningful collections of polymer material.

While semiempirical and *ab initio* quantum mechanical methods have found wide acceptance in the prediction of more direct and stronger chemical phenomena such as bonding arrangements and energies, predictions involving weaker interactions such as those represented by hydrogen bonds are more questionable. It has been generally thought that these types of effects could only be modeled accurately by very high level *ab initio* methods employing extremely large basis sets augmented with special functions (diffuse orbitals). Given the success of SAM1 in predicting ground state properties (see below), we determined to test its performance in these contexts.

The New SAM1 Model in Context

Some of the most successful semiempirical methods such as AM1,[2] MNDO,[3] and MINDO3[4] have been formulated by Dewar and co-workers as part of a continuing 30 year development process. This work culminated in the release of

AMPAC 2.1 in 1985,[5] a comprehensive package containing all of the above semiempirical methods along with a powerful set of tools to study molecular structures and chemical reactions. While SAM1 is the subject of this paper, a few lines will be used to summarize the theoretical basis of AM1, as SAM1 is an expansion on and modification of that basis. AM1 has been shown through extensive trials to perform as well as very high level *ab initio* methods,[6, 7] so any improvement in its performance is a worthwhile endeavor.

AM1 is a representative of the NDDO (neglect of diatomic differential overlap) family of methods. As this model is applied to AM1, there are seven basic parameters that must be considered for each atom. These are augmented with up to three (four in the special case of carbon) Gaussian corrections to adjust the core/core repulsion function (CRF). The two parameters with the largest effect are the one-center/one-electron energies, U_{ss} and U_{pp}. These represent the kinetic energy and core-electron attractive energy of single electrons in *s*- and *p*-orbitals. The next pair of parameters are adjustments to the two-center/one-electron resonance integral, $\beta_{\mu\upsilon}$, and are termed β_s and β_p. These parameters are responsible for bonding interactions between atoms. Another approximation within AM1 is the use of Slater functions to describe spatial features of the atomic orbitals. Slater functions are used because they require fewer parameters and are more easily parameterized. The exponents of the *s*- and *p*-orbitals on each atom become parameters and are abbreviated respectively, ζ_s and ζ_p. It should be noted that ζ_s and ζ_p also affect $\beta_{\mu\upsilon}$, as $\beta_{\mu\upsilon}$ is proportional to the overlap integral ($S_{\mu\upsilon}$), which is in turn derived from the Slater exponents.

The CRF combines the Hamiltonian's nuclear terms with core-electron repulsive and attractive effects. The α parameter scales the entire CRF vertically in energy, providing a radial dimension to the atom's core. The Gaussian correction functions added with AM1 operate on the CRF directly to correct it for specific defects. This type of correction can accommodate several chemical phenomena not directly accounted for in the model. One example of this is atoms that expand their octet when bonding, such as sulfur[8] and phosphorous.[9] Atoms with expanded octets are thought to use *d*-orbitals to allow back-donation of electrons for additional bonding. Although, AM1 has no explicit descriptions of *d*-orbitals, some of the effects *d*-orbitals have on bonding can be imitated by the Gaussian functions and hypervalent molecules are described with some accuracy.

Another example of the utility of Gaussian functions within AM1 is the added flexibility of the model. For instance, the single-ζ nature of the AM1 orbital basis set makes the method insensitive to some electronic factors. Specifically, semiempirical approximations tend to yield poor results for molecules where large charges are localized on atoms. The occupied orbitals of negatively charged species tend to expand in space due to electron repulsion and an analogous but opposite effect is noted for positively charged systems. These phenomena are accounted for in *ab initio* methods by using double- or triple-ζ basis sets or even additional special orbital functions (diffuse functions). This lack of flexibility in the AM1 orbital basis can be overcome to some degree by the Gaussian corrections.

The use of Gaussians is an artificial correction to the semiempirical treatment and should be avoided if possible. When Gaussians were first introduced to the model, they were intended as “patches” for very specific problems such as those mentioned above. From our experience, they should only be applied as specific corrections AFTER a suitable set of semiempirical parameters has been located. Each Gaussian function consists of three parts: intensity (in eV), position (in Å) and width (in $Å^{-2}$).

SAM1 is the next step in the development process that produced MNDO and AM1, using a new approach to calculate important portions of the underlying model. SAM1 will also allow the study of new elements (those that require *d*-orbitals) while retaining the traditional computational efficiency of semiempirical methods. The need for a new theoretical approach to semiempirical calculations is evident for two primary reasons. First, the known deficiencies of AM1 and MNDO in certain cases[10] are resistant to removal by parameterization within the present theoretical model. The problems are inherent to the theoretical treatment and the only reasonable solution is a new method. Second, the present formalism for computing bicentric two-electron repulsion integrals (TERIs) is very cumbersome and difficult when applied to *d*-orbitals, an important next step in the maturation and applicability of these methods At present AM1 and MNDO use a modified multipole expansion (ME) for computation of TERIs:

$$\langle \mu\nu | \lambda\sigma \rangle = \sum_{l_1} \sum_{l_2} \sum_{m} \left[M^A_{l_1 m}, M^B_{l_2 m} \right]$$

where M values are multipoles. The TERI values are computed by determining the magnitude of the electrostatic repulsions between the multipoles as described by Dewar, Sabelli, Klopman, and Ohno (DSKO method).[10-14] While this expansion yields only 22 unique repulsion integrals for *s*- and *p*-orbitals, the number of integrals when *d*-functions are included is enormous. Also, integrals of this form are *not* rotationally invariant in the case of *d*-functions[15] and the ME fails to converge to the appropriate one-center integral boundaries at short distances.[16] A number of approaches have been attempted to "patch" the problems with DSKO repulsion integrals, but they either failed due to poor performance or reduced the flexibility of the semiempirical approximation to unacceptable levels.[17, 18, 19]

The mode of calculation of TERIs is extremely important to semiempirical methods of the MNDO/AM1 family as other key quantities depend on their values. These include the CRF (the repulsive force between core electrons), which is simulated by use of an <ss/ss> integral and the core-electron attraction integrals (CEA, the attractive force between core electrons on one atom and nuclei on another). With this in mind, it becomes clear that an alternative method of computing TERIs is a functional definition of a new semiempirical methodology. This new method of computing TERIs is the primary difference between SAM1 and the earlier NDDO semiempirical methods.

The approach chosen in the case of SAM1 is to utilize *ab initio* integrals, which are scaled using adjustable parameters so as to retain the implicit treatment of correlation effects, characteristic of the semiempirical methods. The form of the TERIs used is:

$$\langle \mu\nu | \lambda\sigma \rangle = f(R_{AB}) \langle \mu\nu | \lambda\sigma \rangle^{CGF}$$

where $\langle \mu\nu | \lambda\sigma \rangle^{CGF}$ are *ab initio* integrals evaluated from contracted Gaussian basis functions fit to the Slater type orbitals (STO-3G[20]) using standard methods.[21] STOs were previously used in the AM1 model and are retained in the SAM1 approach. The $f(R_{AB})$ is a scaling function that features adjustable parameters, altered systematically (see below) to reproduce experimental data. It should be noted that Thiel[22] experimented with *ab initio* TERIs previously and found their performance to be equivalent to similar level *ab initio* techniques without correlation corrections.

There are several significant advantages in the use of *ab initio* TERIs with a semiempirical procedure. First, the formulae are well defined and have been used for many years. Second, the gradients are easily computed from the Cartesian forms of the Gaussian functions. This allows for the rapid and efficient construction of the Gaussians and their derivatives simultaneously. The "shell" approximation for computing integrals using the Gaussian basis functions is used. This requires that a single exponent be used for the Gaussian functions. The geometric mean of the two ζ values (ζ_s and ζ_p) resulting from the semiempirical parameterization is used in this context. Preliminary results indicate that this approximation cannot be extended to *d*-orbitals. Third, formulae for *d*-orbitals already exist and these functions are bound by none of the limitations of the ME TERIs discussed earlier. It should be noted that the computation and manipulation of repulsion integrals is the most resource-consuming (both CPU and I/O intensive) part[23] of any *ab initio* calculation. The main reason for this is not the time required for a single integral calculation, but the large number of independent integrals which must be computed. In the SAM1 protocol, the NDDO approximation is still active, limiting the integrals which must be explicitly calculated to bicentric terms only, excluding three- and four-center terms from consideration. Because of the increased time spent on each integration, preliminary data shows that SAM1 is about two times slower than AM1. This is still well within acceptable limits when compared to *ab initio* treatments of similar accuracy.

After determining a new method for computing repulsion integrals, the most successful function for modification of the integrals ($f(R_{AB})$) was found to be one depending on the interatomic distance as an exponent. Several equations were tested,[16] and the following equation provides the best general results:

$$f(R_{AB},\mu,\nu,\delta,\sigma) = 1 - \rho_A^{l_{\mu\nu}} \exp\left(\frac{-R_{AB}}{\lambda_A}\right) - \rho_B^{l_{\delta\sigma}} \exp\left(\frac{-R_{AB}}{\lambda_B}\right)$$

In this expression, λ_X is an atomic parameter and R_{AB} is a quantity computed from the one-center repulsion integrals (g_{ss}, g_{pp}, and h_{sp}). These changes to the basic formalism of the AM1/MNDO family comprise the new method's theoretical basis. On a performance basis, SAM1 provides better results than previous semiempirical methods (Table 1).

Table 1. Error Analysis for Calculated Heats of Formation[24]

Elements	Number of Examples	Type of Error**	Procedure SAM1	AM1	PM3
C, H, O, N	222	MU	4.24	7.46	4.50
		MS	-0.54	1.27	-0.20
		SD	5.82	7.04	4.12
		RMS	5.83	10.24	6.09
F, Cl, Br, I	186	MU	3.66	5.15	6.30
		MS	-0.13	0.43	3.03
		SDS	4.64	4.36	5.29
		RMS	4.62	6.17	8.21
C, H, O, N, F, Cl, Br, I	408	MU	3.97	6.40	5.32
		MS	-0.35	0.89	1.27
		SDS	5.31	6.07	4.77
		RMS	5.31	8.82	7.14

* The total of the number examples is greater than the total due to overlap in the categories.
** MU=mean unsigned error; MS=mean signed error, SD=standard deviation, RMS=root mean square error.

Application to Hydrogen Bonds

Though very weak (10-100 times less than a covalent or ionic bond), the hydrogen bond is a fundamental effect contributing to the chemistry of many systems of importance. Various interactions of this type largely govern biomolecule conformation. As noted above, semiempirical methods are likely to be applied to larger and larger systems in the future and their performance in the context of intramolecular hydrogen bonding is a potential limitation on their applicability. Hydrogen bonds also direct hydration geometries and energies of molecules and ions in aqueous solutions. Reaction modeling and property computation in explicit solvent systems will require proper prediction of these effects.

The proper modeling of hydrogen bonding has always presented a critical challenge to theoretical methods purporting to reproduce chemical properties. This is especially the case with minimal basis set *ab initio* and semiempirical methods. The lack of sufficient orbital flexibility or extent is often blamed for the discrepancies observed in these systems. While this is certainly part of the problem, an additional concern is whether parameterizations using isolated, bonded systems as references can *ever* properly treat weakly bound complexes. The importance of the hydrogen bond to chemistry is indicated by the fact that AM1[2] was largely developed in order to improve on the treatment of hydrogen bonds by MNDO. Any new model presuming to take the place of present methods must demonstrate its capability in this area. We are pleased to report that our data shows that the new SAM1 method appears to be the first semiempirical parameterization to handle a wide variety of these systems correctly.

A number of papers have appeared recently describing the efficacy of hydrogen bond predictions by the various semiempirical methods of the Dewar style. One of these by Dannenberg and Evleth[25] is especially worthy of mention in that it is quite comprehensive and we will repeat several of the calculation sets here for comparison.

The Water Dimer. A great deal of attention has surrounded the structure of the water dimer in the gas phase both experimentally and theoretically.[25, 26] The first AM1 structure reported for the water dimer (linear, **I**) was later found not to be the minimum energy structure on the potential surface (trifurcated, **II**). See Figure 1. The trifurcated structure is 0.3 kcal/mol lower in energy. High level *ab initio*[26] and experimental[27] results favor the linear structure.

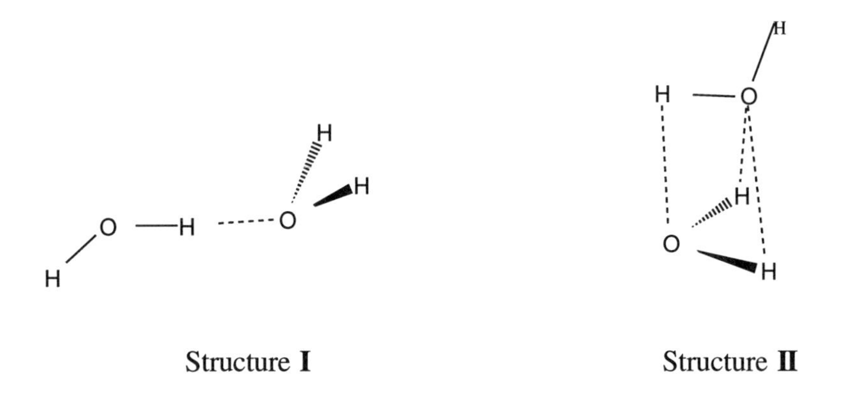

Figure 1. Linear and Trifurcated Structures of the Water Dimer

PM3, however, predicted the water dimer to be linear. SAM1 also prefers the linear Structure **I** by 0.1 kcal/mol. Dannenberg[26] has pointed out that the potential surface of the water dimer system is quite flat, so the very small differences observed are probably indistinguishable from one another. Table 2 lists heat of dimerization by the methods compared to experiment.

Table 2. Heats of Dimerization for Water

Method	ΔH Dimerization*	Reference
Expt'l	-5.2 ±1.5	28
AM1	-3.3	
PM3	-3.5	
SAM1	-4.0	

* kcal/mol

Hydration Studies. The interaction of water molecules with other species is an important and will be examined below.

Hydronium Ion (H_3O^+). The next system examined is the water/hydronium ion hydration series. While AM1 and PM3 both predict trifurcated structures (**IV**), SAM1 predicts a linear arrangement (**III**) to be the most stable (see Figure 2). High level *ab initio* calculations at the SCF level[29] suggest that the linear structure is favored, with a very low barrier to for proton transfer. (This barrier disappears when correlation effects are added *via* MPPT.) Further, experimental data[30] suggests that the bridging proton is moving in a single potential well, supporting **III**.

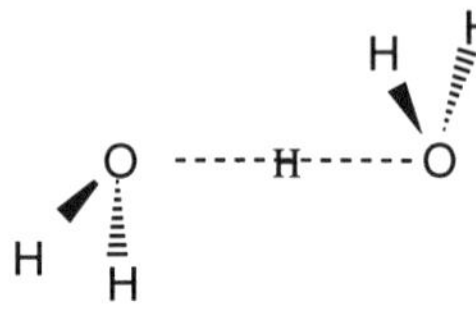

Structure **III** 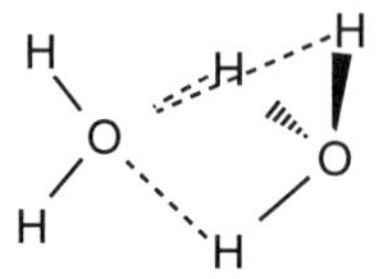 Structure **IV**

Figure 2. Linear and Trifurcated Structures of H_3O^+ / H_2O

Experimental data exists for the stepwise hydration of the hydronium ion.[31] We have carried out these calculations for SAM1 and the results are listed below in Table 3. (AM1 and PM3 were omitted since the incorrect prediction of trifurcated structures disallowed a proper comparison of the computed energies.)

Table 3. Stepwise Hydration Enthalpies* of $H_3O^+ \bullet (H_2O)_n$

n	Expt'l	SAM1
1	32 - 36	36
2	20 - 26	21
3	16 - 18	18
4	13 - 15	12
5	12 - 15	13

* kcal/mol

Ammonium Ion (NH_4^+). The experimental observation that ammonium has virtually no rotational barrier in aqueous solution[32-34] sparked several theoretical studies in an attempt to explain this non-intuitive result. Kassab *et. al.*[35] carried out both AM1 and high level *ab initio* calculations on these systems in an attempt to explain a plausible mechanism for this reaction. The results as compared to SAM1 are listed below on Table 4.

SAM1, PM3 and *ab initio* methods all predict a linear complex for the NH_4^+ / H_2O complex, but AM1 suggests that it is bifurcated (Figure 3). This result is at variance with all other experimental and theoretical findings.

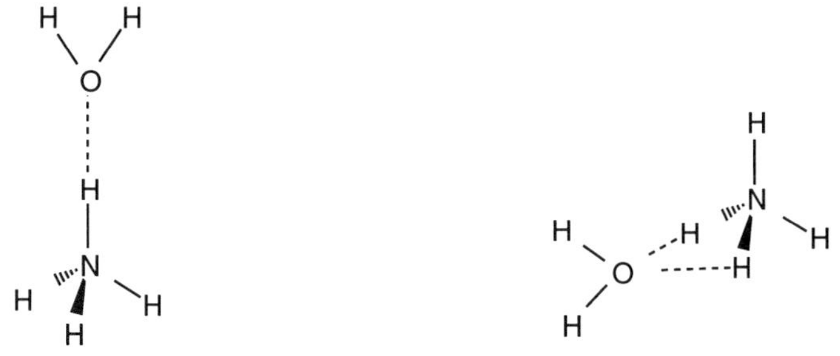

Structure **V** Structure **VI**

Figure 3. Linear and Bifurcated Structures of NH_4^+ / H_2O

Table 4. Stepwise Hydration Enthalpies* of $NH_4^+ \bullet (H_2O)_n$

n	Expt'l	Ref.	AI**	SAM1
1	20.6	36	21.5	18.3
2	16.1	36, 37, 38***	18.2	15.0
3	13.1	36, 37, 38***	15.6	13.4
4	11.5	36, 37, 38***	13.6	11.6

* kcal/mol
** *Ab initio* calculations at the HF/6-31G* level in reference 35.
*** Averaged value.

Hydroxide Anion (OH^-). Another important species in solution is the hydroxide anion. There are two structures, bifurcated and linear, that are generally considered to be likely. MP2/6-31G+* *ab initio* calculations[39] predict an assymmetric structure, but with only a 1 kcal/mol energy barrier for proton transfer. Such a barrier might well be below the V=0 vibrational level and the system would behave spectroscopically as a single well system.[40] SAM1 gives a symmetrical linear (the hydrogen of the hydroxide oriented directly toward the water's oxygen and opposite the water's hydrogens) H-bonded structure close to that obtained at the MP2/6-31G+* level. AM1 and PM3 both predict that the bifurcated geometry (with the hydroxide oxygen bonding symmetrically to both of the water's hydrogens) is the more stable by a few kcal/mol. The complexation enthalpies between water and hydroxide are listed below on Table 5.

Table 5. Complexation Enthalpies* of OH^- and Water

Method	ΔH Complexation*	Reference
Expt'l	27.6	41
AM1	22.2	
PM3	23.5	
SAM1	32.6	

* kcal/mol

Polyatomic Anions. To properly compute reaction energies in solution models requiring explicit definition of water molecules, the ΔH_f (or $\Delta H_{hydr.}$) must be accurately computed. Table 6 contains the hydration/complexation enthalpies of several common polyatomic anions. Both SAM1 and AM1 do very well at the prediction of these values, tending to be slightly below the experimental value. AM1 appears to be marginally better in this application.

Table 6. Hydration Enthalpies* of X • $(H_2O)_n$

X	n	SAM1	AM1	Expt'l	Ref.
NO_3^-	1	12.0	13.1	14.6	42
	2	10.9	12.3	14.3	43
	3	10.0	11.4	13.8	43
CH_3COO^-	1	15.8	12.6	15.8	44
$HCOO^-$	1	15.4	16.2	16.0	43

* kcal/mol

Interaction Between Ammonium and Ammonia. The ammonium/ammonia complex has been investigated experimentally by molecular beam infrared spectroscopy by Y.T. Lee *et. al.*[42] and a symmetric linear hydrogen-bonded arrangement has been deduced from this data. High level *ab initio* calculations[29] indicate a surface behavior close to that of the singly hydrated hydroxide anion in that there is a small (1 kcal/mol) barrier to proton transfer. The experimental value for the interaction is 24.8 kcal/mol.[45] SAM1 predicts the value to be 28.2, AM1 18.7, and PM3 17.3 (all kcal/mol) respectively. SAM1 predicts a symmetrical structure with no barrier for proton transfer, while AM1 and PM3 predict 1.4 and 9.5 kcal/mol, respectively.

Gas Phase Molecule/Molecule Interaction. Gas phase molecule/molecule interactions are somewhat different from ion/ion or ion/molecule interactions in that no strongly charged species are involved. These interactions are generally weaker and more difficult to model. The dimerization of methanol is one of the few such interactions for which experimental data exists. The interaction enthalpy has been determined to be 3.5 kcal/mol.[46] The semiempirical values are 1.4, 2.9, and 2.3 kcal/mol for AM1, PM3, and SAM1 respectively. Dannenberg and Turi[47] have recently completed a study on –C–H • • • • O– type molecular interactions as a method of explaining crystal structures. Their general conclusion is that AM1 is generally a few kcals/mol too low, SAM1 is generally a few kcals/mol too high, and PM3 is highly erratic in predicting these very weak hydrogen bonds.

Conclusions

SAM1 represents a departure from previous NDDO approaches in that the two-electron repulsion integrals are explicitly computed from a set of Gaussian orbitals based on the Slater-type orbitals within the NDDO approximation. The results from SAM1 for virtually every system has improved over AM1 and PM3, fulfilling the criteria for SAM1 to be a reasonable successor to AM1 and PM3 for general purpose semiempirical calculations. This can be seen from both the average error (improved by about 2.5 kcal/mol over AM1 and 1.5 kcal/mol over PM3) and the standard deviation, which has improved substantially. The decrease in standard deviation is important in that the reliability of the method can be directly inferred from this quantity.

A variety of calculations on many different hydrogen-bonding systems indicates that the more rigorous theoretical model that SAM1 is based on produces better results for these types of interactions as well as molecular properties. SAM1 would seem to be the method of choice for modeling these weak interactions with semiempirical methods.

References

1. Using an extended basis set with limited correlation corrections.
2. Dewar, M. J. S.; Zoebisch, E. G.; Healy, E. F.; Stewart, J. J. P. *J. Am. Chem. Soc.* **1985**, *107,* 3902.
3. Dewar, M. J. S.; Thiel, W. *J. Am. Chem. Soc.* **1977**, *99,* 4907.
4. Dewar, M. J. S.; Bingham, R. C.; Lo, D. H. *J. Am. Chem. Soc.* **1975**, *97,* 1285.
5. Dewar, M. J. S. *AMPAC,* Quantum Chemistry Program Exchange: Bloomington, IN.
6. Dewar, M. J. S.; Storch, D. M. *J. Am. Chem. Soc.* **1985**, *107,* 3898.
7. Dewar, M. J. S.; Holder, A. J.; Healy, E. F.; Olivella, S. *J. Chem. Soc., Chem. Commun.* **1989**, 1452.
8. Dewar, M. J. S.; Yuan, Y.-C. *Inorg. Chem.* **1990**, *29,* 3881.
9. Dewar, M. J. S.; Jie, C. *J. Mol. Struct. (THEOCHEM)* **1989**, *187,* 1.
10. Dewar, M. J. S.; Hojvat(Sabelli), N. L. *J. Chem. Phys.* **1961**, *34,* 1232.

11. Dewar, M. J. S.; Hojvat(Sabelli), N. L. *Proc. Roy. Soc. (London)* **1961**, *A264*, 431.
12. Dewar, M. J. S.; Sabelli, N. L. *J. Phys. Chem.* **1962**, *66*, 2310.
13. Klopman, G. *J. Am. Chem. Soc.* **1964**, *86*, 4550.
14. Ohno, K. *Theor. Chim. Acta (Ber.)* **1964**, *2*, 219.
15. Denton, J.; McCourt, M.; McIver, J. W. *THEOCHEM* **1988**, *163*, 355.
16. Ruiz, J. M. "Development and Application of Quantum Mechanical Electronic Structure Methods", *Ph.D. Dissertation*: University of Texas at Austin.
17. Reynolds, C. H. *Ph.D. Dissertation*: University of Texas at Austin.
18. Storch, D. M. *Ph.D. Dissertation*: University of Texas at Austin.
19. Hoggan, P.; Rinaldi, D. *Theor. Chim. Acta* **1987**, *72*, 47.
20. Hehre, W. J.; Stewart, R. F.; Pople, J. A. *J. Chem. Phys.* **1969**, *51*, 2657.
21. Szabo, A.; Ostlund, N. S. *Modern Quantum Chemistry*; McGraw-Hill: New York, 1989; p. 155.
22. Thiel, W. *Theor. Chim. Acta (Ber.)* **1981**, *59*, 191.
23. The advent of "direct" (in core) algorithms for these computations has reduced the I/O dependence of the integral calculation to some extent.
24. Taken from data in Dewar, M. J. S.; Jie, C.; Yu, G. *Tet.* **1993**, *23*, 5003 and A.H. Holder; Dennington, R.D.; Jie, C. *Tet.* **1993**, in press.
25. Dannenberg, J. J.; Evleth, E. M. *Intl. J. Quant. Chem.* **1992**, *44*, 869.
26. Dannenberg, J. J. *J. Phys. Chem.* **1988**, *92*, 6869 and extensive references therein.
27. Dyke, T. R.; Mack, K. M.; Muentner, J. S. *J. Chem. Phys.* **1977**, *66*, 498.
28. Curtiss, L. A.; Blander, M. *Chem. Rev.* **1988**, *88*, 827.
29. DelBene, J. E.; Frisch, M. J.; Pople, J. A. *J. Phys. Chem.* **1985**, *89*, 3669.
30. Yeh, L. I.; Okumura, M.; Myers, J. D.; Price, J. M.; Lee, Y. T. *J. Chem. Phys.* **1989**, *94*, 7319.
31. Keese, R. G.; Castleman, A. W. *J. Phys. Chem. Ref. Data* **1986**, *15*, 1011.
32. Perrin, C. L.; Gipe, R. K. *J. Am. Chem. Soc.* **1986**, *108*, 1088.
33. Perrin, C. L. *J. Am. Chem. Soc.* **1986**, *108*, 6807.
34. Perrin, C. L.; Gipe, R. K. *Science* **1987**, *238*, 1393.
35. Kassab, E.; Evleth, E. M.; Hamou-Tahra, Z. D. *J. Am. Chem. Soc.* **1990**, *112*, 103.
36. Meot-Ner, M.; Speller, C. V. *J. Phys. Chem.* **1986**, *90*, 6616.
37. Payzant, J. D.; Cunningham, A. J.; Kebarle, P. *Can. J. Chem.* **1973**, *51*, 3242.
38. Meot-Ner, M. *J. Am. Chem. Soc.* **1984**, *106*, 1625.
39. Gao, J.; Garner, D. S.; Jorgensen, W. L. *J. Am. Chem. Soc.* **1986**, *108*, 4784.
40. Luck, W. A. P.; Wess, T. *Can. J. Chem.* **1991**, *69*, 1991.
41. Paul, G. J. C.; Karble, P. *J. Phys. Chem.* **1990**, *94*, 5184.
42. Lee, N.; Kesee, R. G.; Castleman, J. *J. Chem. Phys.* **1980**, *72*, 1089.
43. Keesee, R.; Castleman, A. W. *J. Phys. Chem. Ref. Data* **1986**, *15*, 1011.
44. Mautner, M. *J. Am. Chem. Soc.* **1988**, *110*, 3854.
45. Price, J. M.; Crofton, M. W.; Lee, Y. T. *J. Phys. Chem.* **1991**, *95*, 2182.
46. Frurip, D. J.; Curtiss, L. A.; Blander, M. *Int. J. Thermophys.* **1981**, *2*, 115.
47. Dannenberg, J. J.; Turi, L. *J. Phys. Chem.* **1993**, *97*, 7899.

RECEIVED June 2, 1994

Chapter 8

Search for Analytical Functions To Simulate Proton Transfers in Hydrogen Bonds

Steve Scheiner and Xiaofeng Duan

Department of Chemistry, Southern Illinois University, Carbondale, IL 62901

A number of different candidate functions are tested against ab initio calculations for their ability to accurately simulate the energetics of proton transfer. These tests are conducted for H-bonds that have been both stretched and bent from their optimal geometry. A pair of Morse functions appear best, not only in that they closely mimic ab initio results, but also in that their parameters are insensitive to details of the H-bond geometry. A function is devised to handle the interactions that one proton transfer has upon the next in a sequence of consecutive H-bonds.

As a major contributor to the architecture of many molecules, particularly biomolecules, hydrogen bonds have been studied extensively over the years (*1*). The need to develop simple algorithms to predict three-dimensional structure of large molecules, and to simulate molecular interactions, has motivated a continuing search for a computationally tractable means of estimating the H-bond energy for any given pair of groups at an arbitrary relative geometry. This quest has been aided by the ability to conceptually partition the H-bond energy into a set of terms, each with clear physical interpretation (*2*). For example, the Coulombic forces can be modeled by assigning a partial charge to each atom and the dispersion energy simulated by a simple inverse sixth power of intermolecular distance (*3,4*).

Of importance also is the transfer of a proton that may take it from one subunit to the other following the formation of a H-bond. Such transfers are an integral component in the functioning of a large number of enzyme systems (*5-7*). However, there has been much less effort expended in designing a simple analytical function to simulate the energetics of these proton transfer reactions. This task is in some ways more challenging than treatment of interactions between well defined molecules. While in the process of being transferred, the proton cannot be easily assigned to one subunit or the other but must be treated as a third entity. The changes in the subunits that accompany the loss or gain of a proton are large and the use the same set of parameters before, during, and after the transfer becomes less tenable. Moreover, the interactions of the proton with the two subunits will be largely quantum mechanical in origin, and difficult to treat by classical expressions.

0097–6156/94/0569–0125$08.00/0

The present contribution makes a step in this direction. A number of different candidate functions are tested for use in proton transfer reactions by comparison with detailed ab initio quantum mechanical calculations. These tests are conducted not only for the energetically preferred linear H-bonds but also for those that are stretched or bent due to external constraints. The strengths and weaknesses of these functions are assessed and described. The last section goes one additional step and considers the effect on one another of two protons that are parts of consecutive links in a H-bonded chain.

Single Linear H-Bonds

We first define the geometrical parameters to be used in the empirical expressions. Figure 1 illustrates the transfer of a proton between two water molecules (*8*). r refers to the distance of the transferring proton from the midpoint of the O··O axis and R to the interoxygen separation. The two water molecules placed on either end of the chain are there to minimize the "end effects" that would unduly perturb the ab initio energetics of the proton transfer process.

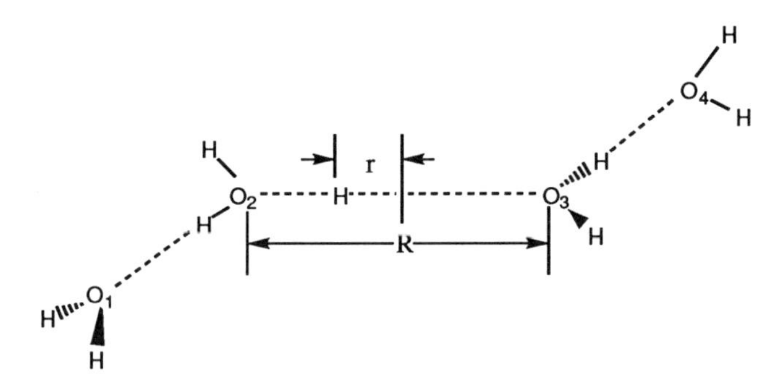

Figure 1. Atomic labeling scheme of protonated water tetramer. All H-bonds are linear. Terminal waters are placed so that O atom lies along O-H axis of neighboring water, with R(OO)=2.54 Å. Internal geometry of each terminal water is r(OH)=0.95 Å; θ(HOH)=112°. (Reproduced with permission from ref. 8. Copyright 1992 Elsevier.)

Morse functions have a successful history of approximating bond stretches and dissociations (*9,10*). One can consider the proton transfer from A to B as the stretch away from group A at the same time as a contraction of the H··B distance occurs. In the case where A and B are identical groups (water in this case), the Morse functions for the two bonds have equal parameters so one can write the total energy as

$$E = D\{1 - \exp[-\alpha(R_e/2 + r)]\}^2 + D\{1 - \exp[-\alpha(R_e/2 - r)]\}^2 - D\{1 - 2\exp[-\alpha R_e]\} \quad (1)$$

The first two terms in equation 1 refer to the bond stretching and contracting, respectively; the last term sets the value of the energy to zero in the two minima of the potential (*8*). The α parameter is related to the stiffness of the O-H bond while the energetic strength of the bond is expressed by D. R_e is defined as R - $2r_e$ where r_e refers to the equilibrium O-H bond length.

A simple polynomial has come into common usage for proton transfer potentials (*11,12*). The following form is generally termed a Φ^4 function

$$E = \varepsilon\{(r/u_o)^2 - 1\}^2 \quad (2)$$

but is quite similar to a more general fourth-order polynomial (*13,14*):

$$E = Ar^4 + Br^2 + C \tag{3}$$

Another approach involves the use of Gaussian functions. When inverted, a pair of such functions can be used to approximate the stretching and contracting A-H and H··B bonds:

$$E = D\{(1 - \exp[-\alpha(R_e/2 + r)^2]) + (1 - \exp[-\alpha(R_e/2 - r)^2])\} + C \tag{4}$$

or a single Gaussian may be combined with a function that produces the steep rise as the proton approaches too closely to either A or B.

$$E = D\{\exp(-\alpha[r^2 - (R_e/2)^2]) + \exp(-\alpha[(R_e/2)^2 - r^2]) - 2\} \tag{5}$$

Another expression that incorporates a Gaussian function is due to Flanigan and de la Vega (*15*); it uses a quadratic to model the repulsive walls.

$$E = D\exp(-\alpha r^2) + ar^2 + a/\alpha[\ln(a/\alpha D) - 1] \tag{6}$$

The final term in equation (6) sets the entire function equal to 0 at the two minima which occur at $\pm[-(1/\alpha)\ln(a/\alpha D)]^{1/2}$.

The Lippincott-Schroeder function (*16*) has found some use in modeling H-bonds over the years. It differs from simpler Gaussians in that the exponent is dependent upon the length of the H-bond.

$$E = D\{1 - \exp[-n(R_e/2 + r)^2/2(R/2 + r)] + 1 - \exp[-n(R_e/2 - r)^2/2(R/2 - r)]\} + C \tag{7}$$

Another possibility simulates the double-well nature of the proton transfer potential through a simple sinusoidal function (*11,12*).

$$E = U_b[\cos(4\pi\alpha r/R) + 1] \tag{8}$$

Within this form, the transfer barrier is equal to $2U_b$ and the width of the barrier is characterized by the α parameter. A principal disadvantage of this function is that it is incapable of treating the energy rise as the transferring hydrogen crowds in on the A or B group at either end of the transfer potential.

Proton transfer potentials were computed by ab initio methods and the results fit by the various functions above (*8*). The distance R between the two central O atoms in Figure 1 was first fixed to one of a set of different values and r varied in uniform increments, with the remainder of the molecular geometry held fixed. The parameters in each empirical function were then fit by least-squares methods to the energies computed at the HF/4-31G level.

The optimal values of the empirical parameters in equation 1 are listed in Table I for a series of progressively longer H-bond lengths, R. One can see from the correlation coefficients, R_c, in the last column that the Morse functions are capable of fitting the ab initio potentials quite well, with R_c exceeding 0.997 at any H-bond length. It is also important to note that the optimized values of the parameters are rather insensitive to the H-bond length. D, for example, remains in the vicinity of 76 kcal/mol. α varies only between 2.6 and 2.7 and r_e does not stray from the 0.963 - 0.971 domain. The near constancy of these parameters permits one to advance the concept of a "mean" set of values, listed in the last row of Table I and evaluated as the simple algebraic average of those in the preceding rows. When these mean parameters are substituted into equation 1, the empirical potential fits all of the data generated, with a correlation coefficient of 0.996. The quality of fit is also evident in Figure 2 where the ab initio energies are shown explicitly and the curves are those obtained from equation 1 with the parameters below.

Table I. Fit of parameters of equation 1 to ab initio proton transfer potential in the protonated water tetramer

R, Å	D, kcal/mol	α, $Å^{-1}$	r_e, Å	R_c
2.55	75.947	2.699	0.9634	0.9999
2.65	76.115	2.688	0.9677	0.9999
2.75	76.044	2.665	0.9707	0.9991
2.85	75.904	2.618	0.9714	0.9986
2.95	75.562	2.614	0.9699	0.9971
mean	75.91	2.66	0.969	0.996

SOURCE: Adapted from ref. 8.

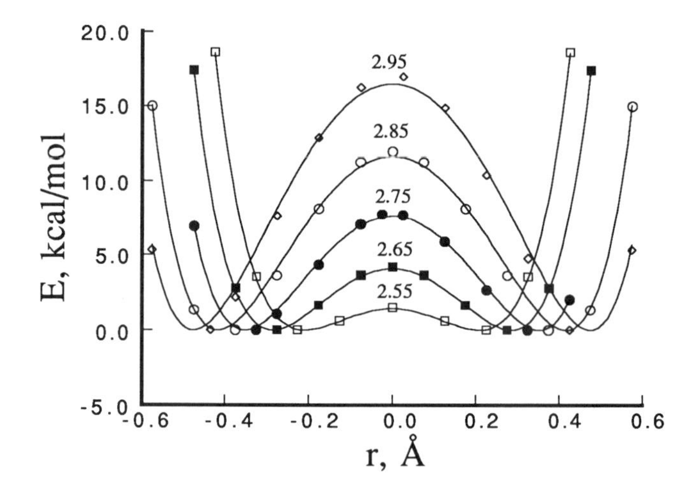

Figure 2. Proton transfer potentials calculated for protonated water tetramer. Curves are drawn from equation 1 with parameters listed in Table I; HF/4-31G energies are represented by points. Numerical labels indicate R in Å. (Reproduced with permission from ref. 8. Copyright 1992 Elsevier.)

The quality of fitting arising from application of the Φ^4 function in equation 2 may be seen in Table II to also be quite good. The correlation coefficients in the last column are all 0.997 or better. On the other hand, the values of the parameters are quite sensitive to the interoxygen separation. Since the height of the energy barrier is known to increase quickly as the H-bond is lengthened, it is not surprising to see the sharp variation in ε. The changes in u_0 result primarily from the drift apart of the two minima as R increases. Although the sensitivity of the Φ^4 parameters to H-bond length make the direct incorporation of this function into molecular simulations awkward, it is possible to derive a relationship such that these quantities can be readily approximated. For example, both u_0 and ε can be fit reasonably well to a linear function of the interoxygen separation (*8*).

Table II. Fit of parameters of equation 2 to ab initio proton transfer potential

R, Å	u_o, Å	ε, kcal/mol	R_c
2.55	0.207	1.787	0.9997
2.65	0.275	4.391	0.9993
2.75	0.342	7.843	0.9991
2.85	0.397	12.274	0.9969
2.95	0.462	17.389	0.9991

SOURCE: Adapted from ref. 8.

Equation 3 is a fourth-order polynomial in r like equation 2 but differs in containing three parameters rather than two. However, as illustrated by the last column of Table III, the fitting of equation 3 is not significantly superior to that of equation 2. Nor are the A, B, and C parameters less sensitive to distance than the u_o and ε quantities of the Φ^4 function. Indeed, stretching R from 2.55 Å to 2.95 Å drops A to less than half its value, while C undergoes a tenfold increase.

Table III. Parameters that fit equation 3 to ab initio potential

R, Å	A, kcal mol^{-1} $Å^{-4}$	B, kcal mol^{-1} $Å^{-2}$	C, kcal mol^{-1}	R_c
2.55	997.72	-86.58	1.678	0.9999
2.65	787.47	-120.31	4.398	0.9996
2.75	577.04	-134.93	7.853	0.9992
2.85	507.34	-160.43	12.359	0.9976
2.95	379.44	-162.07	17.365	0.9991

SOURCE: Adapted from ref. 8.

Table IV reports on the quality of fitting of the dual Gaussian function in equation 4. While this empirical function mimics the ab initio calculations very well for the extremes of R=2.55 and 2.95 Å, the fit is of poorer quality for intermediate H-bond lengths. On the other hand, the values of the parameters are somewhat more stable than in some of the aforementioned cases, but not nearly so constant as the Morse functions. Whereas D, r_e, and C are constant to a first approximation, α is much more variable.

Table IV. Fitting parameters for equation 4

R, Å	D, kcal mol^{-1}	α, $Å^{-2}$	r_e, Å	C, kcal mol^{-1}	R_c
2.55	180.95	4.594	0.922	156.10	0.999
2.65	179.00	3.990	0.935	158.02	0.995
2.75	177.01	3.419	0.943	158.98	0.961
2.85	176.06	3.093	0.954	161.82	0.982
2.95	175.25	2.462	0.934	162.51	0.996

SOURCE: Adapted from ref. 8.

The situation improves, but only marginally, when one uses a single Gaussian to model the barrier plus another function that rises quickly on either end of the potential. As is evident from Table V, D remains between 58 and 59 kcal/mol for the entire range of R but α is still subject to strong variation.

Table V. Fitting parameters for equation 5

R, Å	D, kcal mol^{-1}	α, Å^{-2}	r_e, Å	R_c
2.55	58.800	3.880	1.065	0.994
2.65	58.453	3.579	1.042	0.995
2.75	58.699	3.200	1.036	0.996
2.85	58.497	2.900	1.028	0.997
2.95	58.494	2.498	1.011	0.998

SOURCE: Adapted from ref. 8.

The quality of fitting is good also for the de la Vega function (*15*) which combines a Gaussian with a parabola. On the other hand, all parameters are subject to strong variation upon stretching the H-bond. Both α and a in Table VI are reduced by half upon increasing R from 2.55 to 2.95 Å. None of the parameters in the Lippincott-Schroeder potential undergo such drastic changes as the bond is stretched. Yet Table VII reveals that all four of these parameters remain moderately sensitive to the interoxygen distance.

The cosine function of equation (*8*) fits most poorly of all. In the first place, the inability of this sort of function to model the steep potential walls forced the fitting to ignore those parts of the surface where the energy was higher than that of the transfer midpoint. Even so, the correlation coefficients in the last column of Table VIII are less than 0.98. The U_b parameter is particularly sensitive to the length of the H-bond.

Table VI. Fitting parameters for equation 6

R, Å	D, kcal mol^{-1}	α, Å^{-2}	a, kcal mol^{-1} Å^{-2}	R_c
2.55	170.19	4.039	579.55	0.998
2.65	178.33	3.516	489.98	0.996
2.75	187.91	2.914	395.51	0.996
2.85	201.01	2.754	368.79	0.989
2.95	210.66	2.278	297.67	0.998

SOURCE: Adapted from ref. 8.

Table VII. Parameters for Lippincott-Schroeder function, equation 7

R, Å	D, kcal mol^{-1}	n, Å^{-1}	r_e, Å	C, kcal mol^{-1}	R_c
2.55	127.06	10.42	0.937	-92.94	0.999
2.65	124.04	9.71	0.949	-95.96	0.998
2.75	121.11	8.81	0.951	-97.89	0.995
2.85	119.88	8.36	0.961	-100.12	0.993
2.95	118.06	7.70	0.956	-101.94	0.999

SOURCE: Adapted from ref. 8.

Table VIII. Fit of the cosine function in equation 8

R, Å	U_b, kcal mol^{-1}	α	R_c
2.65	2.350	2.684	0.961
2.75	4.349	2.284	0.938
2.85	6.601	1.933	0.958
2.95	9.286	1.668	0.979

SOURCE: Adapted from ref. 8.

Bent H-Bonds

Of course, not all H-bonds are fully linear as presumed in Figure 1. This nonlinearity is particularly true for H-bonds that occur within the confines of proteins since the three-dimensional structure of such a macromolecule is a compromise between a large number of different considerations. For example, the local (ϕ,ψ) rotational preference around each peptide unit must be balanced against longer range inter-residue interactions, with H-bonding being only one among them. Indeed, surveys of H-bond geometries in proteins confirm that linear H-bonds are the exception rather than the rule (*18*).

It is therefore critical for analytical functions that model proton transfer processes to accommodate various types of angular deformations. Since the Morse and Φ^4 functions (fourth-order polynomial) were found in the prior section as most suitable for linear H-bonds, it seems most appropriate to examine their ability to be adapted to the nonlinear situations. As in the linear cases described above, the ab initio calculations were performed at the HF/4-31G level (*19*).

The angular deformations were introduced into the $H_2O{\cdot\cdot}H^+{\cdot\cdot}OH_2$ system as illustrated in Figure 3 where R again represents the interoxygen separation. The bisectors of the donor and acceptor water molecules were turned by angles of α_1 and α_2, respectively, relative to the O···O axis. The bridging proton was allowed to follow its lowest energy path between the two rotated water molecules. r_x measures the progress of this proton along the x-direction parallel to the O···O axis and r_y refers to its perpendicular deviation from this axis (*19*). One type of deformation investigated keeps the donor molecule optimally oriented (α_1=0°) while the acceptor is turned. A conrotatory type of distortion was imposed by turning the donor and acceptor equal amounts in the same direction (α_1=α_2 as shown in Figure 3) while a disrotatory mode rotates them in opposite directions (α_1=-α_2).

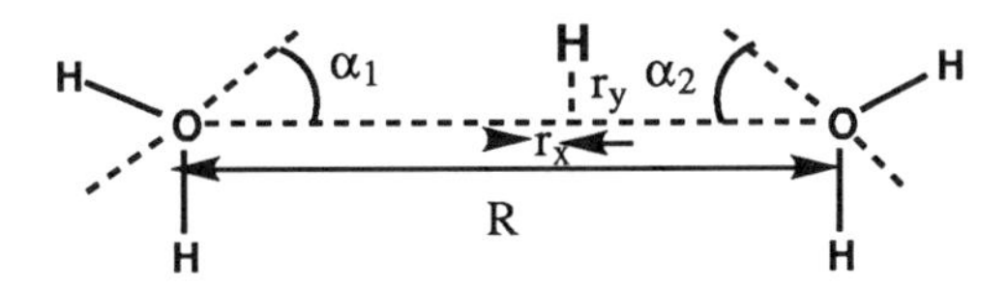

Figure 3. Definition of parameters used to describe angular deformation of H-bond. r_x is measured from center of O··O axis.

The potentials computed for each of the variously distorted $H_2O{\cdot\cdot}H^+{\cdot\cdot}OH_2$ complexes is presented in Figure 4 as a function of r_x. Both the conrotatory and disrotatory motions leave the proton transfer with left-right symmetry, albeit with a higher barrier to proton transfer. The barrier rises as the amount of distortion increases from 20° to 40°.

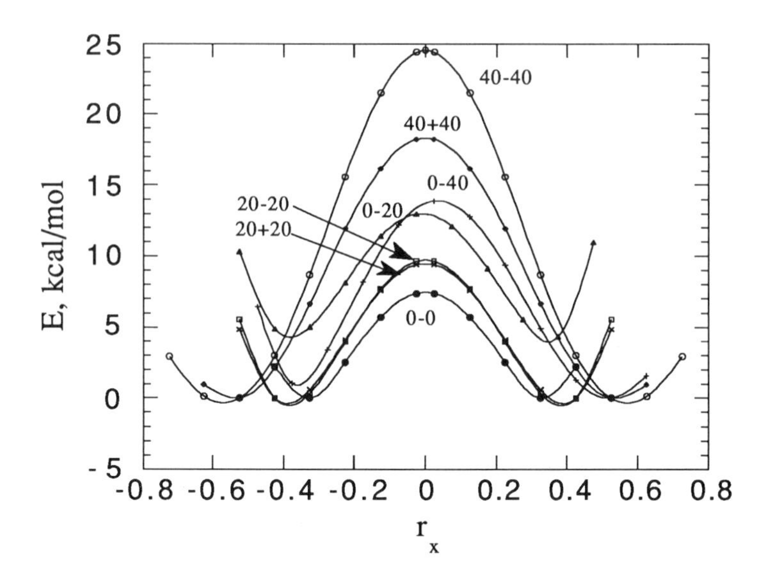

Figure 4. Proton transfer potentials calculated for protonated water dimer with indicated amounts of angular distortion with R(OO)=2.75 Å. Labels on each curve refer to (α_1,α_2) values. All energies are shown relative to minimum in (0,0). (Reproduced with permission from ref. 19. Copyright 1993 Wiley.)

A small modification of equation 1 permits the Morse function to be tested with regard to proton transfers in bent systems. The r distance from equation 1 is taken as the projection of the particular O-H vector onto the O···O direction or x-axis. Hence, R_e is now taken as R-2r_{ex}, where the latter quantity is the aforementioned projection of r_e. A similar modification is required in the Φ^4 function in equation 2; substitution of r_x for r.

Since the conrotatory and disrotatory distortions retain the left-right symmetry of the proton transfer potential, it is possible in principle for the Morse or Φ^4 functions to accurately simulate the ab initio potentials. The results of such a fitting for the Morse function are reported in Tables IX and X. Comparison with Table I reveals that the fit suffers as a result of introducing the bending. Whereas correlation coefficients were consistently above 0.99 in the linear cases, this quantity drops down to the 0.97-0.99 range when angular distortions are involved. Nonetheless, these fits are quite respectable. It is encouraging to note that the D and α parameters do not need to change by very much in order to account for the bending, even by as much as 40°. Most of the influence of the bending is incorporated by the r_{ex} parameter, which drops from 0.97 in the linear case down to as little as 0.75. This change is not unreasonable as the turning of the water molecules would diminish the projection of the O-H vector along the x-direction.

Table IX. Fitting parameters for symmetric Morse function for conrotatory deformation, $\alpha_1=\alpha_2$

R, Å	D, kcal mol^{-1}	α, Å^{-1}	r_{ex}, Å	R_c
		(20,20)		
2.55	75.414	2.608	0.9289	0.9836
2.65	75.241	2.576	0.9320	0.9688
2.75	75.133	2.478	0.9171	0.9871
2.85	75.564	2.512	0.9199	0.9906
2.95	75.887	2.560	0.9227	0.9932
mean	75.448	2.547	0.9241	0.9847
		(40,40)		
2.55	73.126	2.232	0.7813	0.9878
2.65	73.288	2.322	0.7910	0.9847
2.75	73.511	2.404	0.8018	0.9888
2.85	72.443	2.204	0.7502	0.9752
2.95	73.979	2.260	0.7530	0.9785
mean	73.269	2.284	0.7754	0.9830

Table X. Fitting parameters for symmetric Morse function for disrotatory deformation, $\alpha_1=-\alpha_2$

R, Å	D, kcal mol^{-1}	α, Å^{-1}	r_{ex}, Å	R_c
		(20,-20)		
2.55	75.404	2.752	0.9403	0.9854
2.65	74.871	2.666	0.9414	0.9763
2.75	74.409	2.660	0.9318	0.9816
2.85	74.663	2.619	0.9285	0.9918
2.95	75.019	2.599	0.9312	0.9937
mean	74.873	2.654	0.9346	0.9858
		(40,-40)		
2.55	74.380	2.595	0.7829	0.9966
2.65	74.392	2.602	0.7965	0.9837
2.75	74.860	2.635	0.8001	0.9853
2.85	74.261	2.589	0.7657	0.9751
2.95	74.441	2.577	0.7557	0.9785
mean	74.441	2.599	0.7802	0.9838

It was noted earlier that the u_0 and ε parameters of the Φ^4 function are quite sensitive to interoxygen distance. Table XI illustrates a strong dependence upon angular features of the H-bond as well. The correlation deteriorates somewhat as the bond is bent but not as much as in the case of the double Morse function.

A more severe test of the suitability of these functions to model proton transfers would apply them to H-bonds with a different amount of angular deformation in the donor and acceptor groups. In such a case, the transfer potential would not have a left-right symmetry. When a configuration was studied in which $\alpha_1=0°$ and $\alpha_2=20°$, the Φ^4 function in equation 2 yielded correlation coefficients in the range 0.91 - 0.98 for R distances between 2.55 and 2.95 Å (*19*). This poor fitting results despite the fact that the degree of asymmetry in the potential is rather small. Adding first or third-order terms to the fourth-order polynomial does not produce much improvement.

Table XI. Fitting parameters for Φ^4 potentials for conrotatory deformation, $\alpha_1=\alpha_2$

R, Å	u_0, Å	ε, kcal mol^{-1}	R_c
	(20,20)		
2.55	0.255	2.503	0.9969
2.65	0.329	5.066	0.9880
2.75	0.399	9.281	0.9975
2.85	0.461	14.251	0.9994
2.95	0.520	19.790	0.9999
	(40,40)		
2.55	0.428	8.233	0.9940
2.65	0.489	12.562	0.9957
2.75	0.545	17.803	0.9971
2.85	0.635	22.382	0.9909
2.95	0.689	28.225	0.9936

One simple means to allow asymmetry into a dual Morse function is to simply permit the two subunits to have different binding strengths to the proton. This can be accomplished by using two different values of the D parameters:

$$E = D_1\{1-\exp[-\alpha(R_e/2+r_x)]\}^2 + D_2\{1-\exp[-\alpha(R_e/2-r_x)]\}^2 - C \quad (9)$$

However, only a modest improvement is noted, as may be seen by the data in Table XII. Correlation coefficients do not exceed 0.97 and can be as low as 0.92. The last entry in the table of 0.957 suggests that the use of a single set of parameters for all values of R may be questionable with this function. It is interesting that, although the two D parameters are free to vary independently, there is surprisingly little difference in the final values of D_1 and D_2.

Table XII. Fitting parameters for Morse function, equation 9, for asymmetric deformation (0,20)

R, Å	D_1, kcal mol^{-1}	D_2, kcal mol^{-1}	α, Å^{-1}	r_{ex}, Å	C, kcal mol^{-1}	R_c
2.55	74.726	74.223	2.537	0.9353	-47.744	0.9600
2.65	73.562	74.277	2.585	0.9498	-52.718	0.9669
2.75	72.347	74.361	2.455	0.9348	-55.989	0.9215
2.85	72.940	74.512	2.502	0.9372	-60.604	0.9601
2.95	72.619	73.820	2.562	0.9394	-63.808	0.9782
mean	73.239	74.239	2.529	0.9393		0.9573

A better fitting can be attained if the true r(OH) distances are used in the Morse functions and not just their projection along the O··O axis. Taking r_y as the normal distance of the bridging hydrogen from this x-axis (see Figure 3), the true distances from the left and right subunits can be expressed as

$$r_l = ((R/2+r_x)^2+r_y^2)^{\frac{1}{2}} \qquad r_r = ((R/2-r_x)^2+r_y^2)^{\frac{1}{2}} \quad (10)$$

Rather than simply permit two different values of D, one can introduce asymmetry into the function by changing the α and r_e parameters. This additional freedom was accomplished (*19*) by incorporating two new parameters, $\Delta\alpha$ and Δr_e, into the second term of equation 1, which corresponds to the subunit which has been rotated.

$$E = D\{1-\exp[-\alpha(((R/2+r_x)^2+r_y^2)^{\frac{1}{2}} -r_e)]\}^2 \\ +D\{1-\exp[-(\alpha+\Delta\alpha)(((R/2-r_x)^2 +r_y^2)^{\frac{1}{2}}-(r_e+\Delta r_e))]\}^2 + C \quad (11)$$

Only $\Delta\alpha$, Δr_e, and C of equation (11) were fit to the ab initio data while the remaining parameters D, α, and r_e were taken as the average value in the undistorted system as fit by equation (1). The fitting was carried out using a NonLinear program (*20*); the accuracy is evaluated in terms of ΔE, the average variation between the fitted and ab initio energies.

The first set of entries in Table XIII indicates that only small perturbations are required in the α and r_e parameters when the deformation amounts to 20°. The analytical function mimics the ab initio energies of each potential between R=2.55 and 2.95 Å to within less than 0.1 kcal/mol. The fitting is worse for a 40° distortion. Energies are predicted to within about 0.2 or 0.3 kcal/mol on average, and sizable perturbations are required in α.

TABLE XIII. Fitting parameters for asymmetric Morse function, equation 11

R, Å	$\Delta\alpha$, Å^{-1}	Δr_e, Å	C, kcal mol^{-1}	ΔE, kcal mol^{-1}
		(0,20)		
2.55	0.122	-0.035	-50.979	0.028
2.65	0.119	-0.022	-56.592	0.013
2.75	0.103	-0.006	-60.264	0.037
2.85	0.085	-0.001	-63.713	0.054
2.95	0.076	-0.000	-66.518	0.088
		(0,40)		
2.55	0.931	-0.054	-61.751	0.102
2.65	0.801	-0.037	-64.097	0.201
2.75	0.707	-0.027	-66.339	0.279
2.85	0.646	-0.020	-68.306	0.295
2.95	0.629	-0.016	-69.978	0.255

[a]The other parameters in the function are taken from the average value of the undistorted system, i.e., $\alpha_1=\alpha_2=0°$; D=76.069 kcal mol^{-1}; α=2.682; r_e=0.9728.

Multiple Transfers

One of the more interesting characteristics of proton transfers is that they may occur along a chain of more than one H-bond. Indeed, proton shuttling mechanisms are known to be operative in proton conduction within liquid or solid water. Multiple transfers have also been linked to biological proton pumps like bacteriorhodopsin and H^+-ATPase, as well as the functioning of certain enzymes (*21,22*). The work described above has addressed itself only to a single H-bond and the proton within it. No attempt was made to understand how the progress of one proton in a chain can affect the transfer of the next proton. This section deals explicitly with this problem.

Imagine a short segment of a chain of H-bonds, containing only two of these bonds. Transfer of H_1 can precede that of H_2,

$AH_1\cdots BH_2\cdots CH \rightarrow A\cdots H_1BH_2\cdots CH \rightarrow A\cdots H_1B\cdots H_2CH$

can follow it,

$AH_1\cdots BH_2\cdots CH \rightarrow AH_1\cdots B\cdots H_2CH \rightarrow A\cdots H_1B\cdots H_2CH$

or the two can transfer at approximately the same time.

$AH_1\cdots BH_2\cdots CH \rightarrow A\cdots H_1B\cdots H_2CH$

One can expect that the potential for transfer of, say H_2, will be sensitive to the position of H_1; i.e. the three cases above are associated with very different energy profiles.

This problem is considered here as a two-dimensional problem

$$E(r_a,r_b) = E(r_a) + E(r_b) + f(r_a,r_b) \qquad (12)$$

where r_a and r_b refer to the extent of transfer of the two protons individually. Each of the $E(r_a)$ and $E(r_b)$ functions are of one dimension and refer to a single transfer. The coupling function $f(r_a,r_b)$ contains the physics of the interaction between the two protons.

A chain of five water molecules, much like that in Figure 1 was set up to interrogate the problem (*23*). Each r again represents the deviation of the hydrogen from the midpoint of the O···O axis. The third of the five water oxygens was taken as a point of reference; negative values of r_a or r_b indicate closer proximity to this central oxygen and positive to further separation. We focus our attention on R(O··O) distances of 2.95 Å between these water molecules. Energies were computed by the HF/4-31G method as a function of the motion of the two protons transferring between molecules 2 and 3, and between 3 and 4, respectively (*23*). These energetics are presented as a countour plot in Figure 5 where the minima in the upper left and lower right corners represent HAH_1^+···BH_2···CH and HA···H_1B···H_2CH^+ respectively; the excess charge on the central molecule HA···$H_1BH_2^+$···CH corresponds to the lower left corner.

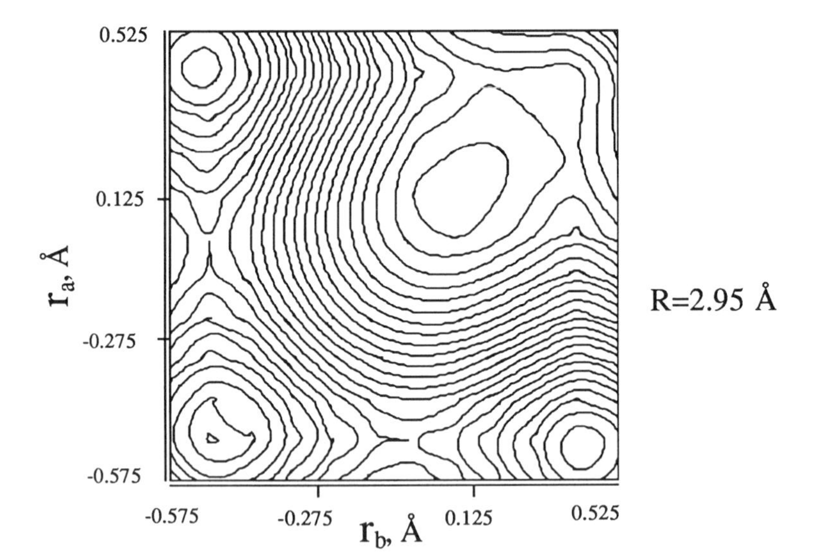

Figure 5. Ab initio potential energy surface of protonated water tetramer. Spacing between contours is 1.95 kcal/mol. (Reproduced with permission from ref. 23. Copyright 1992 Wiley.)

Attempts were made to reproduce the energetics in Figure 5 as closely as possible using equation 12. Morse functions were used for the one-dimensional E(r) functions. The remaining discrepancy was fit by a number of different choices for a coupling function $f(r_a,r_b)$. These included the following (*23*).

$$f(r_a,r_b) = k\,(c + r_a + r_b)^2 \qquad (13)$$

$$f(r_a,r_b) = k\,(c - r_a - r_b)^{-1} \tag{14}$$

$$f(r_a,r_b) = k\,\exp[c(r_a + r_b)] \tag{15}$$

Equation 13 permits the two transferring protons to interact through a harmonic potential while equation 14 incorporates a coulombic interaction between their partial charges. The next function permits exponential growth of the coupling as the two protons move further away from the central water molecule. None of these functions proved very useful. After fitting to the data, variances remained of 2.7, 6.7, and 3.6 kcal/mol, respectively. In other words, these coupling functions leave errors of this magnitude that can contaminate the use of equation 12.

Better results were obtained by equation 16 which expresses the coupling as a linear function of the displacement of the protons away from their equilibrium positions.

$$f(r_a,r_b) = k\,(r_a + R_e/2)\,(r_b + R_e/2) \tag{16}$$

When a least squares fitting was applied to this function, within the context of equation 12, against the ab initio data, k was found to be 52.40 kcal mole^{-1} Å^{-2}. The variance of the empirical energies vs. the 4-31G results is 1.01 kcal/mol. A map of the deviations is illustrated in Figure 6 where the largest errors are located in the region where both r_a and r_b are large. It is in these regions that the large charge separation of the $HAH_1^+\cdots B^-\cdots{}^+H_2CH$ configuration yields a high energy so the system would sample this area to only a limited extent. Restricting the fitting to that part of the map that excludes this region leads to improved fitting, and a variance of only 0.69 kcal/mol.

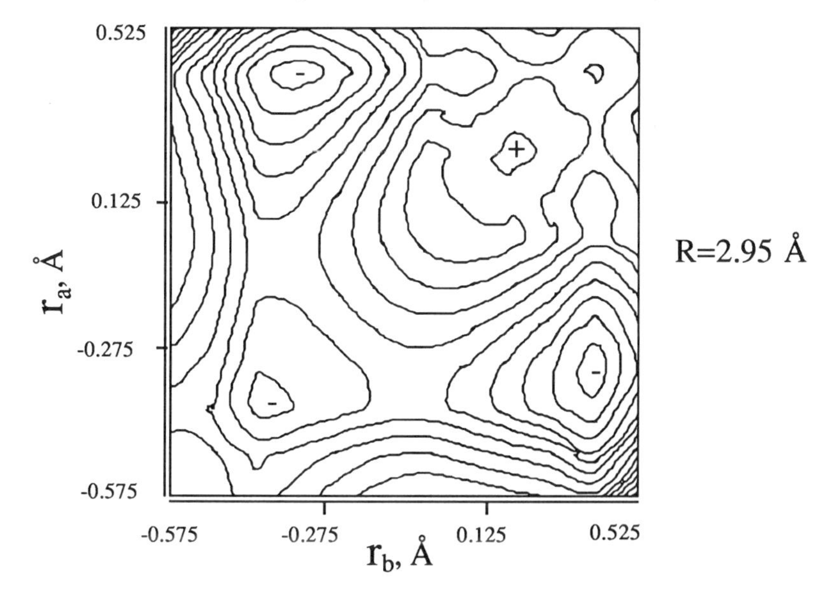

Figure 6. Errors in fitting equations 12 and 16 to ab initio energetics of Figure 5. Positive deiviations are indicated by (+) and negative by (-). Contour spacing is 0.46 kcal/mol. (Reproduced with permission from ref. 23. Copyright 1992 Wiley.)

Acknowledgments

This work has been supported financially by the National Institutes of Health (GM29391).

Literature Cited

1. *The Hydrogen Bond. Recent Developments in Theory and Experiments*; Schuster, P.; Zundel, G.; Sandorfy, C., Eds.; North-Holland: Amsterdam, 1976.
2. Umeyama, H.; Morokuma, K. *J. Am. Chem. Soc.* **1977**, *99*, 1316.
3. Stone, A. J.; Price, S. L. *J. Phys. Chem.* **1988**, *92*, 3325.
4. Dinur, U.; Hagler, A. T. In *Reviews in Computational Chemistry*; Boyd, D. B.; Lipkowitz, K. B., Eds.; VCH: New York, NY, 1991, Vol. 2; pp 99-164.
5. Farnum, M. F.; Klinman, J. P. *Biochemistry* **1986**, *25*, 6028.
6. Howell, E. E.; Warren, M. S.; Booth, C. L. J.; Villafranca, J. E.; Kraut, J. *Biochemistry* **1987**, *26*, 8591.
7. Rose, I. A.; Fung, W.-J.; Warms, J. V. B. *Biochemistry* **1990**, *29*, 4312.
8. Duan, X.; Scheiner, S. *J. Mol. Struct.* **1992**, *270*, 173.
9. Kryachko, E. S. *Chem. Phys.* **1990**, *143*, 359.
10. Sato, N.; Iwata, S. *J. Chem. Phys.* **1988**, *89*, 2932.
11. Pnevmatikos, St.; Savin, A. V.; Zolotaryuk, A. V.; Kivshar, Yu. S.; Velgakis, M. J. *Phys. Rev. A* **1991**, *43*, 5518.
12. Halding, J.; Lomdahl, P. S. *Phys. Rev. A* **1988**, *37*, 2608.
13. Somorjai, R. L.; Hornig, D. F. *J. Chem. Phys.* **1962**, *36*, 1980.
14. Jaroszewski, L.; Lesyng, B.; McCammon, J. A. *J. Mol. Struct. (Theochem)* **1993**, *283*, 57.
15. Flanigan, M. C.; de la Vega, J. R. *J. Chem. Phys.* **1974**, *61*, 1882.
16. Lippincott, E. R.; Schroeder, R. *J. Chem. Phys.* **1955**, *23*, 1099, 1131.
17. Pnevmatikos, St. *Phys. Rev. Lett.* **1988**, *60*, 1534.
18. Baker, E. N.; Hubbard, R. E. *Prog. Biophys. Molec. Biol.* **1984**, *44*, 97.
19. Duan, X.; Scheiner, S.; Wang, R. *Int. J. Quantum Chem., QBS20* **1993**, (in press).
20. Brenstein, R. J. *NonLin for Macintosh,* Robelko Software, Carbondale, IL, **1992.**
21. Miller, A.; Oesterhelt, D. *Biochim. Biophys. Acta* **1990**, *1020*, 57.
22. Braiman, M. S.; Mogi, T.; Marti, T.; Stern, L. J.; Khorana, H. G.; Rothschild, K. J. *Biochemistry* **1988**, *27*, 8516.
23. Duan, X.; Scheiner, S. *Int. J. Quantum Chem., QBS19* **1992**, 109.

RECEIVED June 23, 1994

Chapter 9

Effective Fragment Method for Modeling Intermolecular Hydrogen-Bonding Effects on Quantum Mechanical Calculations

Jan H. Jensen[1], Paul N. Day[1,4], Mark S. Gordon[1], Harold Basch[2], Drora Cohen[2], David R. Garmer[3,5], Morris Kraus[3], and Walter J. Stevens[3]

[1]Department of Chemistry, Iowa State University, Ames, IA 50011
[2]Department of Chemistry, Bar Ilan University, Ramat Gan 52100, Israel
[3]Center for Advanced Research in Biotechnology, National Institute of Standards and Technology, 9600 Gudelsky Drive, Rockville, MD 20850

The effective fragment potential (EFP) method is introduced as a way to model the effect of intermolecular hydrogen bonds on molecules described by standard quantum mechanical (QM) methods. The chemical system of interest is divided into two regions: an "active region" (AR) described by QM, and a "spectator region" (SR) that influences the AR via hydrogen bonding. The SR is replaced by an EFP which describes the interaction by three terms: electrostatics, polarization, and exchange repulsion. *The potentials are derived from separate* ab initio *calculations on the prototypical interactions represented by the spectator region.* The method is currently being implemented in the quantum chemistry code GAMESS. Some applications involving water in the SR are presented.

Hydrogen bonding is one of the most important forms of intermolecular interaction. It is a critical component of biomolecular structure, molecular recognition, and protic solvent effects to name a few. Efficient computational models that describe hydrogen bonding accurately are thus essential for studies of such topics. One such model, the effective fragment potential (EFP) method, is introduced here.

Philosophy Behind The EFP Method

Initial Assumptions. The wavefunction of a chemical system of interest is divided into an "active region" (AR) and a "spectator region" (SR). The AR is the region in which chemical changes (e.g. bond breaking/making) occur. The chemistry in the AR is influenced by intermolecular hydrogen bonds to the SR. Thus, no covalent bonds

[4]Current address: WL/MLPJ Building 651, 3005 P Street, Suite 1, Wright-Patterson Air Force Base, OH 45433
[5]Current address: Department of Physiology and Biophysics, Mount Sinai Medical Center, New York, NY 10029

0097–6156/94/0569–0139$08.00/0

connect the AR and SR. If one initially neglects the overlap of the two regions (errors introduced by this neglect will be discussed later), it is possible to relate the inter-region interaction to the properties of the isolated regions, in a general way.

Following Buckingham *(1)*, the total hamiltonian is defined as the sum of the AR and SR hamiltonians plus an interaction term, *V*:

$$H' = H_{AR} + H_{SR} + V. \tag{1}$$

When overlap, and hence electron exchange, is neglected one can treat the electrons as belonging to one or the other of the two regions. The wavefunction of the un-perturbed system, in which neither region is perturbed by the presence of the other, can then be written as the product of the isolated AR- and SR-wavefunctions. This wavefunction is an eigenfunction of $H_{AR} + H_{SR}$ whose eigenvalue is the sum of the energies of the isolated AR and isolated SR. The energy due to the interaction, and resulting perturbation, of the two wavefunctions can be obtained through perturbation theory with *V* as the perturbation, and is then

$$E' = E_{AR}^{(0)} + E_{SR}^{(0)} + E_{AS}^{(1)} + E_{AS}^{(2)} + \dots \tag{2}$$

In this case *V* describes purely Coulombic interactions and classical interpretations can be given to each energy term. The first order energy corresponds to the electrostatic interactions of the static AR and SR charge distributions. The second order energy is comprised of two polarization energies (AR polarizing SR and SR polarizing AR) and a dispersion energy. The total interaction energy correct to second order is therefore

$$E' = E_{AR}^{(0)} + E_{SR}^{(0)} + E_{AS}^{Elec} + E_{AR}^{Pol} + E_{SR}^{Pol} + E_{AS}^{Disp}. \tag{3}$$

At smaller inter-region distances, where electron exchange becomes important, the total un-perturbed wavefunction must be antisymmetrized and is no longer an eigenfunction of $H_{AR} + H_{SR}$. Hence, it is not obvious how to relate the exchange repulsion energy, that must be added to the total energy, to the properties of the individual regions.

Further Assumptions. The following points are particular to the EFP implementation. (1) The wavefunction of the SR is replaced by an EFP comprised of effective potentials that simulate SR influence on the AR wavefunction. The AR wavefunction is described with standard *ab initio* MO theory. (2) The internal structure of the SR does not change, and the SR hamiltonian (and resulting energy, $E_{SR}^{(0)}$) can thus be ignored. (3) The internal energy of the AR includes E_{AR}^{Pol}, since the AR wavefunction automatically responds to the presence of the EFP in the course of the energy evaluation,

$$E_{AR} = E_{AR}^{(0)} + E_{AR}^{Pol}. \tag{4}$$

(4) The dispersion term (E_{AS}^{Disp} in equation 3) is presumed to have negligible effects on the AR-electronic structure, based on the R^{-6}-distance dependence *(1)*. When exchange repulsion becomes important (at small inter-region separations such that the charge distributions overlap), an additional term, E^{Rep}, must be added. The total energy of the system is then

$$E' = E_{AR} + E_{AS}^{Elec} + E_{SR}^{Pol} + E^{Rep} \quad (5)$$
$$= E + E^{Electrostatics} + E^{Polarization} + E^{Exchange\ Repulsion} \quad (5')$$

The effective fragment potentials are added to the one-electron part of the AR hamiltonian, so the total energy in the AO basis may be rewritten as

$$E' = E + \sum_{\mu,\nu} P_{\mu\nu} V_{\mu\nu} + V_N^{(efp)}. \quad (6)$$

The second and third terms describe the interactions of the EFP with the electrons and nuclei of the AR, respectively:

$$V_{\mu\nu} = \int d\mathbf{r}_1 \phi_\mu^*(1) V^{(efp)}(1) \phi_\nu(1), \quad (7a)$$

$$V_N^{(efp)} = \sum_A Z_A V^{(efp)}(A). \quad (7b)$$

The EFP can further be divided into electrostatic, polarization, and exchange repulsion contributions, cf. equation 5.

$$V^{(efp)}(1) = \sum_k^K V_k^{Elec}(1) + \sum_l^L V_l^{Pol}(1) + \sum_m^M V_m^{Rep}(1), \quad (8)$$

where K, L, and M are the total number of reference points associated with the respective potentials. The first term is the molecular electrostatic potential (MEP) of the isolated SR. The second term represents the change in this MEP induced by the AR wavefunction. The third term is a repulsive potential that describes the exchange repulsion between the AR and SR. The nuclear part of the EFP consists only of the first two terms since the exchange repulsion is a purely electronic effect.

A key feature of the EFP approach is that these potentials are derived from separate *ab initio* calculations. The previous discussion stated that the electrostatic and polarization terms can be rigorously derived from separate calculation of SR properties. This is not rigorously possible for the exchange repulsion term. The next section describes how each component of the potential is obtained, and how the potential is used during the derivation of the AR wavefunction.

Constructing An Effective Fragment Potential

Electrostatic Interactions. The electrostatic interaction dominates the hydrogen bond energy. Buckingham *(1)* has shown that this interaction potential can be related to the properties of the free molecules by expanding the molecular electrostatic potential (MEP) of one charge distribution in a multipolar expansion about an expansion point, k. Thus the electrostatic interaction potential of the AR and SR can be expressed as

$$V_k^{Elec}(1) = \frac{q_k}{\mathbf{r}_{1k}} - \sum_a^{x,y,z} \mu_a^k F_a(\mathbf{r}_{1k}) - \tfrac{1}{3} \sum_a^{x,y,z} \sum_b^{x,y,z} \Theta_{ab}^k F'_{ab}(\mathbf{r}_{1k}) - \tfrac{1}{15} \sum_a^{x,y,z} \sum_b^{x,y,z} \sum_c^{x,y,z} \Omega_{abc}^k F''_{abc}(\mathbf{r}_{1k}) + \ldots \quad (9)$$

Here, q_k is the net charge of the SR charge distribution, μ , Θ, and Ω are the dipole, quadrupole, and octupole, respectively, of the SR, and F, F', and F'' are the electric field, field gradient, and field second derivative operators, due to the AR, at point *k*. As with the perturbative analysis described above, this expansion is only rigorous if the molecules have non-overlapping charge distributions.

In general, an infinite number of terms is required to get an exact expansion of the MEP. However, by choosing several expansion points (*K* in equation 8) for a given molecule, the expansion's convergence can be greatly accelerated. Numerous schemes (*2-4*) have been developed to efficiently describe the MEP. The efficiency is usually determined by comparing the accuracy of the fitted MEP, relative to the quantum mechanical MEP, to the number of terms in the expansion. The electrostatic part of the EFP can be any expansion, but a compact expansion obviously reduces computational expense.

The distributed multipolar analysis (DMA) of Stone (*5-6*) has been found to give well-converged multipolar expansions for several small test molecules (*7*). This permits (but does not require) truncation at the quadrupole term at expansion points at the atom centers and bond midpoints, the expansion centers recommended by Stone et al. (*6*). Multipolar expansions of each gaussian product density element are evaluated at the expansion centers closest to the density element. Thus, the best expansion points coincide with large concentrations of gaussian product centers, e.g. atoms and bond mid-points. This is an approximation to the method of Rabinowitz, *et al.* (*8*) in which each of the N(N+1)/2 gaussian product centers in the basis set is used as an expansion point. While this yields finite expansions at each point, it results in an unwieldy number of points that are basis set dependent.

Charge Penetration. Typical hydrogen bonded distances between two atoms are generally shorter than the sum of their van der Waals radii, indicating that the atomic charge distributions are overlapping to a non-negligible extent. As mentioned previously, the form of the interaction potential in equation 9 is rigorous only for non-overlapping charge distributions. The multipolar expansion is not an accurate representation of the exact quantum mechanical MEP inside the region of significant charge density [one definition of this region is the 0.001 au charge density envelope (*9*)]. As the charge distributions interpenetrate, the MEP seen by one molecule due to the charge density on another molecule is significantly altered, due to the overlap of the two charge densities. Since nuclei generally are outside the overlap region they are effectively deshielded, leading to an effective increase in nuclear charge and thus an effective increase in electron-nuclear attraction. Charge penetration effects are therefore always attractive. Neglecting this charge penetration effect can result in serious errors.

The penetration effects are included in the EFP model by fitting the multipolar expansion of the MEP to the exact quantum mechanical MEP of the isolated spectator molecule. This is done by adding a penetration potential to each multipolar expansion, and optimizing penetration parameters to obtain the best fit to the accurate quantum mechanical MEP of the isolated SR. Preliminary test calculations on neutral atoms (*7*) indicate that the penetration effects decay rapidly with distance, and can be modeled with a single gaussian. The gaussian form facilitates easy implementation in integral evaluation and derivative schemes. Thus, by introducing adjustable parameters α_k and β_k and making the substitution

$$V_k^{Elec}(1) \rightarrow (1-\beta_k e^{-\alpha_k r_{1k}^2})V_k^{Elec}(1), \tag{10}$$

in the electrostatic part of the EFP, intermolecular electrostatic interactions were consistently reproduced to within 5% or less of *ab initio* values at van der Waals distances (*7*).

Polarization. As indicated by equation 3, a part of the intermolecular interaction energy arises from the change in electronic structure in one molecule due to the presence of another, i.e. polarization. This interaction can be expressed in terms of properties of the isolated molecules, i.e. molecular multipolar polarizabilities, (*1*) in an expression similar to that for the electrostatic interaction,

$$E^{Pol} = -\tfrac{1}{2}\sum_{a}^{x,y,z}\sum_{b}^{x,y,z}\alpha_{ab}F_aF_b - \tfrac{1}{3}\sum_{a}^{x,y,z}\sum_{b}^{x,y,z}\sum_{c}^{x,y,z}A_{a:bc}F_aF'_{bc} - \tfrac{1}{6}\sum_{a}^{x,y,z}\sum_{b}^{x,y,z}\sum_{c}^{x,y,z}\sum_{d}^{x,y,z}C_{ab:cd}F'_{ab}F'_{cd} - \ldots \quad (11)$$

Here $\boldsymbol{\alpha}$ is the dipole polarizability tensor, and **A** and **C** are dipole-quadrupole and quadrupole-quadrupole polarizability tensors, respectively. The field and field gradient terms (F and F') are similar to those in equation 9. In the EFP methodology, several expansion points (L in equation 8) are used to describe the polarizability of the SR. This leads to accelerated convergence and allows the individual expansions to be truncated after the first term while still maintaining some description of the higher order effects. Thus, the polarization term in the EFP is given by

$$\begin{aligned} V_l^{Pol}(1) &= -\sum_{a}^{x,y,z}\sum_{b}^{x,y,z}F_a(\mathbf{r}_{1l})\alpha_{ab}^{l}\langle F_b(\mathbf{r}_{1l})\rangle \\ &= -\sum_{a}^{x,y,z}F_a(\mathbf{r}_{1l})\Delta\mu_a^{l}, \end{aligned} \quad (12)$$

where $\Delta\boldsymbol{\mu}$ is referred to as the induced dipole moment at point l. Since $\langle F_b(\mathbf{r}_{1l})\rangle$, the expectation value of the field due to the AR at point l, depends on the final wavefunction, the polarization term is non-linear with respect to the wavefunction. This is solved by obtaining an initial guess at the induced dipole, e.g. calculated by using the current electric field, and iterating to self-consistency. Since the distributed polarizabilities within an EFP are derived from fully-coupled SCF calculations (see below), the induced dipoles within an EFP molecule are not required to interact directly. This is an approximation if they arise from a non-uniform field.

A few methods have been developed to obtain distributed polarizabilities (*10-13*). The method most compatible with the EFP methodology is the one due to Garmer and Stevens (*13*) in which the molecular polarizability is decomposed in terms of localized molecular orbital (LMO) contributions. Each LMO polarizability is given by the (numerical) first derivative of the LMO dipole ($\boldsymbol{\mu}^l$) with respect to a uniform field:

$$\alpha_{xy}^{l} = \lim_{F_y \to 0}\frac{\mu_x^l(F_y) - \mu_x^l(0)}{F_y}. \quad (13)$$

Thus, L in equation 8 is the number of LMOs in the SR, and each point, l, is at the position of the LMO centroid of charge. Summing α_{xy}^{l} over all LMOs gives the xy component of the total molecular polarizability. It is important to note that the molecular polarizability tensor is symmetric, i.e. the *sum* of, say, the *xy*- and the *yx*-components of the LMO-α's are equal, but that this is not necessarily true for each individual LMO-α.

Several tests have been performed (*7*) to compare the distributed polarizability model to the standard molecular polarizability and to Hartree-Fock SCF results. It is found that the distributed model generally reproduces the SCF results better than the single polarizable point model. The average error in energy for the former is around 10-20%, thus the majority of the polarization energy in equation 11 can be modeled through the first term.

Exchange Repulsion. Exchange repulsion can dominate the intermolecular interaction energy at distances where the charge density of two molecules overlap significantly. It arises in part from the fact that charge density in the overlap region is depleted, leading to a decrease in electron-nuclear attraction and thus a net repulsive interaction (*14*). As pointed out previously, no rigorous way to describe this repulsion in terms of properties of the isolated molecules exists. Thus, approximate methods for implementing this effect in EFP calculations must be tested on molecular systems for which this exchange repulsion energy has been calculated explicitly. In the EFP method reported here, the exchange repulsion energy representation is chosen to be as simple as possible, requiring only one-electron integrals and depending only on the density of the AR. The general approach used here is to calculate the exchange repulsion energy for a variety of intermolecular geometric arrangements for a complex (A···B) of interest, and fit the resulting energy surface to some functional form. The exchange repulsion energy, EXO, is calculated by using the energy decomposition scheme of Morokuma and Kitaura (*14-15*). Alternatively, it can be redefined as the difference between the total *ab initio* energy and the electrostatic plus polarization energy, which implicitly includes any charge-transfer and dispersion effects.

Repulsive effective potentials (REPs) are chosen here as a way to implement the exchange repulsion interaction in the EFP methodology. Here, a REP consists of several (*M* in equation 8) linear combinations of gaussians,

$$V_m^{Rep}(1) = \sum_j^J \beta_{m,j} \mathbf{r}_{1m,j}^n e^{-\alpha_{m,j} \mathbf{r}_{1m,j}^2} . \tag{14}$$

where the coefficients β and α have been optimized to reproduce the exchange repulsion energy surface (ERES) of a pair of molecules (A and B), for a given *M*, *J*, and *n*. This fitting of the ERES is accomplished by minimizing the following error function,

$$\Delta = \sum_p^P \frac{\left[\left\langle \Psi_A \left| \sum_m^M V_m^{Rep} \right| \Psi_A \right\rangle_p - EXO_p \right]^2}{EXO_p^2}. \tag{15}$$

Here, *P* is the number of points on the ERES, corresponding to various orientations of A and B; EXO_p is the exchange repulsion energy at point *p*, and $\left\langle \Psi_A \left| \sum_m^M V_m^{Rep} \right| \Psi_A \right\rangle_p$ is the exchange repulsion energy due to the REP of molecule B interacting with the unperturbed wavefunction of A at point *p*. Molecule B is then the molecule to be replaced by an EFP, and molecule A is the molecule in the AR to be described by quantum mechanics. It will be shown below that the repulsive part of the EFP can be used in calculations where the AR is a molecule other than A, and still give reasonable results. This transferability is not necessarily a given, so it has been established through testing. Alternatively, new repulsive potentials must be obtained for each molecular species used in EFP calculations.

Locating Stationary Points

While the internal structure of the of the EFP is fixed, its position relative to the AR is not. So, to obtain the optimum interaction energy, the overall structure must be optimized. This is achieved through a standard Newton-Raphson procedure. Since the internal EFP-geometry is fixed, each EFP adds six degrees of freedom to the system. The six degrees of freedom chosen are the three Cartesian components of the overall translation of the EFP relative to an arbitrary origin and the three Cartesian components of the rotation vector around the EFP center of mass (COM). The corresponding energy derivatives, depicted schematically in Figure 1, are the Cartesian components of the net force and total torque around the COM.

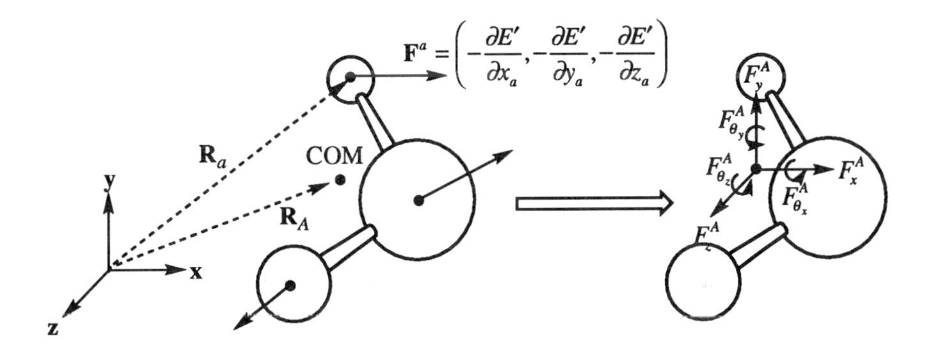

Figure 1. Schematic representation of the transformation of the Cartesian gradient components on a fragment to internal coordinate components defined relative to the center of mass (COM) of the EFP.

These new derivatives are obtained through the following transformation.

$$F_x^A = \sum_a^{a\in A}\left(-\frac{\partial E'}{\partial x_a}\right), \tag{16a}$$

$$F_{\theta_x}^A = \sum_a^{a\in A}\left[(\mathbf{R}_a - \mathbf{R}_A)\times\mathbf{F}^a\right]_{x_a} + \sum_a^{a\in A}\tau_{x_a}^{(\varepsilon)}. \tag{16b}$$

Here, F_x^A and $F_{\theta_x}^A$ are the x components of the total translational force and torque due to all terms on fragment A, respectively and $[\mathbf{v}]_x$ is the x component of vector **v**. Similar equations apply for the x and y components. The last term in equation 16b describes the torque induced on the fragment multipoles by the electric field of the *ab initio* system. The detailed expressions for the energy derivatives and torques are given elsewhere (*17*).

Applications

In the applications of the EFP method described in this section the SR region is taken to be a water molecule. The EFP used for the water molecule is described by five multipolar expansion points (corrected for charge penetration), four polarizable points, and three exchange repulsion points. All terms in the EFP are evaluated at the RHF/CEP-31G* (*18*) level of theory (d orbital exponent=0.85) using a fixed water geometry with bond lengths of 0.957Å and a bond angle of 104.52°. The multipole expansion points are located at the three nuclei and at the two bond midpoints. Multipoles through octupoles are included in the evaluation of the energy and its derivatives. Multipoles have been evaluated from *ab initio* calculations on the water molecule by the method described by Stone (*5-6*). Four effective polarizability points are used, located at the centroids of the four valence localized molecular orbitals. These polarizabilities are obtained from *ab initio* calculations carried out under the influence of an electric field. Three exchange repulsion points are included in the effective fragment potential, one located at each nucleus. The contribution to the interaction potential from each of these points is given by a sum of two spherical gaussians. The potential was fit to the ERES of the water dimer. All calculations were performed with a local version of the GAMESS (*19*) program.

Water Dimer. The water dimer system is chosen as an initial test case for the EFP method. The effect of replacing either the hydrogen bond donor or acceptor water molecule with an EFP-water is compared to all-*ab initio* calculations. Properties of interest include the dimer structure, interaction energy, and vibrational frequencies, evaluated at the RHF/CEP-31G* level of theory. In addition, the effect of polarization functions on the interaction energy is studied.

Table I lists the optimized structure of the water dimer. The most important geometric parameters in the effective fragment calculations are the internal coordinates of the *ab initio* molecule. These values are underlined in Table I. The values marked with an asterisk are fixed in the effective fragment method. Table I indicates that the internal structure of the *ab initio* water molecule is predicted quite accurately by the effective fragment method. In both effective fragment calculations, the bond lengths in the solute molecule agree with those from the full *ab initio* calculation to within 0.001 Å, and the bond angle agrees to better than 0.1°. As for the relative positions of the two molecules, the hydrogen bond length is off by at most 1% (0.022Å for EFP=acceptor). The orientational angle θ is off by as much as 11° when the EFP acts as the acceptor. This also represents the largest difference in structure between the two EFP calculations.

Table II gives the interaction energies for the two water molecules at the equilibrium structures obtained from each of the three types of calculations with three basis sets. In the calculations with the CEP-31G* basis set the interaction energies predicted by the H-donor EFP calculation and by the H-acceptor EFP calculation are less than the 5.0 kcal/mol predicted in the all-*ab initio* calculation by 0.2 kcal/mol and 0.3 kcal/mol, respectively. This is virtually unchanged when the basis set quality is increased by adding p polarization functions (p orbital exponent=1.0) on the hydrogens in the *ab initio* water molecule. Larger discrepancies, 1.2 and 2.6 kcal/mol, arise when the oxygen polarization functions are removed. The source of these discrepancies is the fact that the EFP models a CEP-31G* water molecule and the EF calculations therefore resemble calculations with one CEP-31G water and one CEP-31G* water. Such all-*ab initio* calculations result in optimized interaction energies of 6.9 kcal/mol and 5.6 kcal/mol for d functions on only the donor or acceptor water, respectively. These more sophisticated full *ab initio* calculations are in better agreement with the EFP calculations.

Table III gives the harmonic vibrational frequencies and vibrational zero-point energy (ZPE) changes obtained from hessian calculations on the dimer geometries in

Table I. RHF/CEP-31G(d) Optimized Geometries. The * Marks Frozen EFP Coordinates, While The Underlined Numbers Refer To The Internal Water Structure

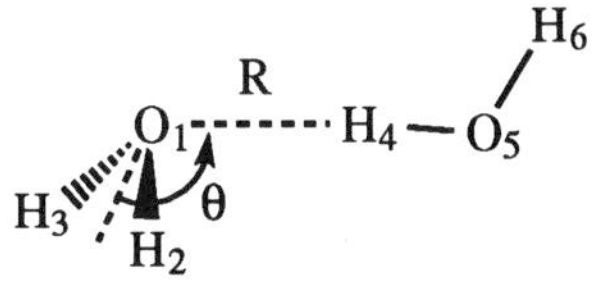

	All-*ab initio*	EFP=Donor	EFP=Acceptor
$r(O_1\text{-}H_2)$	0.952	0.951	0.957*
$a(H_2\text{-}O_1\text{-}H_3)$	106.4	106.4	104.5*
R	2.04	2.05	2.07
θ	41	40	52
$r(O_4\text{-}H_5)$	0.955	0.957*	0.955
$r(O_4\text{-}H_6)$	0.950	0.957*	0.950
$a(H_5\text{-}O_4\text{-}H_6)$	105.9	104.5*	105.8

Table II. Interaction Energies For The Water Dimer In Kcal/mol

Basis Set	All-*ab initio*	EFP=Donor	EFP=Acceptor
CEP-31G	7.6	6.4	5.0
CEP-31G(d)	5.0	4.8	4.7
CEP-31G(d,p)	4.9	4.7	4.6

Table III. Harmonic Frequencies (cm^{-1}) Of The RHF/CEP-31G(d) Water Dimer

Frequency	All-*ab initio*	EFP=Donor	EFP=Acceptor
1. A"	138	134	130
2. A'	145	163	144
3. A"	151	187	204
4. A'	173	233	210
5. A'	342	346	414
6. A"	605	559	642
7. A'	1808	1809	ff
8. A'	1831	ff	1836
9. A'	4074	ff	4093
10. A'	4113	4113	ff
11. A'	4214	ff	4221
12. A"	4234	4236	ff
ΔZPE (kcal/mol)	2.2	2.3	2.5

Table I. Only numerical hessians are available in effective fragment calculations. To ensure accuracy, the maximum component of the gradient of each geometry was reduced to less than 10^{-6} Hartree/Bohr, and the symmetrical displacements around the minimum were reduced to 0.001 Bohr. For the water dimer this generally leads to frequencies that are within 4.1% of analytical results. The harmonic analysis in the effective fragment calculations shows an overestimation of the frequencies associated with the internal coordinates of the *ab initio* molecule by 0.06%, 0.0%, and 0.05% when the H-donor is replaced with a fragment and by 0.3%, 0.5%, and 0.2% when the H-acceptor is replaced with a fragment. For the frequencies associated with the relative motion of the two waters, the H-donor effective fragment calculation agrees quite well with the full *ab initio* calculation, except for frequencies #3 and #4, for which the fragment results (187 and 233 cm^{-1}) are 19% and 26% greater than the *ab initio* frequencies. In the H-acceptor EF calculation, frequencies #3-6 are up to 26% (for #3) higher than in the all-*ab initio* calculation. However, these deviations translate to only minor (≥0.3 kcal/mol) errors in the zero point energy (ZPE) correction to the interaction energy.

Water-Formamide. In order to evaluate the more general usefulness of the effective fragment potential for the water molecule, we need to study its interaction with AR molecules other than water. The interaction between the formamide molecule and the water molecule is of interest in biochemistry because formamide is the simplest prototype for a peptide linkage. In an *ab initio* study by Jasien and Stevens (*20*) four stationary points were found on the RHF/DZP (*21*) formamide-water potential energy surface, within the constraint of C_s symmetry. We have carried out *ab initio* geometry optimizations in C_1 symmetry on this system with the water molecule replaced by an effective fragment. In addition we have done full *ab initio* geometry optimizations in C_1 symmetry. Both lead to three C_1 minima similar to three of the C_s structures. The fourth structure was a C_s transition state at the all-*ab initio* level of theory.

Figure 2 depicts the three C_1 minima, labeled I-III, that were located by both all-*ab initio* and EFP optimizations. Selected structural parameters are listed for the full *ab initio* (bold) and EFP calculations. In the effective fragment calculations on the three minima, the length of the hydrogen bonds between the two molecules is longer than predicted by the *ab initio* calculations by just 0.04 (structure III; 2%) to 0.17 (structure I; 8%) Å. The orientational angles obtained in the effective fragment calculations do not agree exactly with the *ab initio* calculations either, but are qualitatively correct.

The geometric parameters of greatest interest, the internal coordinates of the formamide molecule, are compared in Table IV. The first column in Table IV lists the internal coordinates of a lone formamide molecule in its equilibrium configuration, as obtained in an *ab initio* calculation with the DZP basis, and the other six columns list the change in these coordinates caused by the presence of a water molecule. For each of the three minimum energy configurations, results are listed both from full *ab initio* calculations and from effective fragment calculations. While the changes in these coordinates are small, the effective fragment method consistently predicts nearly the same perturbation in these internal coordinates as was obtained in the full *ab initio* calculations.

Table V lists the interaction energies for structures I-III. For the three minima on the potential energy surface, the interaction energies obtained in the effective fragment calculations differ from those obtained from the full *ab initio* calculations by 1.6 (19%), 0.9 (15%), and 0.5 (9%) kcal/mol. Although the effective fragment

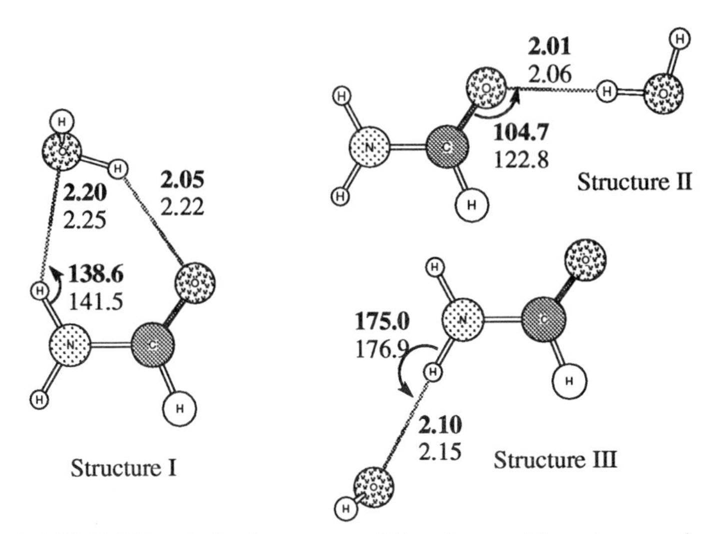

Figure 2. RHF/DZP optimized structures of three formamide-water complexes, with and without the water replaced by an EFP. The bold numbers are structural parameters for all-*ab initio* calculation, and may be compared to the numbers obtained in the EFP calculations. Bond lengths are given in ångstroms and bond angles in degrees.

Table IV. RHF/DZP Internal Coodinates Of The Isolated Formamide Molecule, And The Change In These Coodinates Caused By An *Ab Initio* Or EFP Water In Structures I-III. Bond Lengths In Ångstroms And Angles In Degrees

	Form-amide	Structure I		Structure II		Structure III	
		ab inito	EFP	*ab initio*	EFP	*ab initio*	EFP
r(C-N)	1.353	-0.009	-0.007	-0.006	-0.006	-0.005	-0.004
r(C-O)	1.196	0.009	0.007	0.007	0.004	0.004	0.004
r(C-H)	1.092	-0.002	-0.001	-0.003	-0.002	0.000	0.000
r(N-H)	0.995	0.005	0.005	0.000	0.000	0.000	-0.001
r(N-H')	0.992	0.000	0.000	0.000	0.000	0.005	0.004
a(OCN)	124.9	0.2	0.2	-0.5	-0.3	0.5	0.5
a(HCN)	122.3	-1.0	-0.8	-0.4	-0.2	-0.3	-0.3
a(HNC)	121.3	-0.7	-0.7	-0.1	-0.1	-0.1	-0.1
a(H'NC)	119.1	0.1	0.1	0.1	0.1	-0.5	-0.6
d(OCNH)	0.0	1.6	2.6	0.4	-0.3	0.0	0.3

Table V. RHF/DZP Interaction Energies For Three Water-Formamide Complexes In Kcal/mol

Structure	All-*ab initio*	Water=EFP
I	-8.3	-6.7
II	-6.0	-5.1
III	-5.3	-4.8

method underestimates the interaction energies, it does correctly predict that structure I is considerably more stable (by 1.6 kcal/mol, compared to 2.3 kcal/mol in the *ab initio* case) than structures II or III, and that structures II and III are comparatively close in energy. Clearly, one arrives at the same qualitative picture of the water-formamide interaction based on both methods.

The fact that the EF method does not do quite as well at predicting the formamide-water interaction energy as it did at predicting the water dimer interaction energy is probably due to the exchange repulsion part of the effective fragment potential, which was fit to the water dimer interaction. More sophisticated exchange-repulsion potentials which explicitly take into account the overlap between the AR and the SR may provide improved transferability. Considering the simple form of the potential used here, the effective fragment method does quite well. Since the formamide-water system is small enough to be treated in a full *ab initio* calculation, this system could in principle be used to construct an exchange repulsion potential that might be more accurate in modeling the hydration of larger peptide systems.

Conclusion and Future Directions

The effective fragment potential (EFP) method is introduced as a way to model intermolecular hydrogen bonds and their effect on quantum mechanical wavefunctions. It is shown that the effect of a water molecule on the *ab initio* wavefunctions of water and formamide can be modeled relatively accurately by introducing an EFP in the *ab initio* one-electron Hamiltonian. The potentials are obtained from other *ab initio* calculations on the isolated water molecule and water dimer. Thus for the formamide-water complexes no parameters in the EFP model have been adjusted to reproduce the interaction. It is therefore encouraging to find only relatively modest deviations from calculations in which both the formamide and the water are treated quantum mechanically. The method is in principle extendible to model any intermolecular hydrogen bond.

Current research is focused on a parallelizing the EFP code in GAMESS, as well as including energy and gradient terms that describe EFP-EFP interactions. The latter would allow, for example, to surround an *ab initio* wavefunction with more than one EFP water to approximately model the first solvation shell in aqueous solvation.

Acknowledgments

This research was supported, in part, by grant No. 88-00406 from the United States-Israel Binational Science Foundation (BSF), Jerusalem, Israel. Two of the authors (MSG, JHJ) acknowledge the support of the National Science Foundation (CHE89-11911).

Literature Cited

1. Buckingham, A. D. In *Intermolecular Interactions: From Diatomics to Biopolymers*. Pullman, B., Ed.; Perspectives in Quantum Chemistry and Biochemistry; John Wiley & Sons Ltd.: New York, NY, 1978, Vol. I, pp 1-68.
2. Williams, D. E.; Yan, J. -M. *Adv. Atm. & Mol. Phys.* **1988**, *23*, pp 87.
3. Lavery, R.; Etchebest, C.; Pullman, A. *Chem. Phys. Lett.* **1982**, *85(3)*, pp 266.
4. Etchebest, C.; Lavery, R.; Pullman, A. *Theor. Chim. Acta (Berl.)* **1982**, *62*, pp 17.
5. Stone, A. J. *Chem. Phys. Lett.* **1981**, *83(2)*, pp 233.
6. Stone, A. J.; Alderton, M. *Mol. Phys.* **1985**, *56(5)*, pp 1047.
7. Garmer, D. R.; Stevens, W. J.; Kraus, M.; Basch, H.; Cohen, D. manuscript in preparation.

8. Rabinowitz, J. R.; Namboodiri, K.; Weinstein, H. *Int. J. Quantum Chem.* **1986**, *29*, pp 1697.
9. Wong, M. W.; Wiberg, K. B.; Frisch, M. J. *J. Amer. Chem. Soc.*, **1992**, *114*, pp 1645.
10. Thole, B. T.; Van Duijnen, P. T. *Chem. Phys.* **1982**, *71*, pp 211.
11. Thole, B. T.; Van Duijnen, P. T. *Theor. Chim. Acta*, **1980**, *55*, pp 307.
12. Stone, A. J. *Mol. Phys.* **1985**, *56(5)*, pp 1985.
13. Garmer, D. R.; Stevens, W. J. *J. Phys. Chem.* **1989**, *93*, pp 8263.
14. Salem, L. *Roy. Soc. (London)* **1961**, *A264*, pp 379.
15. Morokuma, K. *J. Chem. Phys.* **1971**, *55*, pp 1236.
16. Kitaura, K.; Morokuma, K. *Int. J. Quantum Chem.* **1976**, *10*, pp 325.
17. Day, P. N.; Jensen, J. H.; Gordon, M. S.; Stevens, W. J.; Garmer, D. R. manuscript in preparation.
18. Stevens, W. J.; Basch, H.; Krauss, M. *J. Chem. Phys.* **1984**, *81*, pp 6026.
19. Schmidt, M. W.; Baldridge, K. K.; Boatz, J. A.; Elbert, S. T.; Gordon, M. S.; Jensen, J. H.; Koseki, A.; Matsunaga, S.; Nguyen, K. A.; Su, S.; Windus, T. L.; Dupuis, M.; Montgomery, J. A. *J. Comp. Chem.*, **1993**, *14*, pp 1347.
20. Jasien, P. G.; Stevens, W. J. *J. Chem. Phys.* **1986**, *84*, pp 3271.
21. Dunning, Jr., T. H.; Hay, P. J. In *Methods of Electronic Structure Theory.* Shaefer III, H. F., Ed.; Plenum press, NY. 1977, chapter 1.

RECEIVED June 2, 1994

Chapter 10

Modeling the Hydrogen Bond with Transferable Atom Equivalents

Curt M. Breneman, Marlon Rhem, Tracy R. Thompson, and Mei Hsu Dung

Department of Chemistry, Rensselaer Polytechnic Institute, 110 8th Street, Troy, NY 12180

Abstract: Hydrogen bonding is recognized as a key feature in determining macromolecular structure, and is often the strongest type of interaction involved in a molecular recognition event.[1] Additionally, recent work has shown that hydrogen bond strengths are affected by their positions within macromolecules.[2] Consequently, in order to predict or evaluate the strengths of hydrogen bond interactions in a proposed struc ture, a method which is sensitive to the local electronic environment must be utilized. The recently developed Transferable Atom Equivalent (TAE) molecular modeling technique has proven to be particularly well suited to this task. The TAE method is based upon the assembly of transferable atomic units of electron density which provide accurate electron density models of large molecules. In addition to some small molecule test cases, TAE modeling of the hydrogen-bonding interaction between TYR 82 of FKBP and the FK506 α-dicarbonyl moiety is included as an example.[3] In these and other test cases, the computational resources required for TAE calculations were found to scale in approximately the same fashion as molecular mechanics calculations.

Introduction: The special considerations required for properly modeling hydrogen bonds have created additional challenges for the theoretical community since the advent of empirical force-field molecular modeling techniques. Since hydrogen bonding has a relatively long range coulombic component, electrostatic terms have been successfully used to account for some of the observed effects.[4] The disadvantage of this method is that it cannot accurately represent the energy of the hydrogen bonding interaction when the donor and acceptor atoms are within bonding distance. Within the electrostatic model, the added complications of exchange repulsion, penetration effects and covalent bonding interactions are lumped together into coulombic and Van der Waals terms. While some of the recent efforts in this area appear promising, the highly parameterized empirical potential functions required to adequately represent hydrogen bonding are cumbersome to implement. Therefore, it would seem that hydrogen bond interactions are best represented by modeling techniques which take into account quantum mechanical phenomena and the effects of changes in the local electron density distribution.

0097–6156/94/0569–0152$08.54/0

Modeling hydrogen bond interactions with quantum mechanics calculations circumvents a number of the problems discussed earlier, but introduces a number of operational concerns. Ideally, high-quality molecular orbital or valence bond quantum mechanical methods should be able to calculate the energy, electronic properties and preferred geometry of just about any array of atoms. One complication associated with comparing the energy of supermolecular arrays to the energies of the individual components is basis set superposition error (BSSE). This effect tends to artificially lower the energy of the complex relative to the isolated partners as a result of apparent basis set inflation. Counterpoise calculations which make use of "ghost" atoms have been used with some success in correcting this error, but they do not allow limits to be placed on the size of the BSSE effect.[5] Another reason why reality falls short of this ideal situation is that a large number of primitive gaussian functions are required to accurately represent the wavefunction of a molecule. When these primitives are used in traditional ab initio calculations, very large numbers of multiple-centered integrals must be calculated and stored or constantly recalculated. This problem limits the size of molecules which can be accurately treated. Several alternative methodologies have been developed to address this problem, including semi-empirical molecular orbital theory, direct SCF and Direct Minimization techniques, and density functional theory (DFT).

Semi-empirical quantum mechanical programs are designed to reduce the impact of integral calculation and storage by using look-up tables of integral values. Such an approach is rapid in execution, but must have tables of integral values parameterized for the bonding situations likely to be encountered. On the other hand, these kinds of semi-empirical methods have reached a high level of refinement in recent years, and such calculations are often the only quantum mechanical method available for use with very large molecules.

Direct SCF and Direct Minimization methods were developed to maximize computational performance on medium-sized molecules without sacrificing ab initio accuracy, but they are only capable of limited performance enhancements relative to the other techniques discussed here. Both Direct SCF and Direct Minimization algorithms rely on fast recalculation of required integrals rather than massive integral storage and retrieval schemes. Aggressive integral cutoff methods are also cleverly used in both schemes to eliminate unnecessary computation. While these techniques allow somewhat larger molecules to be examined, they are still burdened by the n^3-n^4 CPU time dependence characteristic of ab initio methods, where n is the number of non-hydrogen atoms.

Density functional theory (DFT) offers an interesting alternative to molecular orbital or valence bond calculational methods since it attempts to directly determine the best electron density distribution for a given molecular framework.[6] DFT is an attractive approach which has begun to gain favor among the ab initio community since it deals directly with electron density distribution, which is an experimental observable. Since knowledge of the electron density distribution around a molecule and its response to external fields uniquely defines the properties of a molecule, it is highly desirable to have access to such information. While DFT programs such as DMOL[7] are more resource-efficient than molecular orbital calculations for obtaining good electron density representations, this method is also quite cumbersome when applied to very large molecules.

It would appear that the ideal molecular modeling method would have the ability to generate and evaluate molecular electron density distributions with a high degree of accuracy without the associated computational overhead. All interesting molecular properties would then be available by querying the electron density model with appropriate operators. This technique would also use a consistent potential function for all inter- and intramolecular interactions, and would be capable of determining the best molecular geometry. The Transferable Atom Equivalent (TAE) method is being developed to fill this need.

The Transferable Atom Equivalent (TAE) modeling method is a resource-efficient alternative to routine HF/SCF ab initio calculations. The electron density representations created by TAE reconstruction are designed to allow molecular geometries and electronic properties to be quickly assessed with near ab initio accuracy. The properties which can be calculated include geometries, total energies, electrostatic potentials, Laplacian fields, electronic kinetic energy distributions, gradient vector fields and many other scalar and vector quantities. In the current work, the Transferable Atom Equivalent modeling method has been applied to a number of small hydrogen bonded complexes as well as in the large molecule example of FKBP bound to FK506.

Methods: The roots of the TAE approach can be found in the work of Bader and coworkers, who have described the conditions necessary for atoms to be separately defined within molecules.[8] Although the details of Bader's theories are described fully in the literature, several key points will be described here. First, it is important to emphasize that the key to molecular partitioning lies in the requirement that the resulting atomic pieces be valid quantum subsystems. A valid quantum subsystem is a fragment of a larger system which behaves as a separate entity, and gives unique values for all of the operators associated with the larger system. In an atomic quantum subsystem, these values would be things like electronic kinetic energy, electron population, electric multipole moments and electrostatic potentials. Quantum subsystems are said to be valid if the electron density within each separate atomic volume independently obeys the Virial Theorem ($-V/T = 2$, where V is potential energy and T is kinetic energy). In order for this condition to be met, the atomic fragments must be bound by surfaces of zero flux in electron density (ρ), and contain an attractor (a nucleus). In the general sense, a zero flux surface defines a region of space where the net flux of electron density across the surface is zero. A more practical definition states that the flux of electron density across such as surface is zero at all points on that surface. This can be summarized by the expression $\nabla\rho \bullet N = 0$, where N is the surface normal vector.

Determination of atomic zero flux surfaces begins by an analysis of the electronic topology of the molecule. Electronic topology mapping involves finding all of the stationary (critical) points in rho, and identifying them according to their "signatures". A critical point signature is defined as the number of negative curvatures minus the number of positive curvatures in rho at that point in space. For generality, a prefix of "3" designates the dimensionality of the problem. Consequently, atomic nuclei are (3,3) critical points, bond critical points have (3,-1) signatures, ring critical points have (3,1) signatures, and cage critical points have (3,-3) signatures. With the exception of (3,3) nuclear critical points, all of the other critical points exist on interatomic zero flux surfaces. Interatomic surface definition routines make use of the fact that (3,-1) (bond) critical points represent local maxima of electron density on each interatomic surface. Consequently, all points on the path followed by a steepest descent walk originating in any direction perpendicular to the bond path[9] are on an interatomic zero flux surface. Both Bader's original PROAIM program[10] and our later FASTINT program[11] make use of this atomic surface definition algorithm. Once defined, the zero flux surfaces provide boundaries for integration of the electron density properties of each separate atom within a molecule.

Critics of this partitioning method have pointed out that "atomic charges" derived from the integrated electron populations of such atoms do not reproduce the molecular dipole moment or the molecular electrostatic potential field. It is often overlooked that atomic volumes of this sort are not at all spherical. Attribution of integrated atomic populations to nuclear positions as if they were atomic monopoles implicitly asserts the unrealistic spherical atom model. In fact, when dipole components are calculated by combining the atomic basin charge moments [12] with the individual atomic dipole components for all atoms in the molecule, agreement with the HF/SCF results is exact! Energy and population additivity as well as polarizability have also been shown to be exact.[13]

The implications of the electron density partitioning results are clear: It should be possible to construct new molecules as assemblies of flexible atomic electron density fragments.[14] To accomplish this in practice, however, it is necessary to (1) have a supply of appropriate atom types in a form which can transfer electron density properties, and (2) be able to self-consistently adjust the properties of the base atom types to fit new environments. The Transferable Atom Equivalent method is designed around these criteria.

In order to be able to carry electron density information from a database to a new molecular environment, there must exist a library of atom types from which initial atomic electron density representations may be drawn. Secondly, appropriate atom types must be selectable from the library on the basis of molecular connectivity. It goes without saying that the atomic data structure must also be translationally and rotationally invariant, and must be able to react to changes in their local environments. The key to property flexibility in TAEs was found in the observation that the integrated atomic properties of atoms in molecules are intimately coupled with their shape and bonding partners.

On the basis of that observation, a tentative TAE data structure has been defined which makes use of the relationship between atomic shape and atomic properties. Within this definition, the data for each TAE atom type consists of a spherical polar coordinate model of its physical shape and surface properties in a predefined orientation. The orientation chosen for the current TAE modeling paradigm is one in which the shortest ray from the nucleus to an interatomic surface is aligned with the +Z axis, and the next shortest ray is placed in the XZ plane. This procedure allows each position on the external and interatomic surfaces of an atom to be uniquely defined. The data structure also includes a set of coefficients which describe the response of atomic properties (such as interatomic surface electron densities) and atom energies to radial variations of each unique surface position. Together with the recombination algorithm RECON[15], this results in a flexible atomic density representation which is capable of slightly altering its properties in order to fit a new environment. When each new TAE atom is added to a molecule, the RECON algorithm makes use of "perturbation functions" to find the lowest energy position for a new interatomic surface. In addition to the energy criteria, the new surface is constrained to satisfy the zero flux condition $\nabla\rho \cdot N = 0$, where N is the normal vector of the new interatomic surface. These perturbation functions and their coefficients were derived from the ab initio wavefunctions of the model compounds used in TAE definition by examining the effect of all possible incremental ray length changes on the integrated atomic properties and surface properties of each TAE atom. The recombination algorithm relies on the principle that each atom type has a unique "electronic electronegativity", which is defined as the ratio of atomic electronic kinetic energy to integrated atomic population in the region of interest.[16] A simple way to envision this relationship is to note that a more electronegative atom will receive greater energetic benefit from an increment of electron density than a less electronegative one. This relationship, illustrated in the following figures, has been shown to be linear over a broad range of populations for hydrogens and carbons, but behaves in a somewhat more complex way in highly electronegative atoms such as oxygen and nitrogen. This fits chemical intuition, since donation of electron density to an electronegative atom will be favorable up to a point, after which more electron population will tend to destabilize that atom by increased electron repulsion. In the recombination algorithm the problem is more complex, since radial displacement of some points on an interatomic surface will cause more atomic population and energy changes than others. Assignment of perturbation coefficients to each of the atomic surface points describes this variation.

The perturbation functions are then used to compute changes in the electronic kinetic energy, atomic electron population, and the surface electron density during the recombination process. Each original TAE description includes initial ab initio values for all integrated and surface properties of the atom, but once the ray lengths are changed during recombination, these values must be adjusted accordingly. In the interests of efficiency, only the population, surface electron density and electronic kinetic energy values are adjusted during each step of the recombination process. Since the computational resources required for the evaluation process is linear in the number of properties desired in the final molecular assembly, the other properties are recalculated only when the interatomic surfaces have been refined.

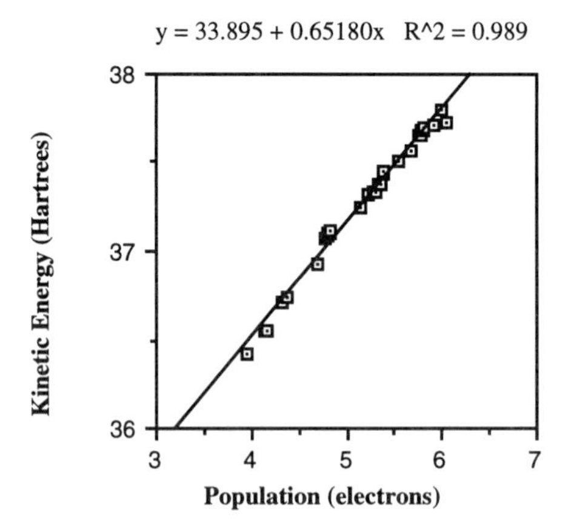

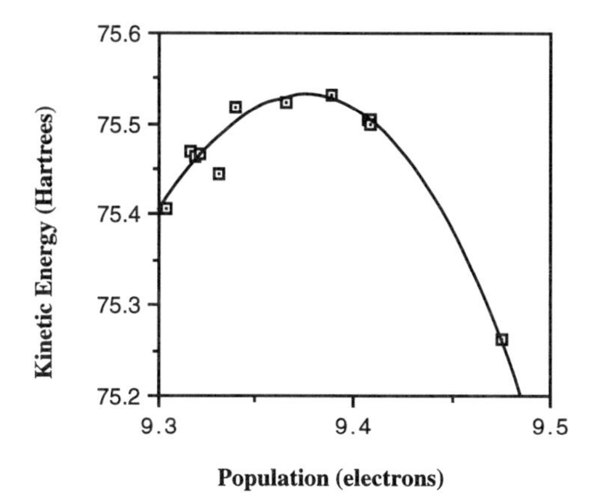

It is important to distinguish between the energetic changes that a particular atom type experiences when faced with changes in its normal electron population (as shown above), and the overall relationship of charge and energy across a large sample of atoms in their normal molecular environments. The figures shown below illustrate the charge and energy ratios of carbons, hydrogens, nitrogens and oxygens taken from large number of FASTINT integrations:

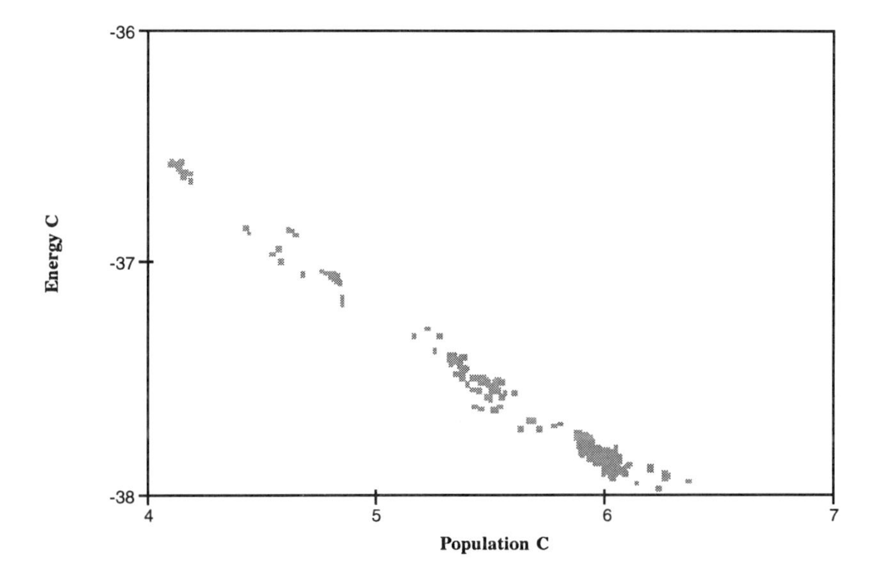

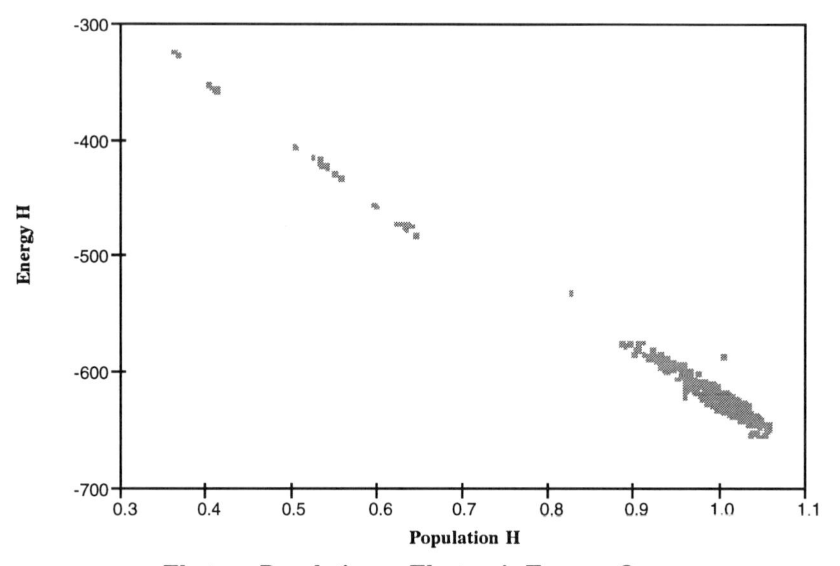
Electron Population vs Electronic Energy: Hydrogens
-300
-400
-500
-600
-700
Energy H
0.3
0.4
0.5
0.6
0.7
0.8
0.9
1.0
1.1
Population H

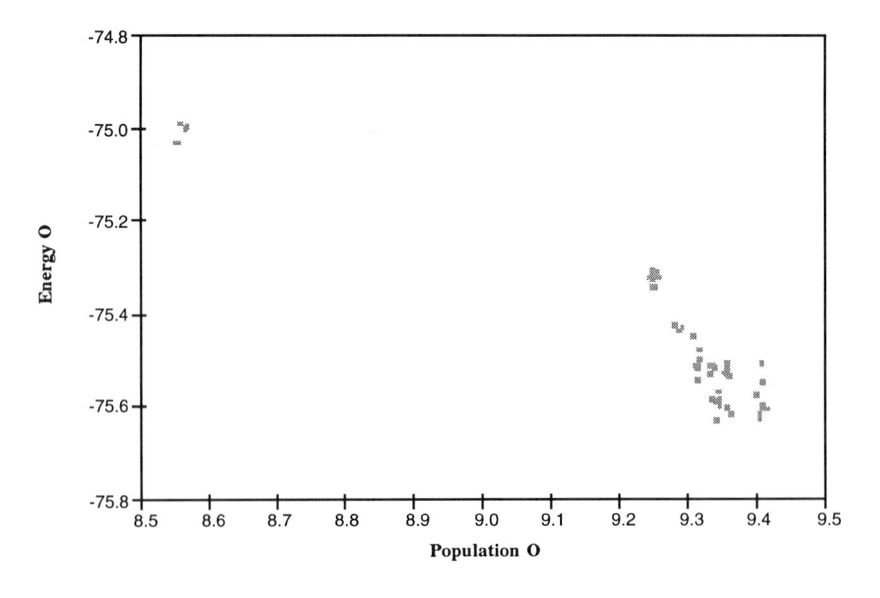
Electron Population vs Electronic Energy: Oxygens
-74.8
-75.0
-75.2
-75.4
-75.6
-75.8
Energy O
8.5
8.6
8.7
8.8
8.9
9.0
9.1
9.2
9.3
9.4
9.5
Population O

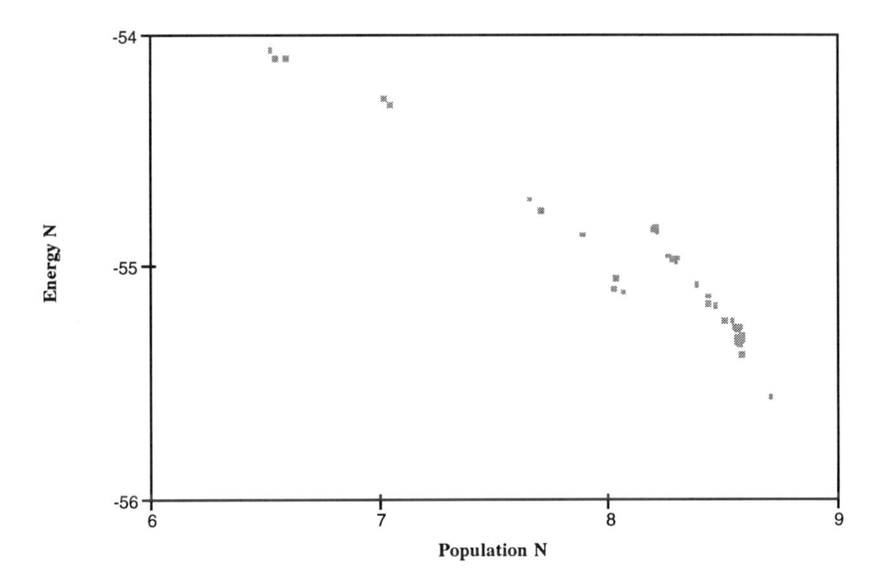

In order to appreciate our current choice of atomic electron density data structures, one must first become familiar with the kind of data which is available during an electron density partitioning event. After the molecular bonding topology has been determined and interatomic surfaces have been generated, the limits of the atomic basin are determined as an ordered list of distances from the atomic nucleus. This list contains distance information at each point in a spherical polar coordinate system using Gauss Quadrature abscissas in both theta and phi. An example of TAE generation is illustrated in Figure 1. In the upper left corner of this graphic, the atomic surface generated during a PROAIMS integration of one of the hydrogen atoms on the sample molecule is shown in raw form. The surface extends outward from the nucleus until it either hits an interatomic boundary or reaches a 10 atomic unit distance limit. If the rays are truncated when the electron density has fallen to the 0.002 e/au^3 level, their endpoints can be used to define the molecular VDW surface.[17] The surface file which contains this information is then mapped onto a regularly spaced spherical polar coordinate system. In order to facilitate comparison with other atomic surfaces, the surface file data is placed with its shortest ray aligned along the +Z axis of the new coordinate system. When more that one valence region exists on an atom, the ray with the highest electron density at its endpoint is used in this capacity. When the data in this regularly spaced surface file is displayed in 2D relief format, the resulting figure becomes known as an atomic Phase Plot. Characteristic Phase Plots are then obtained by rotating the data in each Phase Plot about the new Z axis until a standard orientation is achieved.

The criteria for determining the standard orientation of a surface file depends upon whether the atom has more than one bond critical point. For monovalent atoms, the surface file is rotated until the second shortest ray is placed in the XZ plane. For

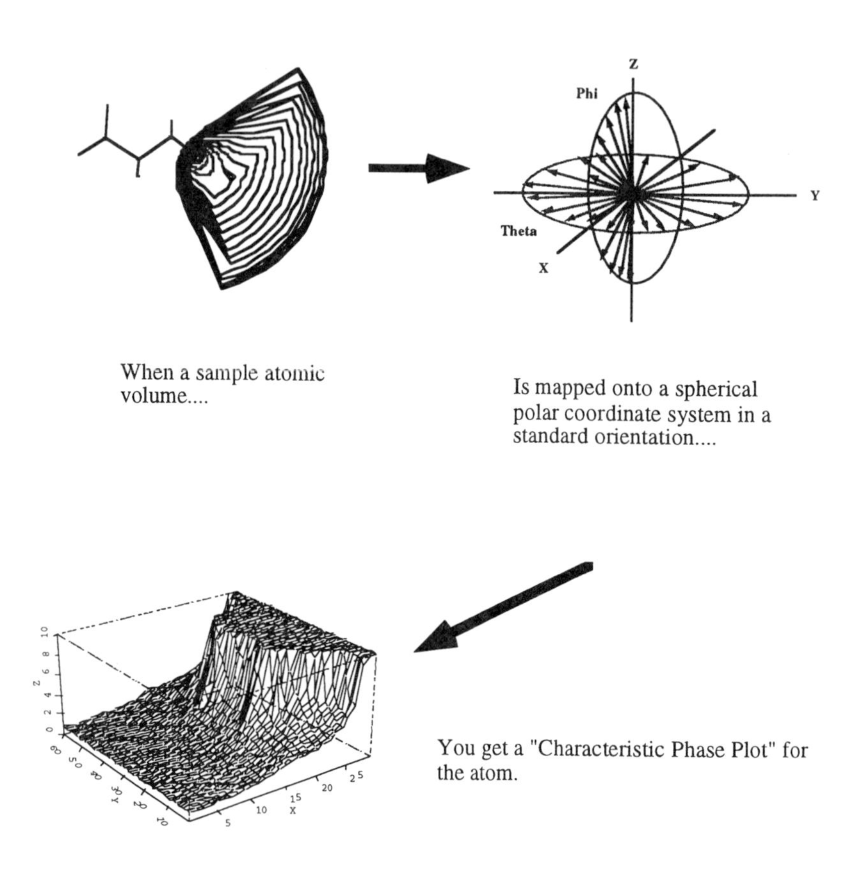

• **All points on the surface are then assigned "perturbation coefficients" which describe how each atomic property is affected by changes in the surface position.**

Figure 1. Origin of the TAE Surface File: A Flexible Atomic Basin Definition.

multivalent atoms, the shortest ray in another valence site is placed in the XZ plane. In either case, it is a simple matter to perform visual or automated comparisons of atomic shapes or surface properties using this representation. This data format is also used in storing the atomic surface shape in the TAE library for use in RECON molecular electron density reconstructions.

TAE Generation: Once the TAE data format and reconstruction paradigm was defined, the integrated properties of a large set of atoms were investigated and divided into a number of groups using the cluster analysis technique. Cluster analysis attempts to define "natural" groupings of objects by measuring the "similarity" or "dissimilarity" between them. Correlation coefficients or Euclidean distances between the variables of each element are typically used as similarity indicators. Each of the available clustering methods is biased towards finding clusters of a specific form. That is, the different methods arrive at cluster groups of a general size, shape, and dispersion inherent to their algorithmic nature. For example, some methods will preferentially find compact spherical clusters of approximately equal size. Other methods will find elongated or irregularly shaped clusters, but fail to find compact forms. In selecting clustering algorithms, one must consider the number of elements being analyzed, the number of variables used for clustering, and also the general "personality" of the data in deciding which method will give the most reliable or "natural" groups.

The algorithms used in defining TAEs were taken from the agglomerative hierarchical clustering methods available in the Statistical Analysis Software (SAS) package from SAS Institute Inc. A hierarchical method allows one cluster to be completely contained within another but does not allow for overlap between clusters. As such, an agglomerative hierarchical method in which each element initially belongs to a separate cluster will require that two elements which are grouped into a cluster stay together. The advantage of a hierarchical method is that it reduces the number of possible clusters and thus is computationally favorable. On the down side, an early wrong decision to group two data elements together is irreversible. One begins with an initial set of single element cluster groups and works toward reducing the number of groups by combining those which have the smallest "variable" distance between them.

The Transferable Atom Equivalent clusters were defined using integrated atomic properties as well as numerous VDW surface properties and integrated VDW surface properties. The integrated atomic properties included electron population, atomic dipole magnitude, electronic kinetic energy, volume, and the three principal quadrupole components. Surface properties included external surface area, valence region surface area and valence critical point electron densities. The following integrated surface properties were also used: $\Sigma(\nabla\rho\cdot N)_i\,\Delta_i$, $\Sigma(\nabla K\cdot N)_i\,\Delta_i$, $\Sigma(\nabla G\cdot N)_i\,\Delta_i$, $\Sigma(K)_i\,\Delta_i$ and $\Sigma(G)_i\,\Delta_i$, where Δ_i is the ith surface element and N is the molecular surface normal vector. These terms are further defined in Table 1. Thus, for an sp^3 carbon atom with four bonds, there are 18 clustering variables for each element. The variables were standardized before clustering was attempted. A total of seven data sets were analyzed: hydrogens, sp^3 carbons, sp^2 carbons, sp^3 nitrogens, sp^2 nitrogens, sp^3 oxygens, and sp^2 oxygens. All the available clustering methods were used on each data set in order to ascertain the most useful method for the given data. The reliability of each method is based on the chemical reasonability of the groups produced and the standard deviations from the mean for each property used in clustering.

Ward's minimum variance method was found to be best for clustering sp^2 nitrogen atom properties. Ward's method utilizes least squares distance criteria and finds spherical clusters of approximately equal size. Pyramidal sp^3 nitrogens were clustered using the complete linkage method. This is a conservative method which requires a group to be defined from maximally complete subgraphs. That is, all pairs of objects within the a given cluster must be related. Moreover, the clusters arising from the complete linkage method are generally small clusters which combine easily to form larger clusters at lower partition levels. The complete linkage method defines the distance between two groups as the maximum distance between a variable in one cluster and a variable in the other cluster.

Table 1. Key to Atomic Properties used in Current TAE Definition and Reconstruction

W, X, Y, Z: valence critical point rho values
A: valence region surface area
B: external surface area (au^2)
N: electron population (electrons)
E: electronic energy (Hartrees)
EP: Electrostatic Potential Surface Integral (au)
D: Dipolc Magnitude (au•e)
V: Atomic Volume within the 0.002 e/au^3 isosurface (au^3)
QAA, QBB, QCC: Diagonalized quadrupole components (au^2•e)
$\nabla\rho\cdot N$: $\Sigma(\nabla\rho\cdot N)_i\,\Delta_i$, The rate of change of electron density perpendicular to the atomic surface
$\nabla K\cdot N$: $\Sigma(\nabla K\cdot N)_i\,\Delta_i$, The rate of change of kinetic energy normal to the atomic surface.
$\nabla G\cdot N$: $\Sigma(\nabla G\cdot N)_i\,\Delta_i$, The rate of change of kinetic energy normal to the atomic surface.
K: $\Sigma(K)_i\,\Delta_i$, The surface integral of the kinetic energy derived from K (au•Bohr)
G: $\Sigma(G)_i\,\Delta_i$, The surface integral of the kinetic energy derived from G (au•Bohr)
Δ_i = Surface area element

K = Kinetic energy operator: $-(\hbar/4m)N\int d\tau' \{\Psi^*\nabla^2\Psi + \Psi\nabla^2\Psi^*\}$

G = Kinetic energy operator: $-(\hbar/2m)N\int d\tau' \nabla\Psi^*\cdot\nabla\Psi$

‡ Kinetic energies K and G are related by K = G + L, where L is the Laplacian of charge density. L vanishes for topologically defined atoms.

Trigonal sp^2 carbon TAEs were also defined by the complete linkage method. However, a small group of these carbons did not cluster well. Instead, this subset consisting of atoms in strained bonding environments and in 5-membered heterocycles with one or two neighboring heteroatoms were clustered in the absence of external or valence surface area variables using the average linkage method. Average linkage utilizes arithmetic averages and defines the distance between two clusters as the average distance between pairs of variables. It is known to find clusters of equal variance. The centroid method was employed for sp^3 carbon atom clustering. It is similar to the average linkage method but defines the distance between two clusters as the squared Euclidean distance between centroids or means instead of using arithmetic averages. Both sp^2 and sp^3 oxygen atoms were clustered using the complete linkage method.

Hydrogen TAEs arose from a single linkage analysis. The single linkage method defines the distance between two groups as the minimum distance between a variable in one cluster and a variable in the other cluster. It is the most liberal of the methods because it does not require every object to be pair-wise linked to every other object in the group. As such, elongated and irregularly shaped clusters can arise. Curiously, there are only six hydrogen TAEs: C-H, N-H, O-H, O-H(Hbonded), S-H, and Si-H. Increasing the number of cluster groups does not afford chemically reasonable subgroups. However, even for the hydrogen atoms bonded to carbon, for which there are 807 in the data set, the group standard deviations are quite small.

The current TAE library contains fifty-eight atom types derived by cluster analysis from the integrated atomic properties of numerous dipeptides, over 500 small acyclic and heterocyclic molecules, and all of the common nucleic and amino acids. More examples are being added to the integrated atomic property database on a continuous basis, which serves to improve TAE modeling performance for unusual bonding situations.

Small Molecule Results: In order to test the performance of the TAE modeling method on hydrogen-bonded systems, a number of small molecules must be constructed in the presence and absence of water complexation. Figure 2 illustrates the process of TAE assembly for a water molecule. In the first step, a hydroxyl hydrogen atom is selected from the TAE library and aligned in its standard orientation along the local -Z axis. In step 2, an appropriate oxygen TAE has been retrieved from the database and placed with its bond critical point vector in the +Z direction. The oxygen atom has also been translated and rotated from its standard orientation to achieve a good fit between its interatomic surface and the interatomic surface of the hydrogen atom. In step 3, the third hydrogen atom has been added, and the interatomic surfaces of the first two atoms have been made coincident using the RECON procedure. Figure 3 illustrates the results of this operation: A completed water molecule with the electrostatic potential encoded on its VDW surface. Note that the surface electrostatic potential is simply one of the properties which can be calculated from the TAE assemblies, but it has been chosen to be the default display property during reconstruction. The slight gaps apparent in the figure between the oxygen and its hydrogen atoms are actually artifacts of the visualization procedure and do not represent a fitting problem. When the electron density distribution of the resulting molecule is contoured to produce a new isosurface, no discontinuities can be seen.[18]

TAE assemblies can only be used to model hydrogen bonding if a number of representative donor and acceptor atoms from hydrogen-bonded complexes are included in the source data used to produce the TAE library. In anticipation of the need to model hydrogen bonds, 27 molecules capable of hydrogen bonding were used as sources for both donor and acceptor atoms. Within each elemental category, hydrogen bonded atom types are distinct from the rest of the 58 TAE groups and cluster together well. As a result of the tight clustering, the relatively low number of examples were considered sufficient to define each hydrogen bonded TAE type. The proof of cluster validity comes during recombination procedures which force each atom to change its surface position by a relatively large amount when compared to the surface shape variation found in each cluster. If the cluster is sufficiently small, the errors in energy and population introduced during the RECON procedure will be small. If the surface motion is excessive, the population and energy sums will not

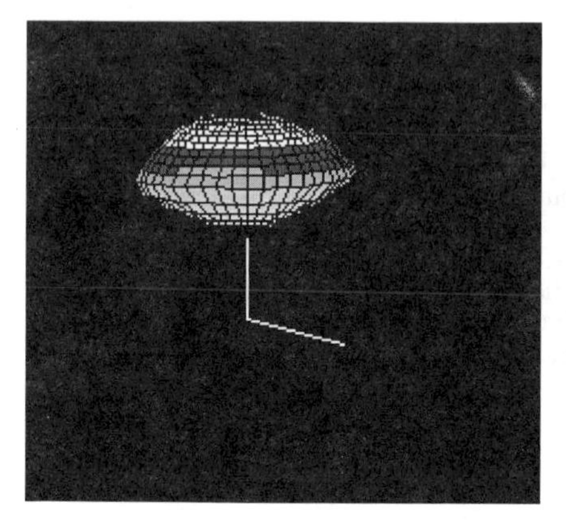

1

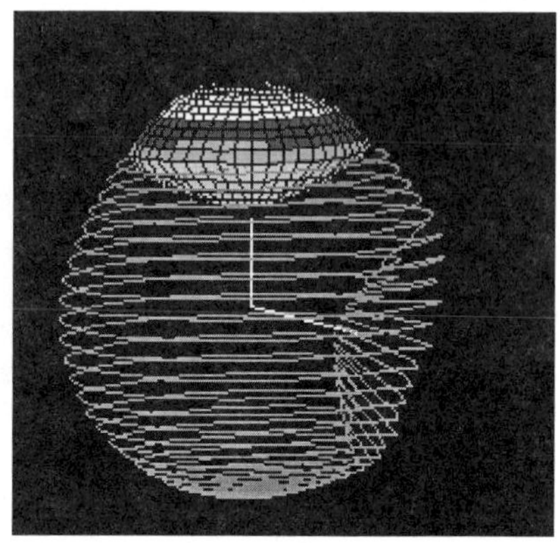

2

- In step 1, the first hydrogen TAE is aligned along the Z axis.
- Next, an sp3 oxygen TAE is positioned for best interatomic surface matching.
- After initial O-H interatomic surface optimization, the second hydrogen TAE is added.
- All atomic properties are slightly altered upon surface optimization.

3

Figure 2. The TAE Assembly Process.

compare well to the ab initio reference results. Using the current TAE library and recombination procedure, the error limits for total energy are usually within 2 kcal/mole.

Figure 4 illustrates the construction process of a water dimer by the TAE method. In the upper figure, the hydrogen donor is shown undergoing polarization by the hydrogen acceptor. In the bottom figure, the acceptor is being polarized by the donor. Figure 5 shows the final results of the dimer reconstruction. The ab initio binding energy (BSSE corrected) was found to be 17.7 kJ/mole using the HF/6-31+G* level of theory, while the TAE reconstruction data indicates 16.9 kJ/mole binding energy for the complex (Table 2.). As expected, the majority of this small error can be attributed to the calculated energy of the water dimer. The energy of the water monomer TAE assembly was found to be only 0.06 kcal/mole different than the HF/6-31+G* energy for water. Figure 6 shows the VDW electrostatic potential and atomic surface envelope of the formaldehyde / water TAE assembly. As in the water dimer case, the TAE results fall quite near the ab initio reference data. Similar agreement is seen for the water / methylamine complex shown in Figure 7, and the water / methanol complex illustrated in Figure 8.

Table 2. TAE vs HF/6-31+G* binding energies

Molecule	TAE (kJ/mole)	HF/6-31+G* (w/BSSE, kJ/mole)
Water Dimer	16.9	17.7
Water / Formaldehyde Complex	16.5	17.1
Water / Methylamine Complex	23.2	24.3
Water / Methanol Complex	16.5	17.0
Water Trimer (Third Water)	13.7	14.2

An interesting test of the TAE method is found in the addition of water to water dimer. In this case, illustrated in Figure 9, the center oxygen of the trimer is seen to be highly polarized by the addition of the third water since it is now a double hydrogen bond donor. Both TAE analysis and HF/6-31+G* geometry optimization picked this structure as a low energy conformation, but no attempts were made to exhaustively explore the potential energy surface of the trimer in this study. It should be noted that while the formation of water dimer initially provides 16.9 kJ/mole in stabilization, the addition of the third water is calculated to provide only 13.7 kJ/mole addition stability. This 3.2 kJ/mole difference compares well to the 3.5 kJ/mole energy difference obtained from the ab initio calculation.

In summary, it should be noted that the TAE method appears capable of providing good results when modeling hydrogen bonded interactions. It has been demonstrated in the comparative results above that the TAE method is capable of representing the loss of hydrogen bonding affinity when a third water molecule is complexed to a water dimer. This responsiveness to the local chemical environment, in this case polarization, is crucial to investigations which attempt to assess changes in the electronic properties of large molecules in response to conformational changes or the onset of non-covalent interactions. In the next section, the energy of an important hydrogen bonding interaction in the FKBP - FK506 crystal structure is examined by the TAE method.

Figure 3. TAE Model of Water.

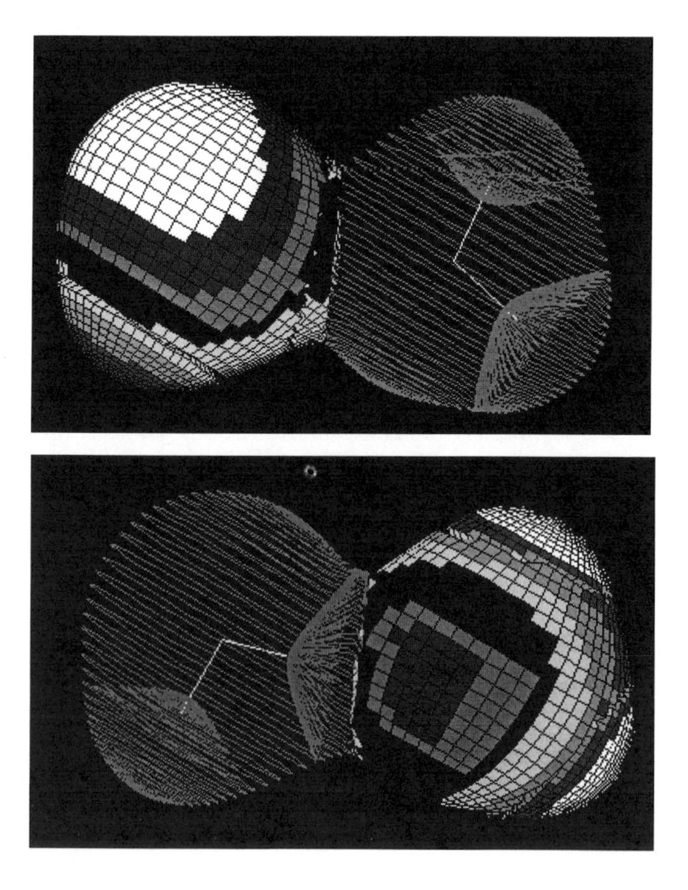

Figure 4. TAE Reconstruction of Water Dimer.

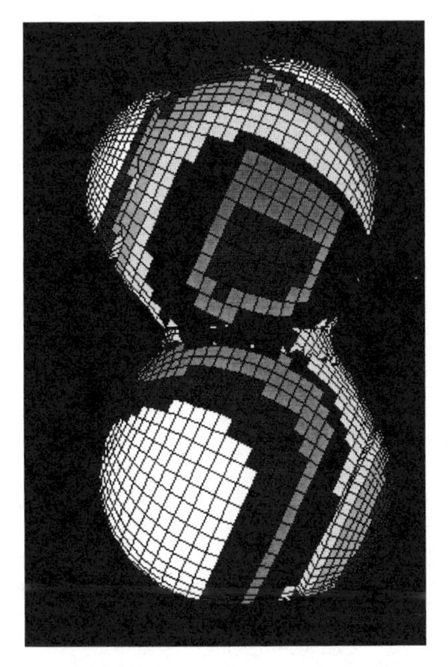

Figure 5. TAE Model of Water Dimer.

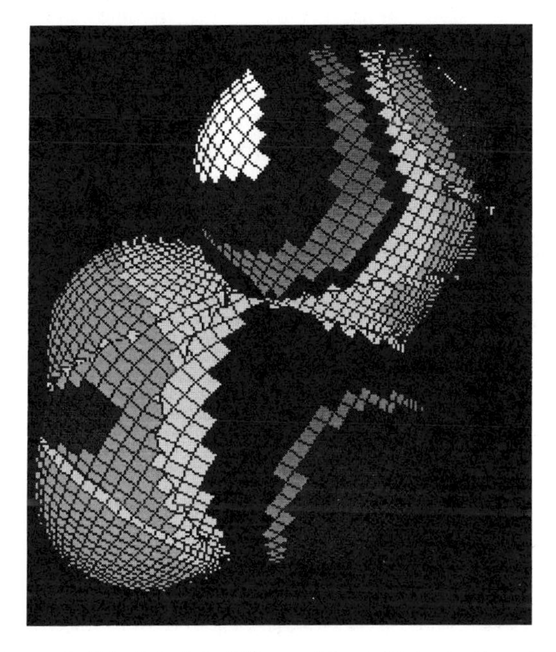

Figure 6. TAE Model of Water / Formaldehyde Complex.

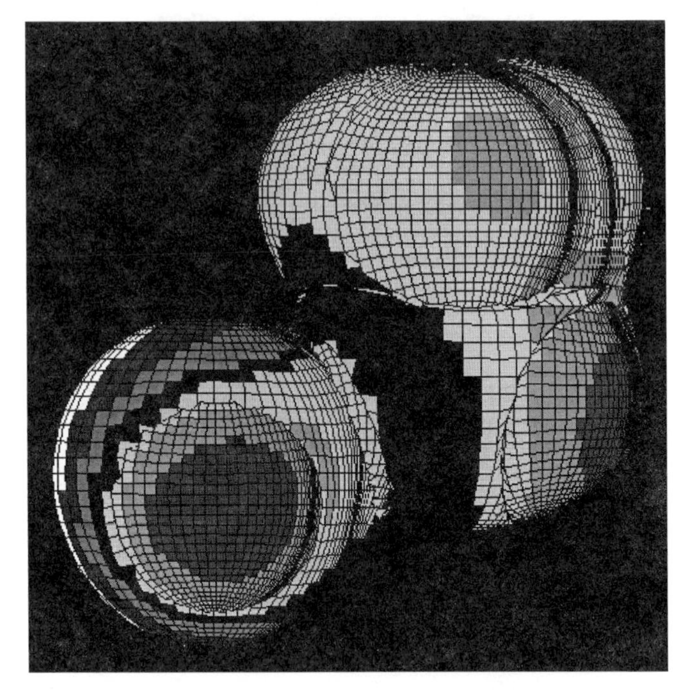

Figure 7. TAE Model of Water / Methylamine Complex.

Figure 8. TAE Model of Water / Methanol Complex.

Figure 9. TAE Model of Water Trimer.

Large Molecule Results: In order to provide an interesting and relevant example of large-molecule TAE modeling, the structure of the complex of immunosuppressant drug FK506 with FKBP protein was examined. The structure was obtained from Protein Data Bank entry PDB1FKF.ENT, which was derived from a 1.7 Angstrom resolution x-ray crystallographic determination of the complex. Figures 10 and 11 show the binding site of FKBP in the presence and absence of the FK506 macrocyclic drug. The interaction of interest in this system involves a hydrogen bond between the hydroxyl hydrogen of tyrosine 82 (TYR 82) shown in Figure 12 and one of the carbonyl oxygens in the critical α-dicarbonyl region of FK506. It is thought that this interaction takes the place of an amide carbonyl - TYR 82 hydrogen bonding interaction which may assist in the isomerase activity of FKBP. The interaction between TYR 82 and FK506 can be more plainly seen in Figure 13, where only this residue and the bound drug are shown as space-filling models. The magnitude of this individual hydrogen bonding interaction can be assessed using the TAE method. First, the decision was made to use the actual crystallographic geometry in the assessment. Using this geometry, the TAE energy of the complex was found to change by 24.2 kJ/mole when interactions between the hydrogen and the two oxygens involved in the hydrogen bonding were turned off. This is accomplished by temporarily replacing the hydrogen bonded atom types with non-hydrogen bonded TAEs and allowing the system to reach electronic equilibrium in the absence of the pertinent interactions. Even though the difference is not large, it should be noted that this binding energy is significantly larger than the model hydrogen bonding energies shown in Table 2.

In order to further study the potential for peptide isomerase activity by FKBP, the positions of both oxygens and the interposed hydrogen atom were extracted and placed into a small model system designed to mimic an amide - TYR 82 interaction. The TAEs in this model system were then allowed to relax electronically, but not geometrically, resulting in the triatomic system shown stereographically in Figure 14. This exercise was completed by the addition of further atoms to generate the methanol - formamide complex shown in Figure 15. The geometry of the model retained the 2.766 Angstrom O-O distance observed in the crystal structure, along with a 120.34 degree C=O - O angle and a 108.8 degree C-C=O - O torsion. The TAE assembly placed the hydrogen atom a reasonable 1.855 Angstroms from the carbonyl oxygen. This hydrogen was found to be 0.943 Angstroms from the hydroxyl oxygen with an O - H - O angle of 162.9 degrees. The TAE energy for that assembly of atoms was found to be 17.8 kJ/mole lower than the sum of the TAE energies of formamide and methanol at their equilibrium geometries. When this system was returned to the FKBP binding site environment, the energy was assessed in a manner similar to that used for determining the TYR 82 - FK506 hydrogen bonding energy. Under these conditions, the binding energy was found to increase to 29.5 kJ/mole. This suggests that the electronic environment found in the hydrated FKBP protein binding site is capable of enhancing hydrogen bond interactions between its residues and a bound peptide. It is unlikely that this is an isolated phenomenon peculiar to FKBP. Undoubtedly this phenomenon will also be observed in other systems when TAE active site modeling is employed. In an effort to expand these kinds of large molecule modeling studies, a version of the TAE modeling program is going to be made available for distribution in the foreseeable future.

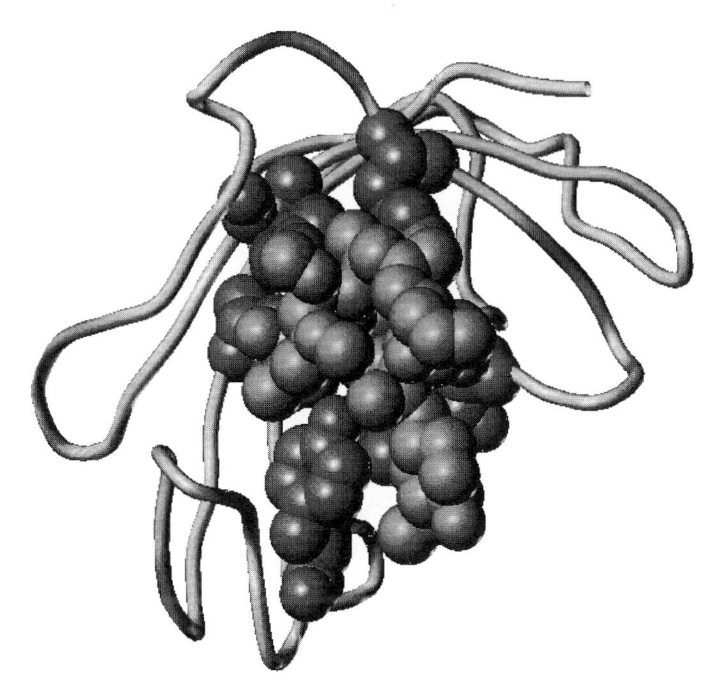

Figure 10. FKBP Binding Site with FK506 Bound.

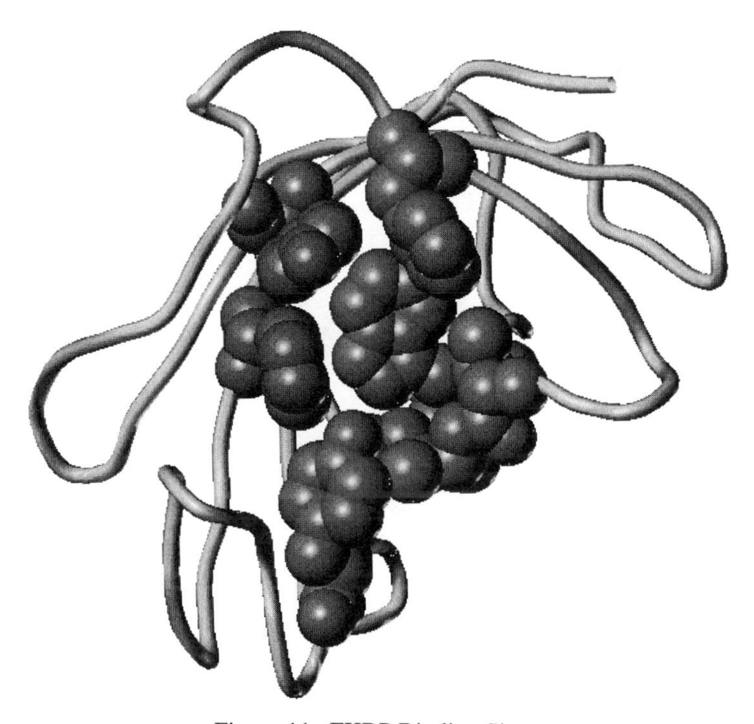

Figure 11. FKBP Binding Site.

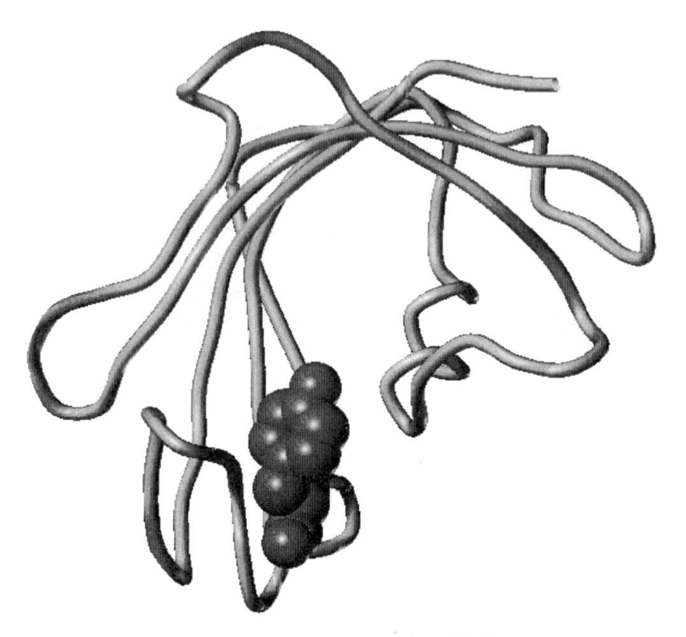

Figure 12. FKBP with Tyrosine 82 Shown.

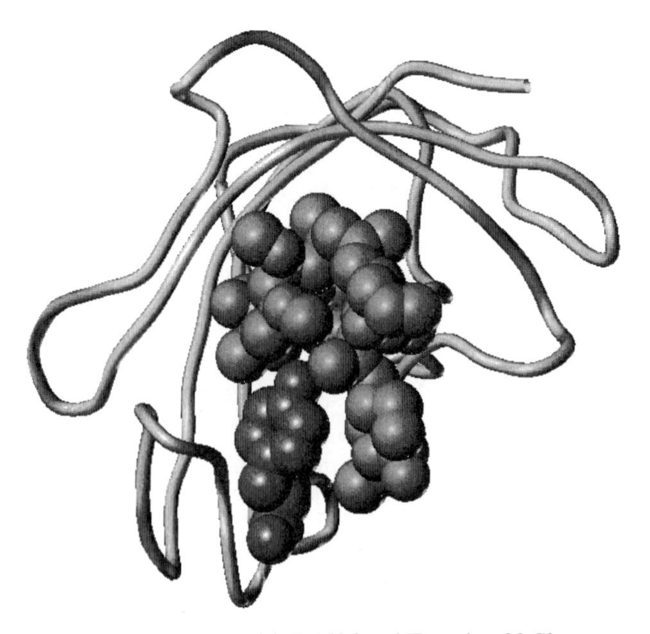

Figure 13. FKBP with FK506 and Tyrosine 82 Shown.

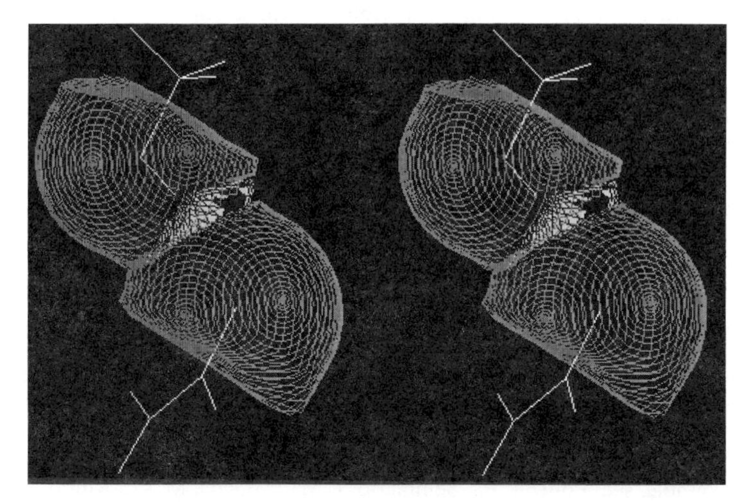

Figure 14. Stereo View of TAE Hydrogen Bonding Interaction.

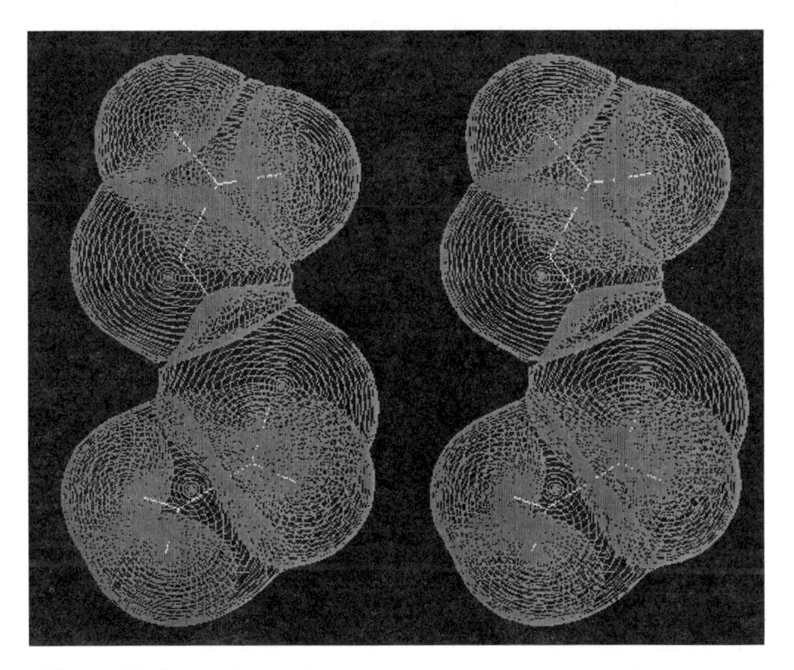

Figure 15. Stereo View of FKBP / FK506 Hydrogen Bonding Model.

Literature Cited

(1) The Hydrogen Bond, Recent Development in Theory and Experiments; Schuster, P.; Zundel, G.; Sandorfy, C., Eds,; North-Holland; Amsterdam, 1976.

(2) Chakrabarti, Pinak; Pal, Sourav "Difference in the energies of interactions at the binding sites in protein structures." *Chem. Phys. Lett.* **1993**, *201,* 24.

(3) Van Duyne, G.D.; Standaert, R.F.; Karplus, P.A.; Schreiber, S.L.; .Clardy, J."Atomic Structure of FKBP-FK506, and Immunophilin-Immunosuppressant Complex" *Science*, **1991**, *252*, 839.

(4) Bhattacharjee, Apurba Krishna; Bose, S. N. "Molecular structure of derivatives of some model anticholinergic and anti-inflammatory compounds: a theoretical conformational and electrostatic potential study." *Indian J. Chem.*, Sect. B **1991**, 30B, 991.

(5) See discussions and references in the Gaussian90 User's Manual: Frisch, M. J.; Binkley, J. S.; Schlegel, H. B.; Gonzalez, C.; Raghavachari, K.; Melius, C. F.; Martin, R. L.; Stewart, J. J. P.; Head-Gordon, M.; Rohling, C. M.; Kahn, L. R.; DeFrees, D. J.; Seeger, R.; Whiteside, R. A.; Fox, D. J.; Fleuder, E. M.; and Pople, J. A. *Gaussian90* , Developmental version, Revision G, Carnegie-Mellon Quantum Chemistry Publishing Unit, Pittsburgh, PA, 1990.

(6) For example, see: Politzer, P., Seminario, J. M., Concha, M. C., Murray, J. S. "Some Applications Of Local Density Functional Theory To The Calculation Of Reaction Energetics" *Theoretica Chimica Acta*, **1993,** *85*, 127.

(7) For examples of the use of DMOL, see: Delley, B "Quantitative Local Density Functional Calculations On Molecular Electronic Properties With Dmol" *New Journal Of Chemistry*, **1992**, *16*, 1103.

(8) Bader, R. F. W. "Atoms in Molecules: A Quantum Theory," Oxford Univ. Press., Oxford, 1990.

(9) As defined by the two negative curvatures and the single positive curvature, which is aligned along the bond path.

(10) Bader, R.F.W. *Acc. Chem. Res.* **1985**, *18*, 9.

(11) FASTINT, C. Breneman and M. Rhem, 1992.

(12) Dipole moments for neutral molecules may be calculated for point-charge (monopole) charge models by summing the atomic charge moments from an arbitrary origin.

(13) Laidig, K. E. *Chem. Phys.* **1992**, *163*, 287; Laidig, K. E. *Chem. Phys. Lett.* **1991**, *185*, 483.

(14) Bader, R.F.W.; Carroll, M.T.; Cheeseman, J.R.; Chang, C. "Properties of Atoms in Molecules: Atomic Volumes" *J. Amer. Chem. Soc.* **1987**, *109*, 7968.

(15) Program RECON: Rhem, M. 1993.

(16) C.M. Breneman and L. W. Weber, "Charge and Energy Redistribution During Rotation About the S-N Bond in Sulfonamide and Fluorosulfonamide. Hybridization and Electrostatic Control of Conformer Stability." Submitted to the *Journal of the American Chemical Society.*

(17) Bader, R.F.W.; Carroll, M.T.; Cheeseman, J.R.; Chang, C. "Properties of Atoms in Molecules: Atomic Volumes" *J. Amer. Chem. Soc.* **1987**, *109*, 7968.

(18) The SCIAN package (Florida State University, SCRI) and an AVS visualization network have been used for smooth visualization of TAE assemblies.

RECEIVED June 2, 1994

Chapter 11

Analysis of Hydrogen Bonding and Stability of Protein Secondary Structures in Molecular Dynamics Simulation

S. Vijayakumar[1], S. Vishveshwara[2], G. Ravishanker[1], and D. L. Beveridge[1]

[1]Department of Chemistry, Wesleyan University, Middletown, CT 06459
[2]Molecular Biophysics Unit, Indian Institute of Science, Bangalore 560 012, India

Molecular Dynamics (MD) simulations provide an atomic level account of the molecular motions and have proven to be immensely useful in the investigation of the dynamical structure of proteins. Once an MD trajectory is obtained, specific interactions at the molecular level can be directly studied by setting up appropriate combinations of distance and angle monitors. However, if a study of the dynamical behavior of secondary structures in proteins becomes important, this approach can become unwieldy. We present herein a method to study the dynamical stability of secondary structures in proteins, based on a relatively simple analysis of backbone hydrogen bonds. The method was developed for studying the thermal unfolding of β-lactamases, but can be extended to other systems and adapted to study relevant properties.

Molecular Dynamics (MD) is a powerful tool that can generate the intricate details of a sequence of molecular events at atomic resolution(*1,2*). The positional fluctuations of individual atoms, average distances between any combination of functional groups, and the dynamical range of motion spanned by any given set of atoms can all be calculated from the trajectory. Knowledge of these parameters can aid in the understanding of specific interactions such as those within the active site of an enzyme or between an enzyme complexed with a substrate or inhibitor. However, the structural characterization of a protein is a very complex task(*3-6*) since the structure is governed by numerous interdependent covalent and non-covalent interactions. Simultaneous consideration of all of these interactions to derive a quantitative measure of the structural stability is possible but tedious. Therefore, obtaining a reduced measure of the stability is desirable, particularly one that is focussed on a particular level in the structural hierarchy, that influences structures of high order.

We present herein a method for analyzing the intrinsic stability of secondary structures in proteins. The method, based on hydrogen bonds, is ideally suited for

0097–6156/94/0569–0175$08.00/0

studying unfolding (or refolding) phenomena or studies involving substantial changes in the secondary structure, induced by mutation and/or complex formation. The method is illustrated here by means of an unfolding simulation of the hydrolytic enzyme, β-lactamase (Figure 1), including explicit waters of solvation and carried out at relatively high temperature to induce denaturation. Application of the new method to the analysis of secondary structures in MD simulations and comparison with other commonly used methods are presented.

Background

Hydrogen bonds are characterized by several unique properties such as proximity, cooperativity, changes in IR and NMR spectra etc. Comprehensive studies and reviews exist on this subject(*7-10*), and extensive studies are still being carried out(*11-13*). Empirical observations made in a number of small peptides, proteins and other biological macromolecules have provided us some insight and led to the development of practical rules for the identification of hydrogen bonds(*14*).

The phenomenon of cooperativity in hydrogen bonding led to the prediction of secondary structures in proteins, even before the first three-dimensional structure of a protein was available(*15*). A distinguished pattern of hydrogen bonding among protein secondary structural regions, was subsequently confirmed by x-ray crystallography. Such patterns serve as a fingerprint and have since been used to identify secondary structural motifs in proteins(*16-18*). Other characteristics of secondary structural motifs, such as backbone dihedral angles, (ϕ, ψ), are also routinely used in the identification process(*19,20*). These methods have been useful as a general tool, but they do not provide a ready measure of the stability of secondary structures in MD simulations.

The stability and extent of secondary structures in MD simulations have been analyzed by several researchers, recently. In a study of helix-coil transition, Daggett et al.(*20,21*) have analyzed the fractional helicity of a twenty residue polyalanine helix based on (ϕ, ψ) angles, wherein a helical region is defined as having at least three residues with (ϕ, ψ) angles within 30° of the ideal values ($\phi = -57°$, $\psi = -47°$). Also, the angular variance of ϕ and ψ angles were employed as a measure of the propensity of a given residue to remain in the helical region. Tirado-Rives et al. have employed the above approach based on (ϕ, ψ) in a study of the unfolding of apomyoglobin in water(*22*). Daggett and Levitt have extended the above approach to sheets in a study of the molten globule state of BPTI(*23*), wherein they calculate percent α-helical and β-structure based on whether a particular residue is within 50° of their corresponding ideal values. Fan et al. have employed a distance criteria to estimate the fractional helicity during the partial unfolding of α-lactalbumin(*24*). Swaminathan et al.(*25*) have analyzed the secondary structure in BPTI using a dimer irregularity function (DIF) based on the "Curves" analysis method(*3*), which measures the local deviation from well defined secondary structure. The above approaches are useful for studying the overall stability of helical regions. However, β-strands pose an additional problem of having to be in close proximity with sequentially distant regions of the protein, in order to facilitate stabilization by inter-strand hydrogen bonds.

Graphical presentation tools have been employed to illustrate the average strength and changes in hydrogen bonding pattern that affect the structure of proteins

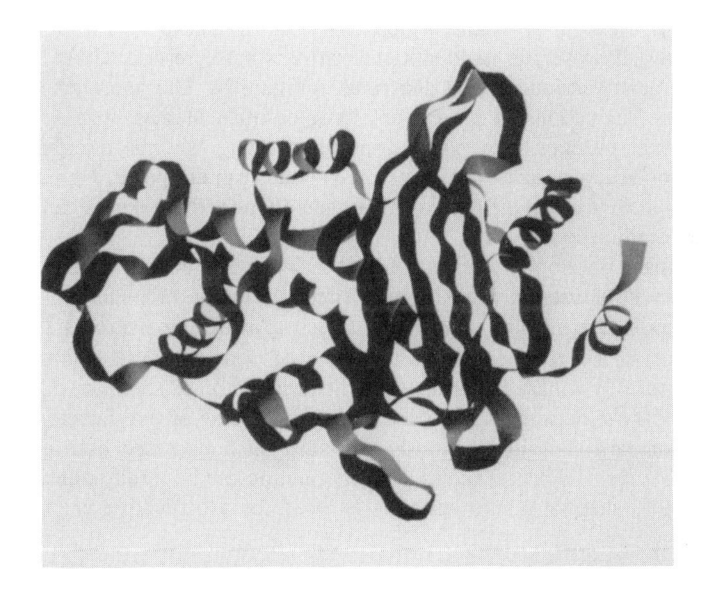

Figure 1: Ribbon drawing of β-Lactamase from S. Aureus(*30*).

and DNA. Analysis of specific hydrogen bonds from MD simulations in DNA simulations have been reported recently(*26*). These methods involve making simple two dimensional plots of the time evolution of one or more types of hydrogen bond based on a set of analysis criteria. Topology diagrams, two-dimensional drawings representing the connectivity and spatial disposition of secondary structural regions relative to each other, have been used as a template to display time dependent changes in hydrogen bonding from MD simulations of proteins(*27,28*). Automatic generation of these graphical templates is now possible through use of computer programs and one such algorithm was presented at this symposium(*31*). Topology diagrams are useful in providing a global view of the backbone hydrogen bonds in small systems, however, in large systems clarity is best preserved only if applied to specific secondary structure(s), as illustrated in Figure 2a. This approach is unsuitable for analyzing simulations of protein unfolding or other studies, wherein substantial conformational changes take place. During unfolding, several transient and new hydrogen bonds appear and existing hydrogen bonds can disappear. Inclusion of all of these dynamical events in the topology diagrams (Figure 2b) makes it clumsy and can lead to mis-leading conclusions.

We have developed herein a comprehensive measure for analyzing the stability of secondary structures in proteins, that circumvents the problems inherent in the above methods. Our method is based on analysis of backbone hydrogen bonds and is illustrated using a simulation of the unfolding of β-lactamase in solution(*29*). The method captures time based information and provides a quick summary of the stability of all secondary structures in a protein, during the course of the trajectory. We have provided comparisons with previously published methods based both on hydrogen bonds and backbone dihedral angles. The information can be readily translated into % helicity and % sheet structure and may be used for a qualitative comparison with experiment.

Methods

MD studies were recently carried out at high temperature (600 K) to study the thermal unfolding of β-lactamase wild type enzyme in solution along with a room temperature (300 K) simulation to serve as a control. Simulation of the native enzyme was started from the published crystal structure(*30*). The simulations were carried out using the GROMOS force field (RT37C4) and the SPC model for water in an NVE microcanonical ensemble using periodic boundary conditions. Full details of the simulation have been reported previously(*29*). The total length of the trajectory for the unfolding study was 200 ps. The total energy and temperature stabilized after a series of initial rescalings and was constant after some 30 ps into the simulation. There was a major rescaling at about 80 ps. Root mean square deviation (RMSD) increased gradually until about 60 ps, then dramatically during the following 20 ps and attained a value of 5.5 Å. Further increase was slow and RMSD reached a maximum value of 6.75 Å and remained so for the last 30-40 ps. Expansion of the enzyme was significant initially but began to level off during last 30-40 ps. The simulation was stopped at this point for further analysis.

The analysis of hydrogen bonds essentially involves two steps; (a) identifying all backbone hydrogen bonds using a distance and angle criteria and filtering out all

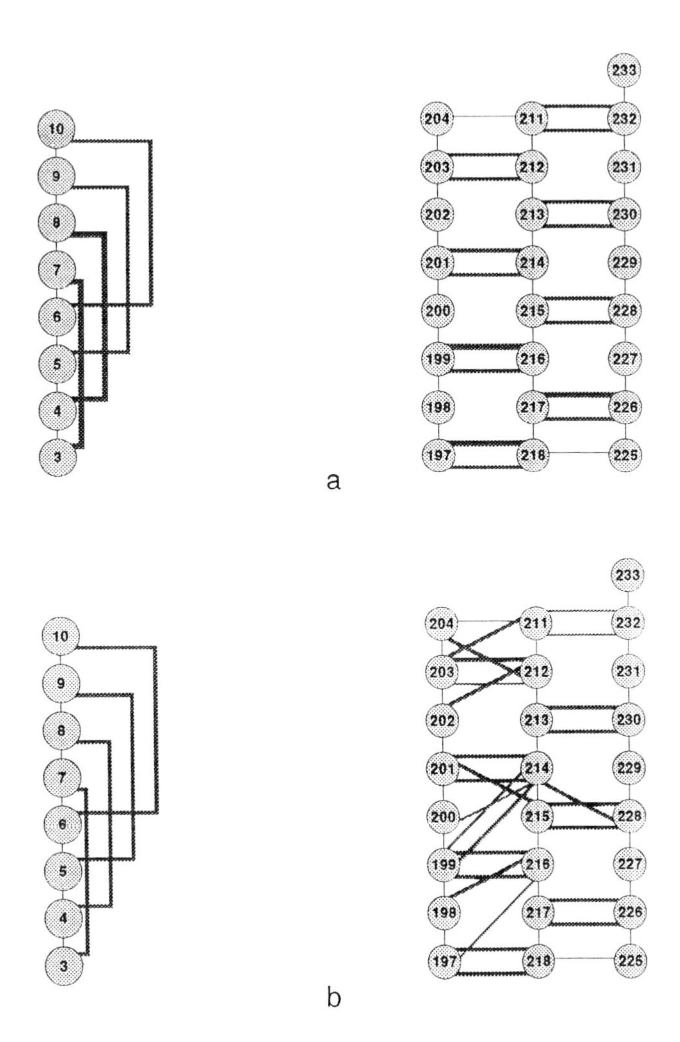

Figure 2: Topology diagram superposed with hydrogen bond information for helix H1 and β-strands S7, S8 and S9 of β-Lactamase. Residue ranges for the above secondary structures are identified in Table 1. Numbered circles represent residues. Vertical lines connecting residues represent chain connectivity and horizontal or curved lines connecting two residues represent hydrogen bonds between those two residues. The thickness of the lines are indicative of the average strength of a given hydrogen bond. a) Data from an MD simulation at room temperature. Dynamical averages indicate strengthening or weakening of a given hydrogen bond. b) Data from an MD simulation of β-Lactamase at high temperature. Diagram illustrates the potential for confusion due to straying lines arising from transient hydrogen bonds and structural rearrangement.

weak hydrogen bonds based on a crude energy calculation and (b) segregating the hydrogen bonds into secondary structural regions and summing up the total number of hydrogen bonds within each region. The segregation of hydrogen bonds into secondary structural regions is based on a user specified list that identifies residue ranges for each secondary structure. In this study, residue ranges for identification of secondary structures were obtained from the published crystal structure(*30*).

The criteria used for identifying the backbone hydrogen bonds are as follows. We used an N-H···O=C distance cutoff of 2.25 ± 0.75 Å, an N-H···O angle cutoff of 150 ± 30°, and an energy filter which discards any hydrogen bond weaker than -1.0 Kcal/mol. The hydrogen bond energy was assumed to arise only from the electrostatic interaction between the atoms involved in the hydrogen bond, namely, N, H, C and O, and calculated based on the partial atomic charges used in the force field. Additional restrictions were imposed on helical and sheet regions as follows; i/i+3 and i/i+4 hydrogen bonds were considered for helices unless specifically stated otherwise, and for sheets any hydrogen bond between backbone atoms of two different β-strands were included. Further, in order to allow for changes in secondary structural regions during dynamics, the segregation algorithm was initially designed to allow for limited extensions to existing secondary structural regions. This extension came out naturally for helices, a maximum of four residues in either direction. For β-sheets, we set an arbitrary limit of two residues in each direction and analyzed for the presence of additional hydrogen bonds stabilizing each sheet. The above criteria provided a semi-quantitative measure of the extent of existing secondary structures, related to the intrinsic stability of each secondary structural element and can be monitored as a function of time. Note that this method sets a limit on the extension of existing secondary structures and lacks the ability to identify newly formed secondary structures during the dynamics trajectory. Further, estimates of percent secondary structure based on hydrogen bond alone may not be meaningful, since intervening residues can form part of a secondary structure without being directly involved in a hydrogen bond.

A modification to the above procedure is subsequently made to provide a better estimate of the extent of different secondary structures in MD simulations. We focus here mainly on helical and sheet regions but the method can also be employed to study other well defined structural forms. The modification involved identifying residues of a given structural type based on backbone hydrogen bonds and rather than summing hydrogen bonds we sum up the number of residues within each type of secondary structure. This modification allows for unlimited extensions to existing secondary structural regions and identification of new secondary structures formed during the course of the trajectory. We employed the criteria of Kabsch and Sanders(*18*) to determine the secondary structural regions within each snapshot, subsequent to identification of hydrogen bonds using our above criteria. For overlap regions we have used the following priority; α-helices > β-sheets > 3_{10}-helices. After an initial secondary structure is assigned for every time point in the trajectory, we employ a filter based on the ideal backbone dihedral angles for all unique secondary structures. The backbone dihedral angles, (ϕ, ψ), for each residue is determined and any residue whose (ϕ, ψ) angles differ by more than 50° of the ideal values(*23*) for a given structural type, is changed to a random type. The number of residues of each type is then summed and expressed as a percentage with respect to that of the crystal form.

Results and Discussion

We have previously discussed the results from the high temperature simulation of β-lactamase in solution with respect to overall stability and protein unfolding(*29*). We focus here on the issue of stability of major α-helices and β-strands, estimation of percent secondary structure and provide comparison with results obtained using other methods. Analysis of the backbone hydrogen bonds of all major α-helices and β-strands were carried out in β-lactamase and the results are summarized in Figure 3. Each panel represents a time evolution of the total number of hydrogen bonds found within the secondary structure identified in the title. Residue ranges for various secondary structural regions are provided in Table 1. The dashed line serves as a reference and indicates the number of hydrogen bonds found in the crystal structure within a given region and the solid line represents the data from the unfolding simulation. Note that none of the secondary structure appears to be intact. Partial and complete loss of secondary structure seems to be the general trend. Helices H2, H7 and H11 appear to have completely melted by about 100 ps while H8 takes longer. However, helices H1, H6, H10 and H14 appear to have melted only partially. In contrast, all of the β-strands appear to be partially intact. Thus, our analysis criteria suggests a difference in the melting of individual secondary structures in the protein. Alternatively, it is possible to estimate the stability of a given secondary structural region by determining the total energetic stabilization provided by all the hydrogen bonds in that region, at least as a first approximation. Implicit in both of these approaches is the assumption that secondary structures are stabilized to a large part by hydrogen bonded interactions and consequently their presence or strength could be interpreted as a measure of the intrinsic stability.

The stability of individual secondary structures, although important knowledge, is hard to determine experimentally. However, the overall percentage of secondary structure can be easily determined and could be used for comparing the results from our analysis. We have examined if the total number of hydrogen bonds within a given motif can be used to determine the percentage secondary structure in the protein. The time evolution of the percentage of hydrogen bonds, retained among α-helical, all helical (both α- and 3_{10} helices) and sheet regions, are shown in Figure 4 with respect to the corresponding data for the crystal form. During the initial stages of the simulation, the fraction of α-helical hydrogen bonds retained appears to be greater than that of sheets. Towards the end of the simulation, the fraction of α-helical hydrogen bonds and that of sheets appear to be the same and attain a value of about 25 % of that observed in the crystal form. This compares with an experimental value of 50 % deformation for helices and no change for β-sheets in the molten globule state of β-lactamase(*32*). Inclusion of the contribution from 3_{10} helices did not change the percentage helicity significantly. It is obvious that the number of hydrogen bonds cannot be used alone to determine the extent of overall secondary structure.

The Kabsch and Sanders' algorithm is widely used for the crystallographic assignment of secondary structures in proteins. We have implemented the criteria of Kabsch and Sanders(*18*) into our approach based on hydrogen bonds in order to

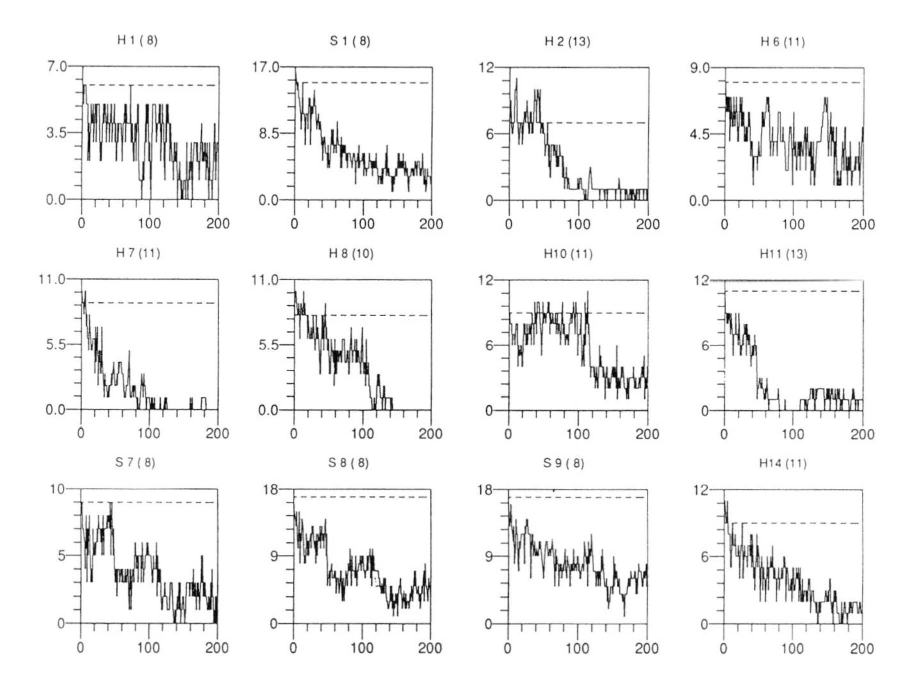

Figure 3: Time evolution of the total number of backbone hydrogen bonds within individual secondary structures. The title on each panel identifies the secondary structure, residue ranges for which are specified in Table 1. The dashed line represents the number of hydrogen bonds within each region in the crystal structure and the solid line represents data from the unfolding simulation. X-axis is time and Y-axis is number of hydrogen bonds.

Table 1. Residue ranges identifying secondary structures

Residue range	Helices	Sheets
3 - 10	H1	
13 - 20		S1
26 - 29		S2
35 - 36		S3
38 - 50	H2	
55 - 57	H3	
61 - 63		S4
66 - 68	H4	
76 - 79	H5	
83 - 85		S5
86 - 96	H6	
99 - 109	H7	
112 - 121	H8	
135 - 137	H9	
147 - 148		S6
150 - 160	H10	
168 - 180	H11	
182 - 184	H12	
188 - 191	H13	
197 - 204		S7
211 - 218		S8
226 - 233		S9
244 - 254	H14	

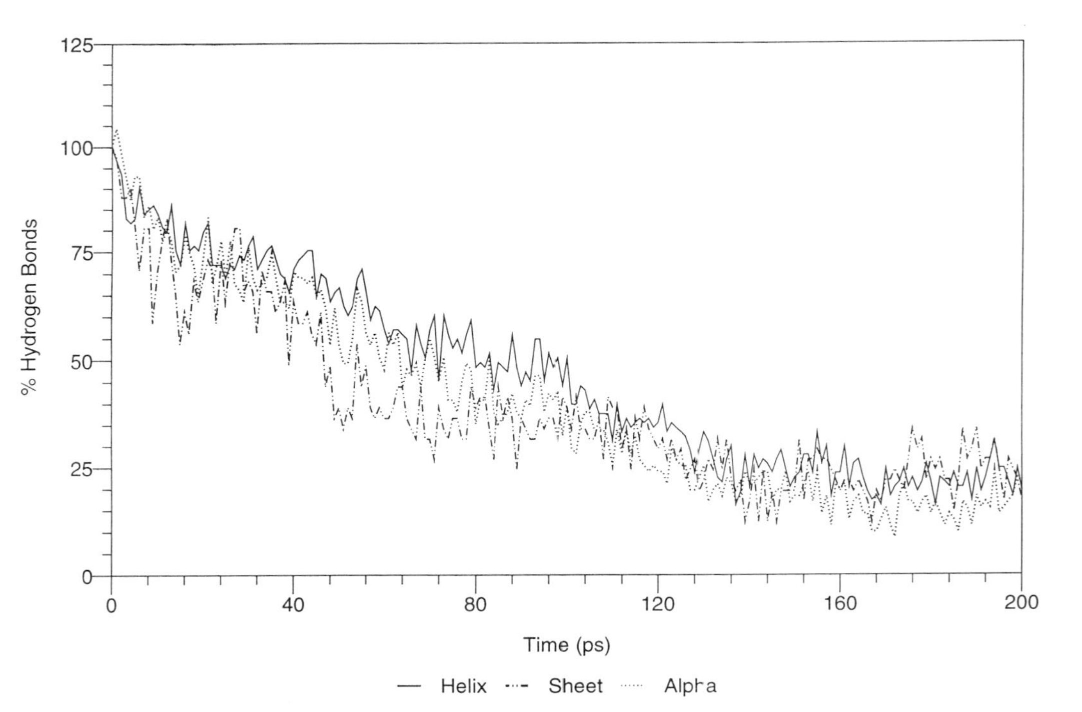

Figure 4: Time evolution of the percentage hydrogen bonds present within α-helical, helical (α- and 3_{10} helices) and sheet regions, respectively. Residue ranges for the secondary structural regions included in the analysis are listed in Table 1. Percentages reported are with respect to corresponding data for the crystal form.

provide a better estimate of the extent of secondary structures in MD simulations, the results of which are summarized in Figure 5. The new results clearly indicate a substantial difference in the extent of helices and sheets retained, a 25 % loss in the α-helical region, while there is a 20 % increase in the sheet region with respect to those found in the crystal structure. When both α-helices and 3_{10}-helices are taken together we observe only a 10-15 % decrease in the extent of helices. Nevertheless, this modification allows us to identify unlimited dynamical changes in the secondary structural regions, which includes the formation of new ones. It must be noted that our implementation correctly identifies almost all of the secondary structures noted in the crystal structure. Nevertheless, the observed increase in the percentage sheet structures and the smaller decrease (compared to experiment) in the percent helicity suggests that the Kabsch and Sanders' algorithm may not be readily applicable to analysis of dynamical structures. So how do methods based on (ϕ, ψ) and others compare ?

The time evolution of the percent helical and sheet residues, derived from analysis of ϕ and ψ angles, is shown in Figure 6 with respect to the corresponding data for the crystal structure. The analysis is based on the methods of Daggett et al.(*20,21*), wherein a given residue is considered helical or sheet like if their (ϕ, ψ) angles are within the dispersion found in the crystal structure for the corresponding region. For helices we employed a range of $-88 < \phi < -40$ and $-66 < \psi < -11$ and for sheets the range included $-166 < \phi < -78$ and $98 < \psi < 173$. Even ignoring the (ϕ, ψ) values for immediately contiguous regions, it is apparent that the number of residues in the helical region drops off quite rapidly. On the contrary, the number of residues in the sheet region appear to increase towards the end of the simulation. The latter point was also noted by Daggett et al.(*23*), however, in that case they employed a tolerance of 50° from ideal angles. Even though the result may appear qualitatively reasonable for the helical region, it is imperative that caution be exercised in interpreting the data for sheets, since the formation of a β-sheet requires not only (ϕ, ψ) criteria to be met, but also requires the presence of another strand in close proximity, in order to allow stabilization by hydrogen bonds.

The RMS fluctuations in ϕ and ψ dihedral angles, calculated with respect to the value in the crystal form, are shown separately for the helical and sheet regions in Figure 7. Note that there is a significant difference between the helical and sheet regions, specifically, that their relative ordering is reversed as unfolding progresses. It is interesting to note that the fluctuation in (ϕ, ψ) space is lower for helical regions during the initial stages (incidentally, this behavior is representative of room temperature simulations and a similar result was obtained from analysis of atomic fluctuations in cartesian space, after removing translation and rotation between internal fragments of the protein and between time points in the trajectory). As the unfolding progresses, the larger fluctuation observed in the helical regions suggests that they become more flexible and this is consistent with the observation that the melting of the helices precede that of sheets.

Another method of analysis called DIF, based on "Curves" (*3*) has been previously employed by Swaminathan et al. to study the dynamical evolution of secondary structure from MD simulations of BPTI(*25*). DIF is a measure of the local deviation from well defined secondary structure and thus would be ideally zero for residues in regular secondary structural regions. We have plotted the DIF function for

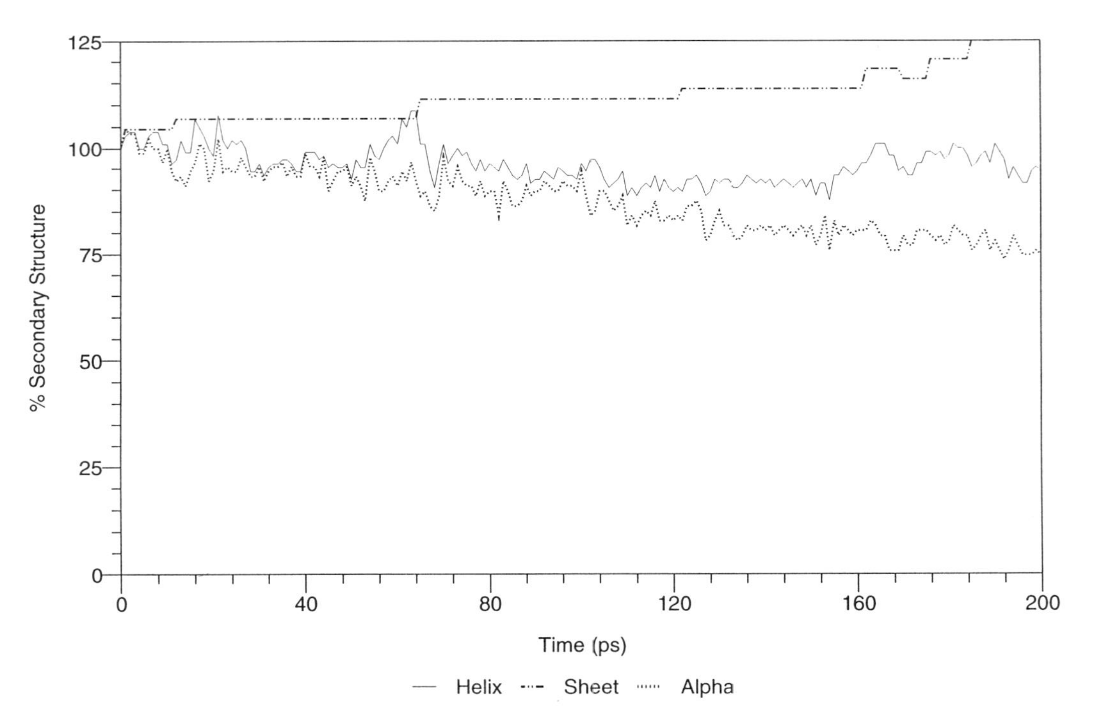

Figure 5: Time evolution of the percentage secondary structure as determined from analysis of backbone hydrogen bonds, imposing the criteria of Kabsch and Sanders(*18*) for α-helical, helical (α- and 3_{10} helices) and sheet regions. Percentages reported are with respect to the values found in the crystal structure (see text for more details).

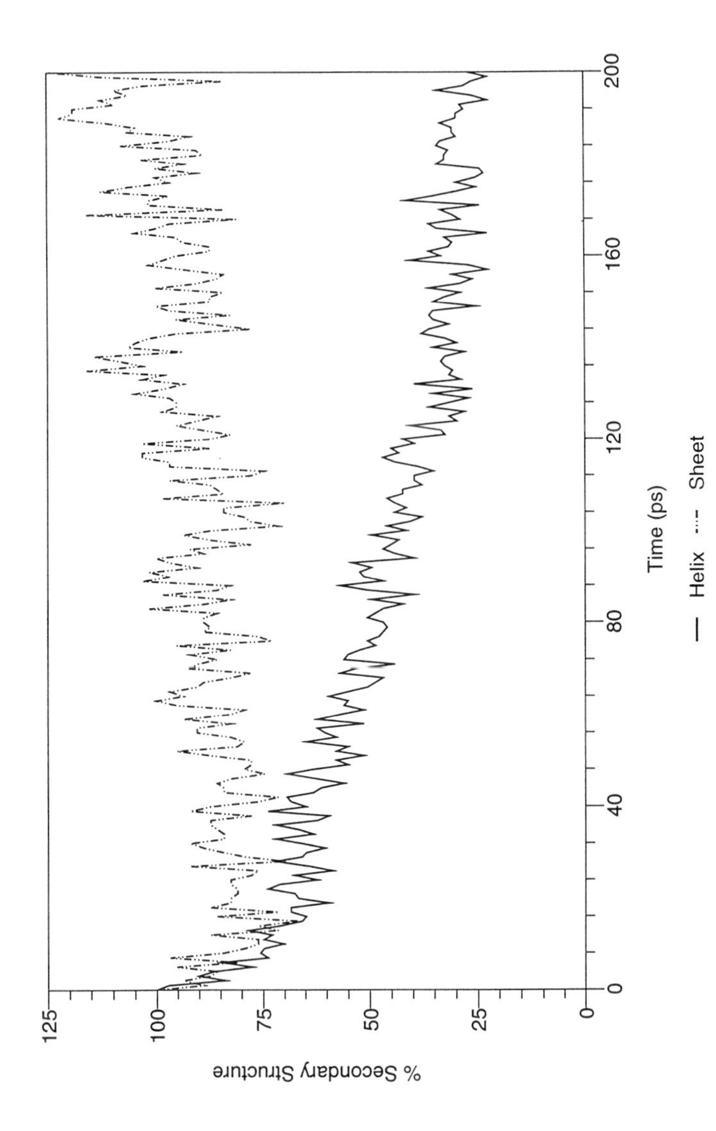

Figure 6: Time evolution of the percentage secondary structure as determined from analysis of (ϕ, ψ) angles. Data is reported as a percentage with respect to crystal form. The criteria used for determining the helical and sheet residues are described in the text.

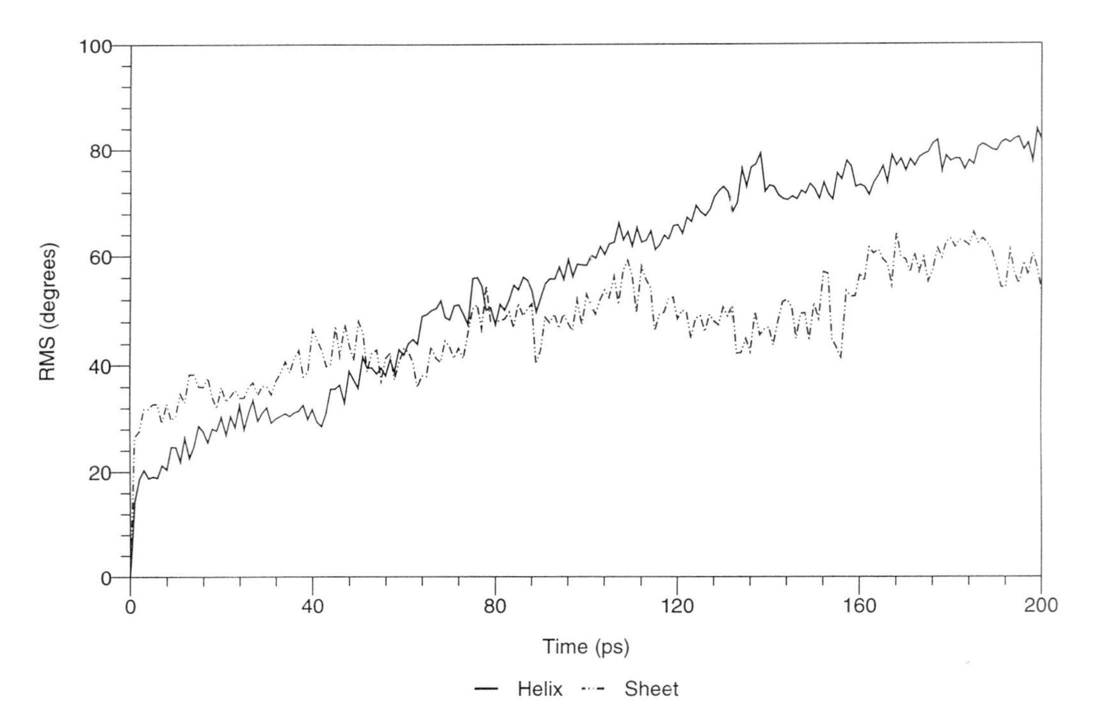

Figure 7: Time evolution of the RMS fluctuations in ϕ and ψ angles averaged over both angles and all residues in the helical and sheet regions, respectively. Residue ranges for the helical and sheet regions included in the analysis are listed in Table 1.

six different points from the MD trajectory (Figure 8). Note that the DIF plot does not produce a value of zero for all secondary structural regions defined in the crystal structure, however, it is able to identify secondary structural regions in a qualitative manner. The more closer a DIF value is to zero the more closer a given residue is to well defined secondary structure, be it helices or sheets. It is clear from the plot that as the MD simulation progresses more of the helical regions are deformed than are sheet regions. The DIF function thus provides a qualitative support to the results from hydrogen bond analysis (Figure 5).

Hydrogen bonds provide a more non-local definition of secondary structures in proteins than do method based on backbone dihedrals, (ϕ, ψ). However, it is obvious from the above results that neither method can independently provide an accurate estimate of the extent of secondary structures, during MD. Both methods are based on empirical observations and perhaps they implicitly include the other to some extent, but most likely they complement each other. We have thus combined structural information from the protein backbone with those from the three-dimensional spatial arrangement in order to obtain a better estimate. The initial assignment of secondary structural regions based on hydrogen bonds was refined by imposing a limited tolerance in the variation of (ϕ, ψ) angles from ideal values for each region, as stated in Methods. The results are presented in Figure 9. The percentage is normalized to the results from the crystal form. The new result clearly brings out the difference in the stability of the helical and sheet regions and is consistent with the analysis of fluctuations, ϕ and ψ dihedral angles and visual inspection of the molecule during the course of the MD trajectory. Note that the addition of the contribution from 3_{10}-helices to that of the α-helices does not make a significant difference in the percent helicity. Towards the end of the simulation there is about 25-30 % drop in the sheet structure and about 50-60 % reduction in the extent of the helical regions, which compares much better with experimental results(*32*). It is important to note that all of these methods can be used only semi-quantitatively, since there is a lot of simplifying assumptions and approximations implicit in the interpretation of both experimental and theoretical results.

Conclusion

We have shown that a method based on the number of hydrogen bonds can be used to monitor the stability of individual secondary structures in proteins, during the course of an MD trajectory, however, it cannot be used to quantitatively estimate the extent of overall secondary structure. A modification has been developed that combines information on backbone dihedral angles as well as hydrogen bonds and allows us to obtain semi-quantitative estimates of the extent of secondary structure in MD simulations. In addition, dynamical changes in existing secondary structures and the formation of new ones can now be accounted. Our analysis has the advantage of considering the backbone structure at spatially and sequentially neighboring residues for a proper evaluation and/or assignment of secondary structural regions, for both helices and sheets. Other methods, such as those based only on backbone dihedral angles, (ϕ, ψ), are limited to local structure definition and as such may not provide a meaningful measure of the extent of secondary structure or yield an accurate assignment of secondary structural regions. Note, however, our quantification may

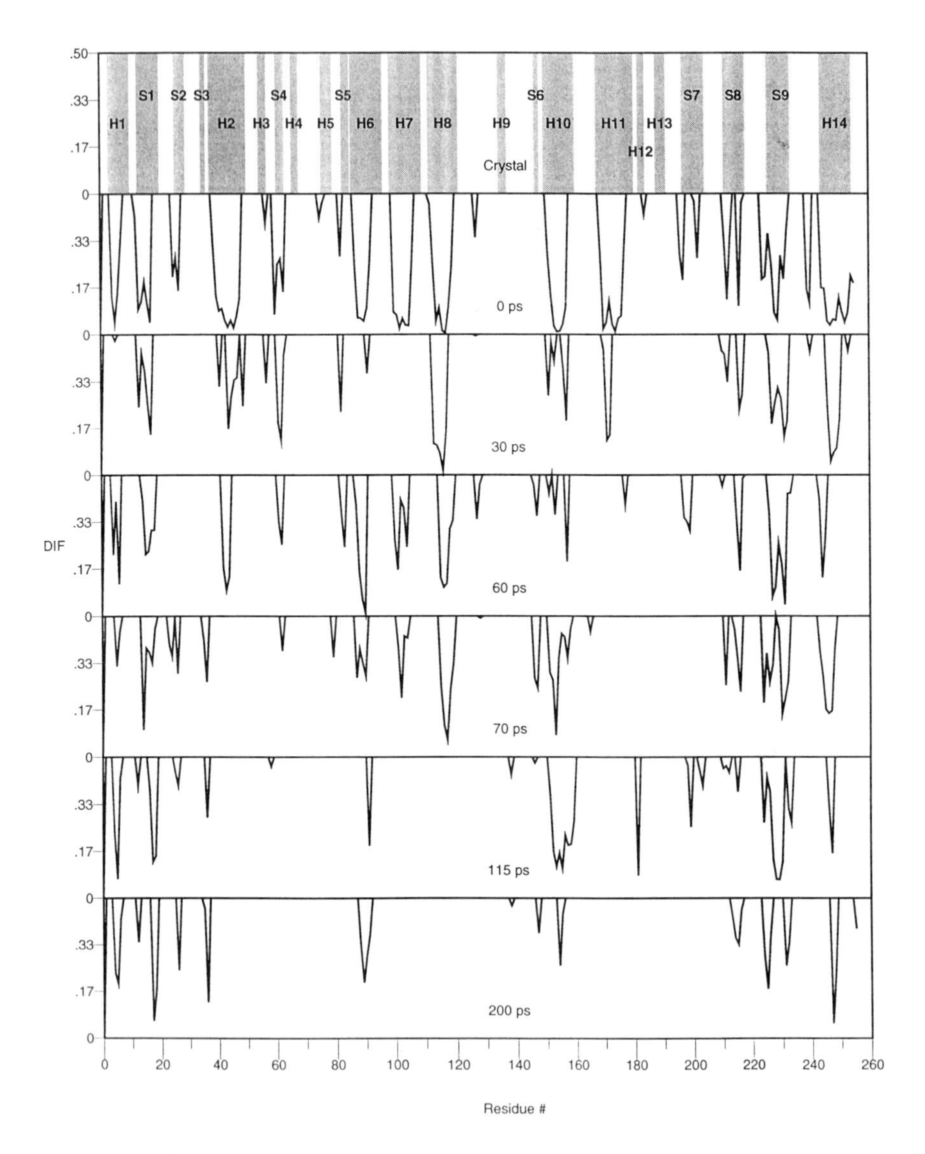

Figure 8: Plot of DIF function(*3,25*) for six different time points from the MD trajectory. DIF values closer to zero represent helical or sheet regions, as identified in the top panel.

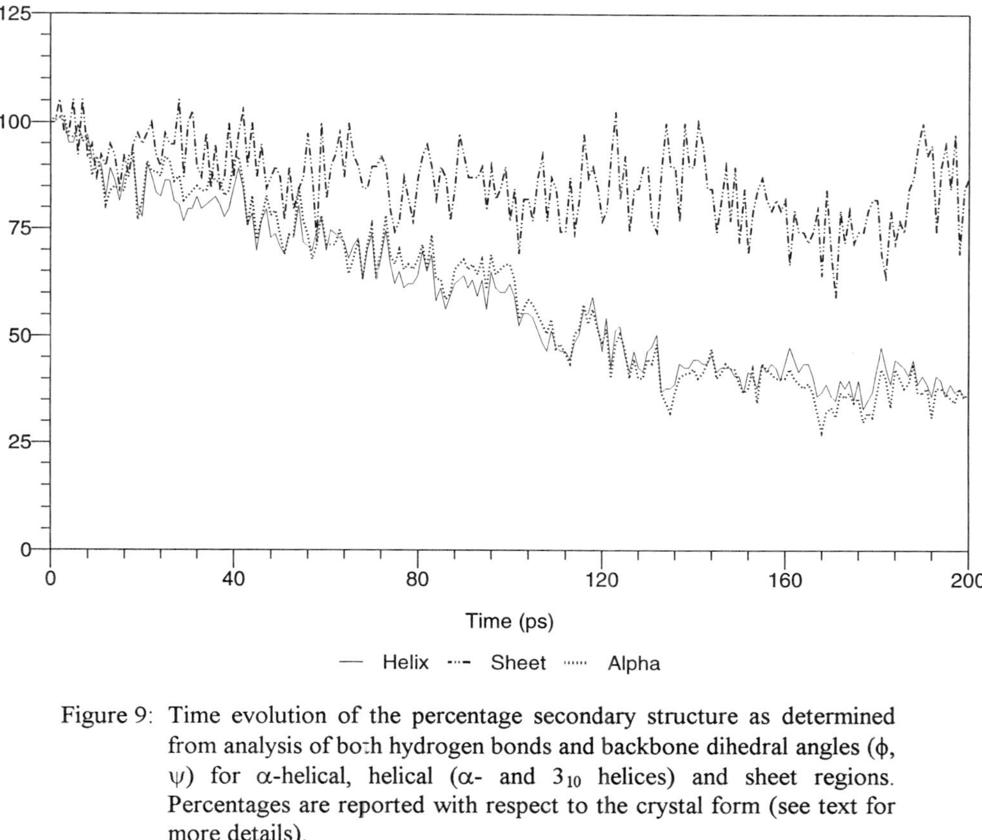

Figure 9: Time evolution of the percentage secondary structure as determined from analysis of both hydrogen bonds and backbone dihedral angles (ϕ, ψ) for α-helical, helical (α- and 3_{10} helices) and sheet regions. Percentages are reported with respect to the crystal form (see text for more details).

not be highly accurate due to the empirical nature of our method. More detailed comparisons on a number of well studied systems are required for further validation.

Acknowledgments

The authors hereby acknowledge funding from the National Institutes of Health, Grant # GM37909 awarded to DLB and Grant # RR07885-01 awarded to DLB and GR and grants of computer time and staff support from the Pittsburgh Supercomputing Center and the Frederick Biomedical Supercomputing Center of the Frederick Cancer Research and Development Center.

Literature Cited

(1) McCammon, J. A.; Harvey, S. C. *Dynamics of Proteins and Nucleic Acids*, Cambridge University Press, Cambridge, 1987.
(2) van Gunsteren, W. F.; Berendsen, H. J. C. *Angew. Chem. Int. Ed. Engl.* **1990**, *29*, 992.
(3) Sklenar, H.; Etchebest, C.; Lavery, R. *Proteins: Struct. Func. Genet.* **1989**, *6*, 46.
(4) Ramakrishnan, C.; Ramachandran, G. N. *Biophys. J.* **1965**, *5*, 909.
(5) Richardson, J. S. *Adv. Protein. Chem.* **1981**, *34*, 167.
(6) Rackovsky, S.; Scheraga, H. A. *Macromolecules* **1978**, *11*, 1168.
(7) Coulson, C. A. *Research* **1957**, *10*, 149.
(8) Kollman, P. A.; Allen, L. C. *Chem. Rev.* **1972**, *72*, 283.
(9) Schuster, P.; Zundel, G.; Sandorfy, C. *The Hydrogen Bond: Recent Developments in Theory and Experiments*, North-Holland, Amsterdam, 1976, Vol. I-III.
(10) Umeyama, H.; Morokuma, K. *J. Am. Chem. Soc.* **1977**, *99*, 1316.
(11) Reiher, W. E., III Ph. D. Thesis, Harvard University, 1985.
(12) Saenger, W.; Jeffrey, G. A. *Hydrogen Bonding in Biological Molecules*, Springer-Verlag, 1991.
(13) Delbene, J. E. *Int. J. Quant. Chem.* **1992**, *26*, 527.
(14) Taylor, R.; Kennard, O. *Acc. Chem. Res.* **1984**, *17*, 320.
(15) Pauling, L.; Corey, R. B. *Proc. Natl. Acad. Sci.* **1951**, *37*, 729.
(16) Levitt, M.; Greer, J. *J. Mol. Biol.* **1977**, *114*, 181.
(17) Lifson, S.; Sander, C. *J. Mol. Biol.* **1980**, *139*, 627.
(18) Kabsch, W.; Sander, C. *Biopolymers* **1983**, *22*, 2577.
(19) IUPAC-IUB Commission on Biochemical Nomenclature, *J. Biol. Chem.* **1970**, *245*, 6489.
(20) Daggett, V.; Levitt, M. *J. Mol. Biol.* **1992**, *223*, 1121.
(21) Daggett, V.; Kollman, P. A.; Kuntz, I. D. *Biopolymers* **1991**, *31*, 1115.
(22) Tirado-Rives, J.; Jorgensen, W. L. *Biochemistry* **1993**, *32*, 4175.
(23) Daggett, V.; Levitt, M. *Proc. Natl. Acad. Sci.* **1992**, *89*, 5142.
(24) Fan, P.; Kominos, D.; Kitchen, D. B.; Levy, R. M.; Baum, J. *J. Chem. Phys.* **1991**, *158*, 295.
(25) Swaminathan, S.; Ravishanker, G.; Beveridge, D. L.; Lavery, R.; Etchebest, C.; Sklenar, H. *Proteins: Struct.Func. and Genet.* **1990**, *8*, 179.

(26) Shibata, M.; Zielinski, T. J. *J. Mol. Graphics* **1992**, *10*, 88.
(27) Harte, W. E.; Beveridge, D. L. *J. Am. Chem. Soc.* **1993**, *115*, 1231.
(28) Harte, W. E.; Swaminathan, S.; Mansuri, M. M.; Martin, J. C.; Rosenberg, I. E.; Beveridge, D. L. *Proc. Natl. Acad. Sci.* **1990**, *87*, 8864.
(29) Vijayakumar, S.; Visveshwara, S.; Ravishanker, G.; Beveridge, D. L. *Biophysical J.* **1993**, *65*, 2304-2312.
(30) Herzberg, O. *J. Mol. Biol.* **1991**, *217*, 701.
(31) Ravishanker, G.; Vijayakumar, S.; Beveridge, D. L. Proceedings of the Symposium on Hydrogen Bonding, Abstract# 29, 206th National Meeting of the American Chemical Society, August 1993.
(32) Carrey, E. A.; Pain, R. H. *Biochim. Biophys. Acta* **1978**, *533*, 12-22.

RECEIVED June 2, 1994

Chapter 12

Unusual Cross-Strand Hydrogen Bonds in Oligopurine•Oligopyrimidine Duplexes

Computer Graphics Presentations of Hydrogen Bonds in DNA Molecular Dynamics Simulation

Masayuki Shibata[1] and Theresa Julia Zielinski[2]

[1]Department of Biophysics, Roswell Park Cancer Institute, Elm and Carlton Streets, Buffalo, NY 14263
[2]Department of Chemistry, Niagara University, Niagara University, NY 14109

In order to obtain better insights into the dynamic nature of hydrogen bonding, computer graphics representations were used in the analysis of molecular dynamics trajectories of the $d(G)_6 \bullet d(C)_6$ and $d(A)_6 \bullet d(T)_6$ duplexes. A schematic representation of hydrogen bonding patterns is generated to reflect the frequency and the type of hydrogen bonding occurring during the simulation period. Various trajectory plots for monitoring geometrical parameters and for the analysis of three-center hydrogen bonding were also generated. For the $d(G)_6 \bullet d(C)_6$ system, three-center hydrogen bonds can be classified as in-plane and major/minor groove types. The in-plane three-center hydrogen bond represents a stable state in which both bonds simultaneously satisfy the relaxed hydrogen bonding criteria for a measurable period. The groove three-center hydrogen bonds behave as a transient intermediate state in a swing hydrogen bonding system. For the $d(A)_6 \bullet d(T)_6$ system, the major groove cross strand hydrogen bond and cross strand middle hydrogen bonds are formed due to large propeller twist. Three center hydrogen bonds were rare in this system. The differences in the hydrogen bonding patterns of the two systems are discussed.

Hydrogen bonding is one of the most common and important molecular phenomena. It plays a vital role in the structure and function of various biomolecules, including carbohydrates, proteins and nucleic acids. Thus it is not surprising that there are numerous books and review articles on the hydrogen bonding of small molecules and biomacromolecules ranging from gaseous to the solution/solid states (*1-10*).

The MD simulation technique is a useful tool for studying the energetics and structures of biological macromolecules that are partly controlled by hydrogen bonding (*11-18*). Analysis of hydrogen bonding from molecular dynamics (MD) simulations have appeared in the literature. For example, Koehler et al. analyzed the occurrence of three-center hydrogen bonding in cyclodextrins (*19,20*); Swanson et al. used the animation technique in their study of hydrogen bonding at the active center of elastin (*21*); Nordlund et al. examined the effect of a modified base on the

0097–6156/94/0569–0194$08.00/0

hydrogen bonding of DNA dodecamers (*22*); and Fraternali studied the dynamics of hydrogen bonding pattern for the linear peptide ahmethicin (*23*). Geller et al. used a bar code type diagram to display normal hydrogen bond occurance in a small peptide/protein complex system (*24*). In this article, the utility of computer graphics is shown by their application to oligopurine•oligopyrimidine duplexes, $d(G)_6$•$d(C)_6$ and $d(A)_6$•$d(T)_6$. The molecular basis for the observed difference in the hydrogen bonding pattern between these duplexes is discussed.

METHOD

Exactly the same method was used for both hexamer duplexes (*17,18*). Initial structures were constructed from standard B-DNA coordinates with a base pair step height of 0.34 nm, a helix twist of 36.0° (*25*) and with both the 5' and 3' terminals ending in hydroxy groups. The united atom approximation was used for all carbon hydrogens. All other hydrogens were treated explicitly. Electrical neutrality was achieved by placing octahedrally hydrated Na^+ ions at each of the phosphate positions. The resulting complex was then submerged in a box of SPC water (*26*). The total number of waters around the complex was reduced to 295 for d(A)•d(T) and 292 for d(G)•d(C) systems by discarding any waters beyond 0.5 nm of a solute atom in order to give a reasonable description of the first hydration shell of a mini helix. The central hydrogen bond for the base pairs at the top and bottom of the mini helices were restrained to their initial interatomic distances using a potential force constant of 1000 kJ mol^{-1} nm^{-2} in order to eliminate the possibility of fraying of the helix during the simulation.

The potential energy function of the GROMOS programs was used without modification (*27*). GROMOS partial charges were also used with a 1.0 dielectric constant. The cutoff range for nonbonded pair interactions was set at 0 8 nm and evaluated every 10 MD steps. The electrostatic interaction was calculated using a 1.8 nm cutoff applied on a "group by group" basis from a pair list which was updated every 10 time steps. The complete solute-solvent system was relaxed with conjugate gradient energy minimization until the difference in energy between two successive steps was less than 10^{-4} kJ/mol.

The initial velocities for the dynamics run were taken from a Maxwellian distribution. Constant temperature was maintained by weakly coupling the system to a 300K thermal bath. No pressure effects were included. The MD run extended for 180 ps with a step size of 0.001 ps for the first 20 ps in order to equilibrate the system to the chosen temperature. A step size of 0.002 ps was used thereafter. Chemical bond lengths for the duplex molecules were maintained at their initial values by using SHAKE (*28*). No restrictions were placed on the bond angles or dihedral angles during the dynamics study. The average temperature remained at 296 $\pm$ 4 K for the post equilibration period. Conformation snapshots obtained every 0.05 ps were used to prepare the trajectory plots.

Hydrogen bonding within the solute molecule was analyzed first with the PROHB program (*27*). The donor (D) to acceptor (A) distance and the angle D-H••A are typical geometrical parameters used to define the normal linear (two-center) hydrogen bonds in crystallographic studies in which the position of hydrogens are not easily obtained (*29*). Since hydrogen positions are monitored easily in MD studies the hydrogen (H) - acceptor (A) distance is used as the hydrogen bonding criteria instead of the D••A distance. In order to account for the occurrence of three-center bonds, more relaxed criteria, 0.27 nm for H••A distance and 90° for D-H••A angle, are used instead of the normal criteria of 0.25 nm and 135°.

Since it is usually difficult to obtain overall hydrogen bonding features such as type of hydrogen bond, changes in strength over time, and the identity of donor and acceptor atoms when using numerical tables, we recently presented several computer graphics based approaches to make the analysis of hydrogen bonding easier (*30*). The advantage of graphic representations in the analysis of MD trajectories for other properties are discussed by Ravishanker et al. (*31*). Since detailed methods were presented earlier (*30*), only a brief description is given here.

In one type of representation a schematic diagram is drawn for the systems examined. In order to observe changes in hydrogen bonding patterns over time during the simulation, the final 160 ps of the trajectory was divided into eight 20 ps time spans. Arrows are used to indicate hydrogen bonding with the direction of the arrow pointing from donor to acceptor. The arrow thickness reflects the frequency of occurrence of the hydrogen bond for a particular time span.

In another type of representation the dynamic nature of hydrogen bonding can be appreciated by monitoring specific geometrical parameters as a function of time. Two important parameters, H••A distance and D-H••A angle for example, are plotted against time in the same diagram with a special scaling scheme. When the two parameters at any given time remain within the zone enclosed by suitable upper and lower limit lines, the bond examined satisfies the hydrogen bonding condition. Useing a different shade for each parameter differentiates the two behaviors.

Since nucleic acid bases contain more than two hydrogen bond donors and acceptors, the situation where a single hydrogen is shared by two acceptors, three-center hydrogen bonding, is frequently encountered. Here the parameters for two constituent hydrogen bonds, two H••A distances and two D-H••A angles, the measures of planality, the dihedral angle A1-D-H-A2 the sum of the angles D-H••A1, D-H••A2 and A1••H••A2, are plotted simultaneously in a third type of representation.

In order to clarify the dynamics of formation and breakage of three center bonds, an additional representation portrays the frequency and duration of three-center bonds easily. Numerical values are assigned to the four possible situations for the two hydrogen bond constituents in a three-center hydrogen bond. For the configuration where all the conditions defining a three-center bond are satisfied (H••An < 0.27nm, D-H••An > 90°, a sum-of-three-angles > 340°, and | A1-D-H-A2 | < 15°), the value 1.5 is assigned. If the three-center hydrogen bond criteria are not satisfied, then each hydrogen bond is examined separately by using the normal linear hydrogen bonding criteria (H••A < 0.25nm, D-H-A > 135°). A value of one is assigned when D-H••A1 satisfies the condition or a value of two is assigned when D-H••A2 satisfies the condition. Finally, if there is no hydrogen bond, the value zero is assigned. In order to eliminate transient noise, a filter is introduced to clearly depict the phenomenon of three center hydrogen bonding. When the two consecutive configurations share the same value, a small line is drawn but if the values are not identical, the pen is just moved without drawing a line. The results indicate the how three center bonds are formed and broken during the MD simulation.

The effect of glycosidic angle rotation on the formation of unusual cross strand hydrogen bonding, are examined using the trinucleotide duplexes in the scheme shown here. Starting from the standard B-DNA conformation, the glycosidic angle of the middle base was rotated in two opposite directions by 45° and then the hydrogen bonding parameters for the system were measured for each geometry. The effect of each glycosidic angle rotation was examined separately.

```
 S-G1   C3-S        S-A1   T3-S
 /          \       /          \
P            P     P            P
 \          /       \          /
 S-G2   C2-S        S-A2   T2-S
 /          \       /          \
P            P     P            P
 \          /       \          /
 S-G3   C1-S        S-A3   T1-S
```

RESULTS AND DISCUSSION

The total potential energy as a function of time for both systems is shown in Figure 1. No net drift in total PE for either system is seen. The d(A)•d(T) system does not show any significant energy fluctuation compared to the d(G)•d(C) system.

Schematic Representation of Hydrogen Bonding. The schematic representation of the hydrogen bonding between the $d(G)_6$ and $d(C)_6$ strands for different MD time slices is shown in Figure 2a. The arrows point from donor to acceptor with a thickness proportional to the frequency of occurrence in the MD simulation. In the base plane, the three normal Watson-Crick type hydrogen bonds are shown by horizontal arrows. An additional in plane hydrogen bond, N1(G)-H••O2(T), in most base pairs has a thickness that indicates that during the simulation these additional in plane non-Watson-Crick type hydrogen bonds occur as frequently as the standard Watson-Crick bonds.

There are several hydrogen bonds that are formed between two adjacent base pairs. The cross strand hydrogen bonds are classified according to the grooves. In the major groove, the N4 of cytosine interacts with an O6 of an adjacent guanine while the N2 of a guanine interacts with an O2 of an adjacent cytosine in the minor groove. The diagram clearly indicates that the major groove cross strand hydrogen bonding is more common than the minor groove one. However, the strength of the groove cross strand hydrogen bonds fluctuates from one time span to another and from one base pair to another. When these groove cross strand hydrogen bonds form while maintaining the original Watson-Crick bonds, they become the three center hydrogen bonding discussed below.

The corresponding figure for d(A)•d(T) system is shown in Figure 2b. Compared to the d(G)•d(C) system, very little hydrogen bonding is present during the simulation. Even normal Watson-Crick A•T hydrogen bonds are broken at various places during the simulation. The more exposed N6(A)-O4(T) major groove side Watson-Crick bonds are more frequently broken than the N1(A)-N3(T) middle Watson-Crick bonds. Instead, there are several strong major groove cross strand hydrogen bonds formed between N6 of adenine and O4 of an adjacent thymine. It should be pointed out that this type of unusual hydrogen bonding has been observed in the crystal structures of deoxyoligonucleotide duplexes (*32,33*) as a part of three center hydrogen bonding while maintaining the original Watson-Crick bonds. Less frequently found is another cross strand hydrogen bond formed between N1 of adenine and the N3 of adjacent thymine. The significant difference on the stability of hydrogen bonding observed between d(G)•d(C) and d(A)•d(T) system will be discussed later.

The possible three-center hydrogen bonds observed in our simulations are

```
    N3(Cn)      (Gn-1)O6          O4(Tn)        O2(Cn+1)      (An-1)N1
    /              \              /              /               \
(Gn)N1-H        H-N4(Cn)     (An)N4-H       (Gn)N2-H          H-N3(Tn)
    \              /              \              \               /
    O2(Cn)      (Gn)O6           O4(Tn-1)       O2(Cn)         (An)N1

 a) In-Plane  b) Major Groove  b) Major Groove  c) Minor Groove  d) Middle
```

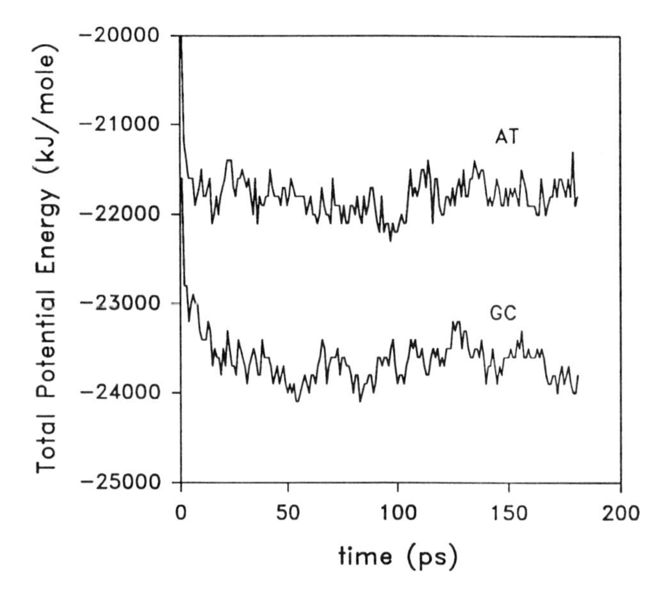

Figure 1. The total potential energy in kJ/mol is plotted against time for the $d(G)_6 \bullet d(C)_6$ and $d(A)_6 \bullet d(T)6$ MD simulations.

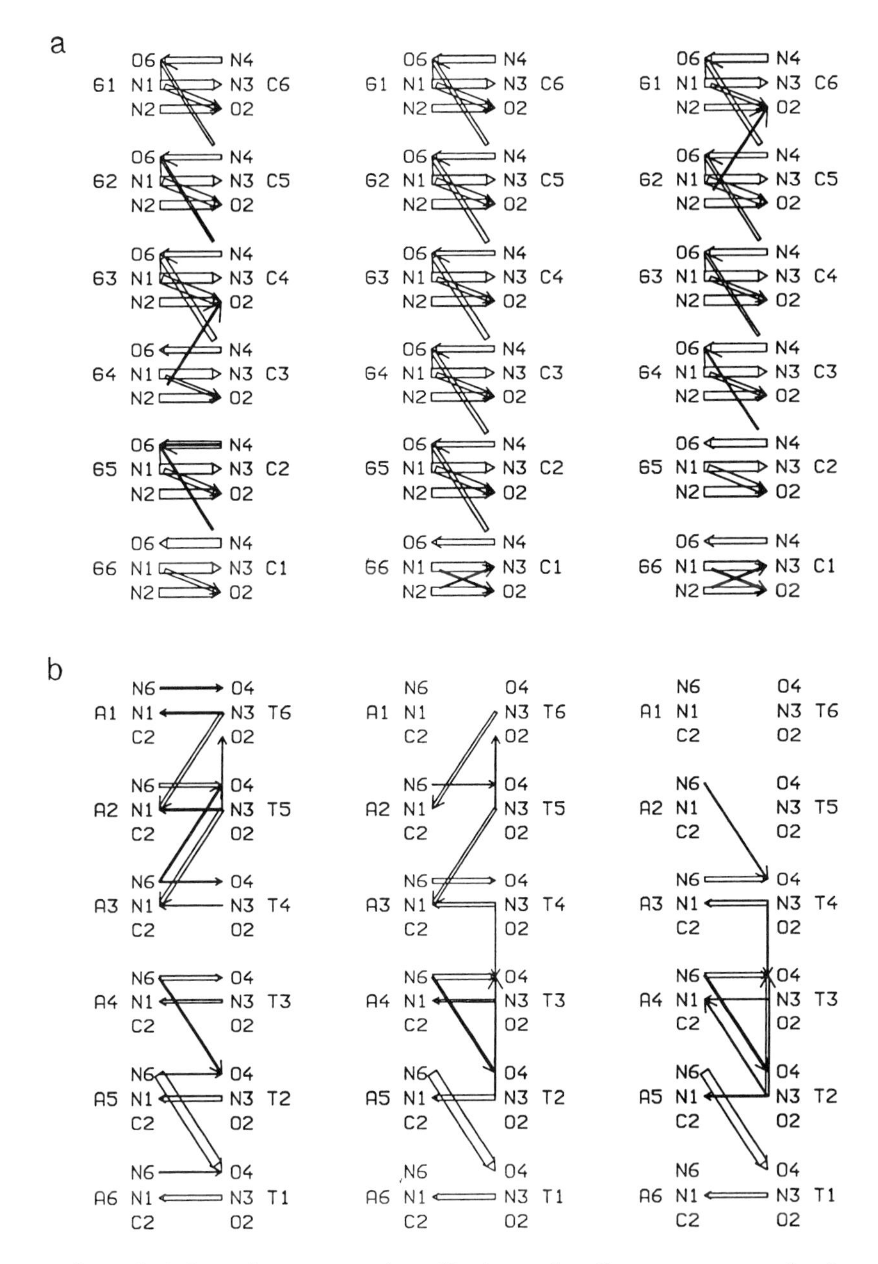

Figure 2. Schematic representation of hydrogen bonding occurrence and pattern for the $d(G)_6$•$d(C)_6$ (Figure 2a) and $d(A)_6$•$d(T)_6$ (Figure 2b) MD simulations. The width of the arrow is proportional to the frequency of occurrence. The unusual hydrogen bonding between neighboring base planes occurring in major and minor grooves as well as in the middle are presented.

grouped into the four types for the the nth base pair in the scheme shown here. The in-plane three center hydrogen bonds observed in d(G)•d(C) system are the most stable while the cross strand minor groove in d(G)•d(C) system and the cross strand middle in the d(A)•d(T) system are least stable. The significance of the major groove cross strand hydrogen bonding will be discussed later.

Hydrogen Bonding Distance and Angle Trajectory Plots. The utility of trajectory plots in representing the dynamic nature of hydrogen bonding is shown with some examples. A typical behavior is shown by the O6(G3)••H••N4(C4) Watson-Crick bond (Figure 3a) for which the angle (gray) remains in an acceptable region but the distance (black) strays out of the acceptable region between 85 to 125 ps. The behaivor of N6(A5)-H••O4(T1) cross strand major groove hydrogen bond is shown in Figure 3b, where dispite some fluctuation between formation and breakage the stability of this particular bond is evident.

Three Center Hydrogen Bonds. Systematic analysis of three-center hydrogen bonding was reported using crystal structures obtained from the neutron diffraction studies (*34*) of pyranoses and pyranosides (*35*), of small organic compounds (*36*), of amino acids (*37*), of nucleosides and nucleotides (*38*), and barbiturates, purines and pyrimidines (*39*) and proteins (*40*).

Trajectory plots of parameters defining the three-center hydrogen bonds were created for various types and some examples are shown to present the utility of this diagram for examining their dynamic behavior. Figure 4 shows the cross strand major groove three center hydrogen bonding of N4(C2)-H hydrogen bonding with the O6s of G4 and G5. No hydrogen bonding is detected near the beginning of the simulation according to the cutoff criteria. Only after 40ps do all the parameters simultaneously start to show the presence of three-center bonding. It is clear both from the distance and angle diagrams that the normal Watson-Crick hydrogen bond and the cross-strand adjacent-pair hydrogen bond take turns dominating.

Three Center Hydrogen Bond and Swing Hydrogen Bond. When one hydrogen is shared by two acceptors, there are two types of three center hydrogen bonding possible, a stable one and a swing type. For the former, the hydrogen bonding condition must be satisfied simultaneously for both bonds. In the latter case, one bond satisfies the hydrogen bonding condition while the other cannot i.e., one bond starts to break when the other starts to form, repeatedly swinging back and forth so that only one hydrogen bond exists at any given time. MD studies of carbohydrates (*19,20,41,42*) suggest that three-center hydrogen bonding serves as an intermediate state for flip-flop hydrogen bonding. A pictorial analysis tool was created to determine if a three center bond in our simulations can serve as an unstable intermediate state for swing hydrogen bonding. In this representation the existence of each state is indicated by the horizontal lines at different heights and a filter is used so that only two consecutive occurrences of each state is drawn as a short line. An example of the stability of the in-plane three center hydrogen bonding is shown for N1-H (G2) interacting with N3 and O2 of C5 in Figure 5a. It is clearly seen that its existence is interrupted only briefly. The second example, N4-H (C2) with O6s of G4 and G5 (Figure 5b) shows a representation of the swing hydrogen bonding where the existence of three center hydrogen bonding is interrupted by the formation of two constituent hydrogen bonds alternately. The Watson-Crick bond and the cross strand major groove adjacent pair bond alternate dominance through frequent formation of a three center hydrogen bond that is serving as an unstable intermediate during switches between the two dominant hydrogen bonds.

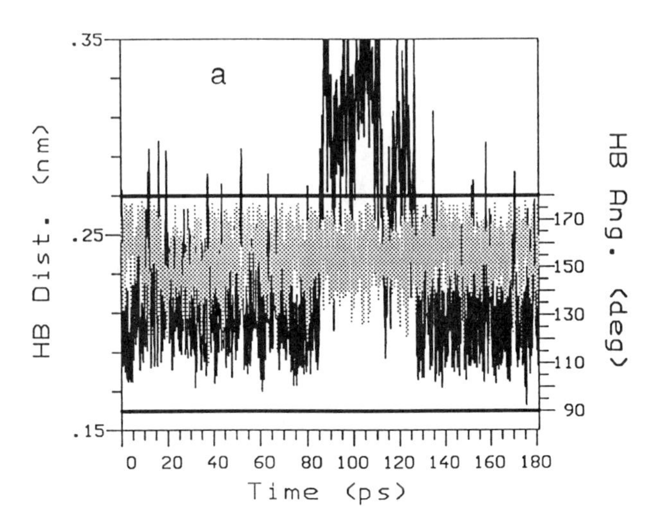

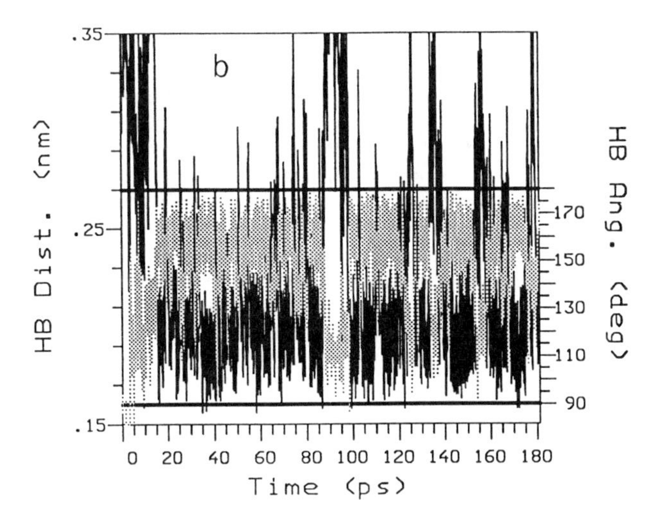

Figure 3. Trajectory plots of hydrogen bond geometrical parameters. The H••A distance (gray) and D-H••A angle (black) are drawn in the same diagram; the horizontal lines indicate the upper and lower limits of these parameters.
a) O6(G3)••H-N4(C4) bond; b) N4(A5)-H••O4(T1) bond.

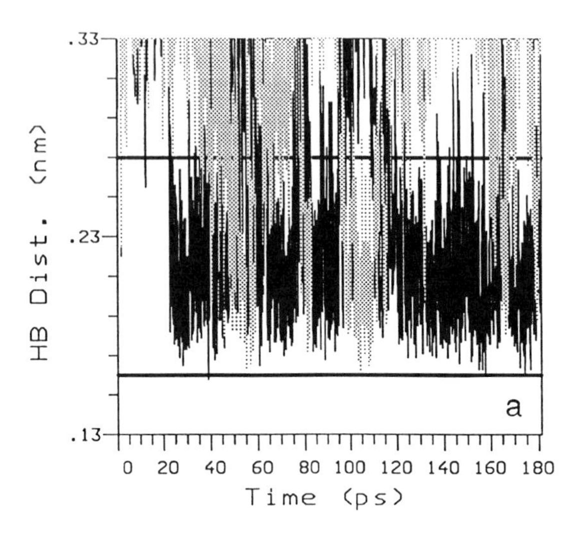

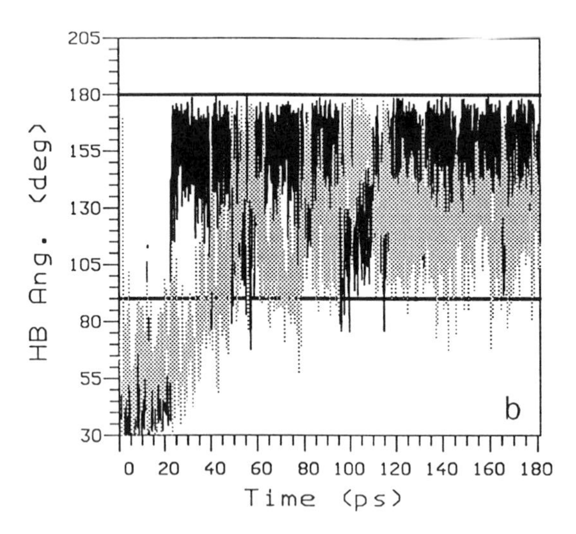

Figure 4. Trajectory plots of geometrical parameters defining three-center hydrogen bonds formed by N4-H of the residue C2 with O6s of residues G4 and G5. The distance diagram shows the H••A distance for both major (black) and minor (gray) components of three-center hydrogen bonds in the same diagram. Similarly the angle diagram shows the D-H••A angles for the major (black) and minor (gray) components of three-center hydrogen bonds.

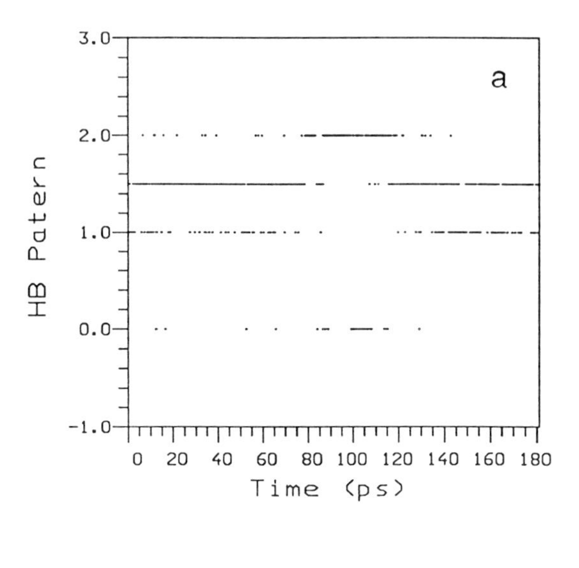

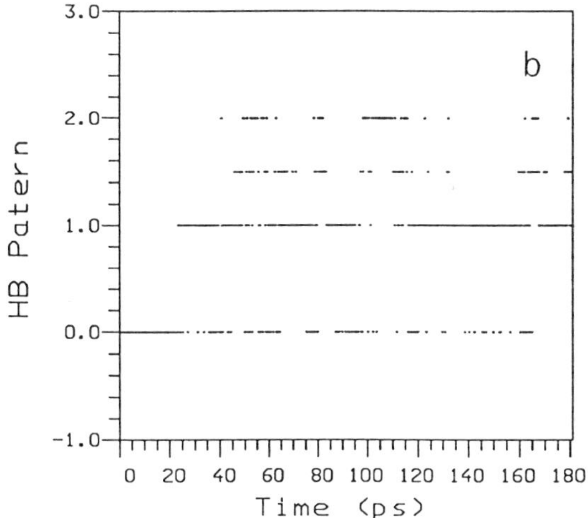

Figure 5. The trajectory of the three-center hydrogen bonding states were analyzed by assigning the values 0, 1, 1.5, and 2 for the states corresponding to no-hydrogen-bonding, a normal Watson-Crick hydrogen bond, a three-center hydrogen bond, and a secondary hydrogen bond, respectively. A filter is applied to plot only the state which persists for more than two consecutive configurations: a) three-center hydrogen bonding formed by N1-H of residue G2 with N3 and O2 of residue C5; b) three-center hydrogen bonding formed by N4-H of residue C2 with O2s of residues G4 and G5.

Difference Between d(G)•d(C) And d(A)•d(T) Systems. The very different patterns of hydrogen bonding stability necessitated an examination of various factors that may affect the hydrogen bonding stabilities. These factors included pseudorotation, propeller twist, and glycosidic torsional angle.

There was no significant difference in the trajectory pattern for the pseudorotation phase angle parameter for the two systems. The sugar ring of purine residues tended to be more stable than those of the pyrimidines residues except for residues A2 and A3 which explore a wide range of conformations. In general, all belong to the O4'endo/C1'exo/C2'endo domain but more concentrated on the C1'exo-C2'endo region. The sugars on the pyrimidine residues tend to explore a larger area including the C3' endo region although only for short periods.

Since the unusual cross strand hydrogen bonds were suggested to stabilize a large propeller twist in crystals (*32,33*), the propeller twist trajectory plots were examined. The results showed steady negative propeller twists for the d(G)•d(C) system and a large fluctuation for the d(A)•d(T) system. Although mainly in the negative regions, the figure (not shown) indicated that the d(A)•d(T) system explored even positive regions.

The dynamic behavior of the glycosidic dihedral angles were also examined from the trajectory plots. For the guanine residues, a majority of trajectory plots indicate that the glycosidic angle remains in the -120° region while adenine residues, especially residues A2, A3 and A4 reach into the -80° region. For the pyrimidine residues, cytosines tend to explore the -80° region more often than guanosine, again, the thymine residues especially the C3, C4 and C5 explore this region much more frequently.

The effect of thc glycosidic rotations on hydrogen bonding systems was examined by measuring various hydrogen bonding geometrical parameters upon the rotation of one glycosidic angle within a fixed DNA trimer duplex. The results for the d(A)•d(T) system are shown in Table I. Since the O4 of thymine lies along the extension of the glycosidic bond of thymine, very little effect was observed for the hydrogen bonding parameters involving O4 of thymine upon the rotation of the T glycosidic bond. The N6(A)-O4(T) Watson-Crick hydrogen bond remained intact. The N1(A)-N3(T) Watson-Crick hydrogen bond, however, was slightly weakened regardless of the sense of rotation. Only a positive rotation assisted in the formation of an additional cross strand middle hydrogen bond N1(A+1)-N3(T). Upon a rotation about the adenine glycosidic bond, the Watson-Crick bond N1(A)-N3(T) was weakened while the N6(A)-O4(T) was only slightly weakened. Meanwhile the magnitude of these weakening effects were independent of the sense of the rotation. The negative rotation assists in the formation of the major groove cross strand N6(A)-O4(T-1) bond while the positive rotation assists in the formation of the cross strand middle N1(A)-N3(T+1) bond. Thus the net result is that the positive rotation may favor one additional cross strand hydrogen bond formation over the negative rotation.

A corresponding examination of the d(G)•d(C) system (not shown) indicates that the major groove side Watson-Crick hydrogen bonding O6(G)-N4(C) is affected only slightly by the rotations while the minor groove N2(G)-O2(C) Watson-Crick bonds are affected much more significantly, especially by the rotation of the guanine glycosidic bond. The formation of the major groove cross strand hydrogen bond is assisted by negative rotation of the glycosidic bonds, regardless of the residue. The minor groove cross strand hydrogen bond is assisted by positive rotation but it seems that there is a problem with satisfying the angle requirement for the three center bond.

The results are consistent with the our trajectory analysis that the major

Table I. Effect of Glycodic Bond Rotations on Hydrogen Bonding

	B-DNA	Thymine Rotations -45.0 (diff)	45.0 (diff)
N6(A-1)-O4(T)	0.41	0.41 (0.00)	0.42 (0.01)
H62(A-1)-O4(T)	0.39	0.38 (-0.01)	0.40 (0.01)
N6(A-1)-H62-O4(T)	99	99 (0.4)	99 (0)
N6(A)-O4(T)	0.28	0.28 (0.00)	0.28 (0.00)
H62(A)-O4(T)	0.18	0.18 (0.00)	0.18 (0.00)
N6(A)-H62-O4(T)	173	174 (1)	171 (-2)
N6(A+1)-O4(T)	0.48	0.49 (0.01)	0.47 (-0.01)
H62(A+1)-O4(T)	0.42	0.43 (0.01)	0.41 (-0.01)
N6(A+1)-H62-O4(T)	120	120 (0)	122 (1)
N1(A-1)-N3(T)	0.48	0.46 (-0.02)	0.56 (0.08)
N1(A-1)-H3(T)	0.42	0.36 (-0.06)	0.57 (0.15)
N1(A-1)-H3-N3(T)	127	171 (44)	84 (-43)
N1(A)-N3(T)	0.30	0.34 (0.04)	0.34 (0.04)
N1(A)-H3(T)	0.20	0.29 (0.09)	0.29 (0.09)
N1(A)-H3-N3(T)	179	109 (-69)	111 (-68)
N1(A+1)-N3(T)	0.40	0.49 (0.09)	0.35 (-0.05)
N1(A+1)-H3(T)	0.36	0.51 (0.15)	0.26 (-0.10)
N1(A+1)-H3-N3(T)	109	70 (-39)	159 (50)
		Adenine Rotations	
N6(A)-O4(T+1)	0.48	0.57 (0.09)	0.45 (-0.03)
H62(A)-O4(T+1)	0.42	0.57 (0.15)	0.36 (-0.06)
N6(A)-H62-O4(T+1)	120	84 (-36)	143 (23)
N6(A)-O4(T)	0.28	0.32 (0.04)	0.33 (0.05)
H62(A)-O4(T)	0.18	0.26 (0.08)	0.27 (0.09)
N6(A)-H62-O4(T)	173	116 (-56)	111 (-62)
N6(A)-O4(T-1)	0.41	0.34 (-0.07)	0.50 (0.09)
H62(A)-O4(T-1)	0.39	0.25 (-0.14)	0.53 (0.14)
N6(A)-H62-O4(T-1)	99	147 (47)	71 (-28)
N1(A)-N3(T+1)	0.40	0.62 (0.22)	0.35 (-0.05)
N1(A)-H3(T+1)	0.36	0.58 (0.22)	0.24 (-0.12)
N1(A)-H3-N3(T+1)	109	110 (1)	146 (37)
N1(A)-N3(T)	0.30	0.41 (0.11)	0.41 (0.11)
N1(A)-H3(T)	0.20	0.33 (0.13)	0.33 (0.13)
N1(A)-H3-N3(T)	179	139 (-40)	141 (-38)
N1(A)-N3(T-1)	0.48	0.41 (-0.07)	0.67 (0.19)
N1(A)-H3(T-1)	0.42	0.32 (-0.10)	0.62 (0.20)
N1(A)-H3-N3(T-1)	127	141 (14)	116 (-11)

groove cross strand hydrogen bond, O6(G)-N4(C-1), is very frequently observed with stable -120° glycosidic angle for guanosine while the cross strand middle N1(A)-N3(T+1) bond is observed with glycosidic angles exploring the -80° range. For the d(A)•d(T) system, stable major groove N6(A)-O4(T-1) cross strand hydrogen bonds are also observed which may be accompanied by N1(A)-N3(T-1) cross strand middle hydrogen bond. Directionality of the bond N1(A) with N3(T+1) or with N3(T-1) depends on the sense of rotation of the glycosidic bond.

CONCLUSIONS

This study demonstrates that both the schematic drawing of hydrogen bonding patterns between base pairs and various trajectory plots are essential tools in analyzing the nature of hydrogen bonding during an MD simulation.

In nucleic acid fragments three-center hydrogen bonds can be classified into four different types: in-plane (in GC system), major groove (both GC and AT systems), minor groove (in GC system) and middle (in AT system). Among these three-center hydrogen bonding types, the in-plane three-center bond is the most stable and the minor groove three-center bond is the least stable. The groove three-center hydrogen bonds behave as transient intermediates connecting two different hydrogen bonding states in a swing hydrogen bonding system.

Although any MD simulation cannot escape the limitations imposed by the particular force field employed, our results are consistent with observations made by others. Mohan and Yathindra systematically examined the formation of the cross strand hydrogen bonding at base-pairs, dimer duplex and trimer duplex levels by molecular mechanics energy minimization study (*43*). They showed that both major and minor groove cross strand hydrogen bond formation are possible for d(G)•d(T) system although they only observed the major groove cross strand hydrogen bonds for the d(A)•d(T) system. MD simulation was performed by Fritsch and Westhof (*44*) and Brahms et al. (*45*) for the $d(A)_{10}$•$d(T)_{10}$ system in their studies of three center hydrogen bonding. Their results indicated that three center hydrogen bonds appear as a consequence of the unusual structural properties of this oligomer rather than from some significant stabilization factor in the system. Our own examination of the effect of the glycosidic bond rotation completely ignored steric hindrance since it was aimed at examining any advantage from glycosidic torsion towards a formation of cross strand hydrogen bonding. A recent systematic study by Hunter of sequence dependent structural features based on base stacking interactions indicated that the steric crush between 3'thymine methyl and 5'neighboring thymine sugar can be removed by a large negative propeller twist (*46*). Thus cross strand hydrogen bonding is a consequence and not the cause of the large propeller twist. In fact from our simulation it was observed that a large propeller twist will break normal Watson Crick hydrogen bonding in order to form a major groove cross strand hydrogen bond.

From our MD simulations, the stability of d(G)•d(C) hydrogen bonding can be attributed to several factors. Of course, G•C pairs have one more Watson-Crick hydrogen bond than A•T pairs. In addition, the ease of the formation of in-plane three center hydrogen bond by the formation of N1(G)-O2(C) is a strong additional stabilization. Furthermore, the major groove cross strand three center hydrogen bond, serving as an intermediate in the swing switch between O6(G)-N4(C-1) and O6(G)-N4(C), also adds to the dynamic stability of this system. In the d(A)•d(T) system, all of the above factors are absent leading to the unstable hydrogen bonding pattern observed in our simulation.

References

1. Pimentel, G.C. and McClellan, A.L. *The Hydrogen Bond.* W.H.Freeman & Co., SF, 1960
2. Hamilton, W.C. and Ibers, J.A. *Hydrogen Bonding in Solids.* W.A.Benjamin, Inc., NY, 1968
3. Vinogradov, S.N. and Linnell, R.H. *Hydrogen Bonding.* van Nostrand Reinhold Co., NY, 1971
4. Toniolo, C. *CRC Crit. Rev. Biochem.* 1980, ,1-44
5. Baker, E.N. and Hubbard, R.E. *Prog. Biophys. Mol. Biol.* 1984, **44**, 97-179
6. Saenger, W. *Ann. Rev. Biophys. Biophys. Chem.* 1987, **16**, 93-114
7. Westhof, E. *Ann. Rev. Biophys. Biophys. Chem.* 1988, **17**, 125-144
8. Legon, A.C. *Chem. Soc. Rev.* 1990, **19**, 197-237
9. Stickle, D.F., Presta, L.G., Dill, K.A., and Rose, G.D, *J. Mol. Biol.* 1992, **226**, 1143-1159
10. Jeffrey, G. A. and Saenger, W., 1991 *Hydrogen Bonding in Biological Structures*, Springer-Verlag, Berlin Heidleberg
11. McCammon, J.A. and Harvey, S.C. 1987 *Dynamics of Proteins and Nucleic Acids*, Cambridge University Press, Cambridge.
12. Levitt, M. *Cold Spring Harbor Symp. Quant. Biol.* 1983, **47**, 251-275.
13. Singh, U.C., Weiner, S.J. and Kollman, P.A. *Proc. Natl. Acad. Sci. U.S.A.* 1985 **82**, 755-759.
14. van Gunsteren, W.F., Berendsen, H.J.C., Geurtsen, R.G., and Zwinderman, H.R.J. *Annals N.Y. Acad. Sci.* 1986, **482**, 287-303.
15. Swaminathan, S., Ravishanker, G. and Beveridge, D.L. *J. Amer. Chem. Soc.* 1991, **113**, 5027-5040
16. van Gunsteren, W.F. and Mark, A.E. *Eur. J. Biochem* 1992, **204**, 947-961
17. Zielinski, T.J. and Shibata, M. *Biopolymers* 1990, **29**, 1027-1044
18. Shibata, M., Zielinski, T.J. and Rein, R. *Biopolymers* 1991, **31**, 211-232
19. Koehler, J.E.H., Saenger, W. and van Gunsteren, W.F. *J. Biomol. Struct. Dyn.* 1988, **6**, 181-198
20. Koehler, J.E.H., Saenger, W. and van Gunsteren, W.F. *Eur. Biophys. J.* 1988, **16**, 153-168
21. Swanson, S.M., Wesolowski, T., Geller, M. and Meyer, E.F. *J. Mol. Graphics* 1989, **7**, 240-242
22. Nordlund, T.M., Anderson, S., Nilsson, L., Rigler, R., Gräslund, A. and McLaughlin, L.W. *Biochemistry* 1989, **28**, 9095-9103
23. Fraternali, F. *Biopolymers* 1990, **30**, 1083-1099
24. Geller, M., Swanson, S.M., and Meyer, E.F., Jr. *J. Biomol. Struct. Dyn.* 1990, **7**, 1043-1052
25. Arnott, S., Smith, P.J., and Chandrasekaran, R. *CRC Handbook of Biochemistry and Molecular Biology: Nucleic Acids, 3rd edition, Vol. 2* ed., G.D. Fasman, CRC Press, Cleveland, Ohio, 1976, pp.411-422
26. Berendensen, H.J.C., Postma, J.P.M., van Gunsteren, W.F. and Hermans, J. *Intermolecular Forces*, ed., B.Pullman, D. Reidel Publishing Co., Dordecht, The Netherlands, 1981, pp.331-342
27. van Gunsteren, W.F. and Berendsen, H.J.C. *Groningen Molecular Simulation System*, BIOMOS b. v., Biomolecular Software, Laboratory of Physical Chemistry, University of Groningen, Groningen, The Netherlands 1986
28. van Gunsteren, W.F. and Berendesen, H.J.C. *Mol Phys* 1977, **34**, 1311-1327
29. Murray-Rust, P. and Glusker, P. *J. Am. Chem. Soc.* 1984, **106**, 1018-1025
30. Shibata, M. and Zielinski, T.J. *J. Mol. Graph.* 1992, **10**, 88-95
31. Ravishanker, G., Swaminathan, S., Beveridge, D.L., Lavery, R., and Sklenar, H. *J. Biomol. Struct. Dyn.* 1989, **6**, 669-699

32. Heinemann, V. and Alings, C. *J. Mol. Biol.* 1989, **210**, 369-381
33. Timsit, Y., Westhof, E., Fuchs, R.P.P. and Moras, D. *Nature* 1989, **341**, 459-462
34. Ceccarelli, C., Jeffrey, G.A., and Taylor, R. *J. Mol. Struct.* 1981, **70**, 255-271
35. Jeffrey, G.A. and Mitra, J. *Acta Cryst.* 1983, **B39**, 469-480
36. Taylor, R., Kennard, O., and Versichel, W. *J. Am. Chem. Soc.* 1984, **106**, 244-248
37. Jeffrey, G.A. and Mitra, J. *J. Am. Chem. Soc.* 1984, **106**, 5546-5553
38. Jeffrey, G.A., Maluszynska, H., and Mitra, J. *Int. J. Biol. Macromol.* 1985, **7**, 336-348
39. Jeffrey, G.A. and Maluszynska, H. *J. Mol. Struct.* 1986, **147**, 127-142
40. Preishner, R. Egner, U. and Saenger, W. *FEBS Letts* 1991, **288**, 192-196
41. Steiner, T. Mason, S.A. and Saenger, W. *J. Amer. Chem. Soc.* 1991, **113**, 5676-5687
42. Ding, J., Steiner, T., Zabel, V., Hingerty, B.E., Mason, S.A. and Saenger, W. *J. Amer. Chem. Soc.* 1991, **113**, 8081-8089
43. Mohan, S. and Yathindra, N. *J. Biomol. Stero. Dyn.* 1991, **9**, 113-126
44. Fritsch, V. and Westhof, E. *J. Amer. Chem. Soc.* 1991, **113** 8271-8277
45. Brahms, S., Fritsch, V., Brahms, J.G., and Westhof, E. *J. Mol. Biol.* 1992, **223**, 455-476
46. Hanter, C.A. *J. Mol. Biol.* 1993, **230**, 1025-1054

RECEIVED June 2, 1994

Chapter 13

STRIPS: An Algorithm for Generating Two-Dimensional Hydrogen-Bond Topology Diagrams for Proteins

G. Ravishanker, S. Vijayakumar, and D. L. Beveridge

Department of Chemistry, Wesleyan University, Middletown, CT 06459

Hydrogen bonding is an important factor in the stabilization of the three dimensional structure of a protein and plays a significant role in the regulation of biological processes. Hydrogen bonds form characteristic patterns among unique secondary structural regions and representing these patterns on the classical three dimensional rendering of a molecule is a major problem, particularly in the superposition of hydrogen bonds in β-sheets. Mapping the hydrogen bond information on two dimensional topology diagrams have proven to be useful in conveying information extracted from computer simulation experiments. So far, the generation of such H-bond topology diagrams have been done manually only for a few selected proteins. We present herein, an algorithm that automates the generation of such topology diagrams given a Brookhaven PDB file.

Hydrogen bonds in proteins stabilize secondary and tertiary structural forms and play a major role in substrate recognition, binding and enzyme catalysis (*1,2*). Hydrogen bonds stabilizing secondary structural regions form characteristic patterns and presenting these patterns in a comprehensible manner for analysis is usually a difficult task due to the three-dimensional complexity of a protein. Representation of the topology and relative disposition of secondary structures is commonly achieved using ribbons (*3,4*), a simplified scheme for presenting protein structure. However, these representations are still too complicated to show the individual hydrogen bonds in proteins and properties associated with them. Detailed presentation can be made by superposing the relevant hydrogen bond information on template drawings, obtained through an optimal placement of residues on a two dimensional plane, that capture the hydrogen bonding patterns characteristic of secondary structures (*5,6*). These diagrams, hereafter referred to as 2-D Hydrogen Bond Topology Diagrams (2-D HBTD), present an alternate view of the protein structure and can be of tremendous help in conveying information about secondary structures in proteins.

0097–6156/94/0569–0209$08.00/0

Molecular Dynamics (MD) simulations of proteins provide a detailed picture of the structure and dynamics of a protein in an assumed force field (*7,8*). Monitoring the hydrogen bonding patterns during an MD simulation can provide very interesting information about the stability and flexibility of secondary structures in proteins. Typical presentation of such data from protein MD involves time series plots showing various hydrogen bond measures such as distance or energy per H-bond etc. In this case, one has to constantly go back and forth between the structure of the protein and the graphs to interpret the results. 2-D HBTD of proteins with average hydrogen bond energies over the MD trajectories superposed on them have been used to present the results from MD very effectively (*9*).

The 2-D topology diagrams generated so far have all been done manually and such generation can be cumbersome. Computational algorithms that automate the generation of 2-D topology diagrams from known secondary structural information will result in consistent and easily usable diagrams. X-ray crystallographic structures from the Brookhaven Protein Data Bank (*10*) is the usual starting point for various structural studies on proteins including MD. These files contain secondary structure information in **SHEET** and **HELIX** records (*11*) that can be used to generate the 2-D topology diagrams. The secondary structure records are typically generated by using computer programs such as DSSP (Dictionary of Secondary Structures in Protein) developed by Kabsch and Sander (*1*). DSSP employs a simple energy function to identify potential hydrogen bonds and uses empirical rules to locate strips of secondary structure based on the list of hydrogen bonds generated.

The generation and presentation of the hydrogen bonding information has been developed as one of several tools in a multipurpose analysis and presentation program suit called the **MD Toolchest** (*12*), developed here at Wesleyan. MD Toolchest is a rich collection of programs that enables the analysis of biomolecular structure in great detail and allows presentation of the results in graphically efficient ways. Analysis of a single X-ray structure, such as from Brookhaven PDB, or comparison of several of them or analysis of MD trajectory are all possible using MD Toolchest. The topology generator tool allows the user to generate the 2-D topology diagram from either a PDB file with or without secondary structure records or from any arbitrary list of hydrogen bonds provided by the user. If the PDB file does not contain the secondary structure records, the program generates those using the Kabsch and Sander algorithm (*1*). The hydrogen bond tool is used to produce the graphical output of the topology diagram with the hydrogen bond information such as energy, distance, angles or combination of those superposed on the diagram.

We present herein a brief description of the algorithm called 'STRIPS', used to produce a protein hydrogen bond topology diagram from the secondary structure records in a PDB file, and some preliminary results from the application of this algorithm to selected protein structures. The detailed description of the algorithm will be published elsewhere (*13*). Topology diagrams generated by STRIPS, for BPTI and HIV-1 protease, are compared with those generated manually, and we show how very simple modifications to the secondary structure records can result in STRIPS generated diagrams which are very close to those manually generated.

STRIPS Algorithm

STRIPS optimally places the amino acid residues of a protein on a two-dimensional grid of columns and rows based on a set of placement rules for secondary structures. The algorithm is called 'STRIPS' because it cuts the amino acid chain into smaller strips of residues forming a secondary structure, before placing them on the grid. The organization of the amino acid residues of a protein on the two dimensional grid will be referred to as the 2-D Hydrogen Bond Topology Diagram (2-D HBTD) since they will be used primarily to show the hydrogen bonding patterns in the protein.

Brookhaven PDB files typically contain HELIX and SHEET records that specify residue ranges for a given type of secondary structure and an index for the residues that form a β-strand. This index identifies the sheet, of which the β-strand forms a part. The directionality of a given β-strand (parallel or anti-parallel) with respect to the one prior to it in sequence is provided by an additional direction index. By definition, the first sheet in a bundle has a direction of zero. Complete description of these records are provided by PDB (*11*).

STRIPS algorithm works with a 2-D grid of arbitrary dimension. Each point in the grid is assigned an occupancy value of '0' (zero) initially and is set to '1' upon the placement of a residue. If the occupancy is 1, then no other residue can be placed on the same grid. Between each pair of horizontal and vertical grids, subgrids or lanes are provided to draw lines connecting the backbone chain in sequential order, as well as to show the hydrogen bond connectivities. Subgrids are also assigned occupancy to avoid overdrawing and the number of such subgrids are user-defineable. STRIPS scans through the PDB file to locate all the secondary structures listed and stores them internally. STRIPS uses two simple placement rules; if the secondary structure is a helix or zero-direction sheet from a bundle, we place the corresponding residues vertically on the next available grid. If it is directional sheet, we place it in the column next to the sheet prior to it, such that the correct hydrogen bonding partners are on the same row and that the direction is maintained with respect to the sheet prior to it.

Secondary structure strips are first sorted in increasing order of residues that begin the strip. The algorithm tries to place all the residues in a protein and since the secondary structure records contain only a subset of residues, the algorithm extends each sorted strip by including residues following the previous strip and preceding the current strip. If the last sorted strip does not end on the last residue of the protein, a new strip is generated to include all residues remaining till the end of the protein. Thus, each extended strip will end exactly at the same residue that the original secondary structure ended, but may begin at different residues. It is important to note that the initial sorting is temporary and only used to determine proper extensions. The original order of secondary structures and their residue ranges as read from the PDB files are retained everywhere.

STRIPS now processes one extended strip at a time in the exact same order they were read from the PDB file. The residues in each record are placed using the rules described above, begining at the left hand corner of the grid. Each strip is placed beginning in the next available column (vertically up or down depending on the nature of the secondary structure and direction), thus introducing a vertical bias for placement. The next available column is advanced to the right by one after placing each of the strip. If the strip is a helix, it is simply placed in the next available column with the residues going upwards. If it is a sheet with a direction, by design the sheet prior to it should have been already placed. Hydrogen bonding partners are placed on the same row and the rest are then filled in, up or down, based on direction. Each residue is represented by a split circle showing the N-terminus with a single color or shade of grey and the other half of the circle could be colored according to the nature of the secondary structure, such as the type of helix or sheet.

The line tracing the protein backbone and hydrogen bonding patterns are superposed on the diagram using the subgrids or lanes. Since each residue has a distinct N- and C-terminal end in the drawing, connectivites are drawn to show the C=O and N-H ends of the hydrogen bond correctly. The rules for drawing these are based on occupancy of the predefined path connecting any two residues which are discussed in great detail in reference 13. If any lane on such a path is occupied, the next path is searched and so on. The drawing of these lines are as important as the placement of the residues and thus proper placement is ensured. The STRIPS algorithm offers a choice of colors and/or line thickness to show a variety of information such as the average strength of the hydrogen bonds or distances or angles. It can also show the information as a time series. The examples shown here are restricted to displaying just the average hydrogen bond strength. More display options are discussed in reference 13.

Results and Discussion

The first example discussed here is a 58 residue protein, bovine pancreatic trypsin inhibitor (BPTI), the coordinates for which were obtained from Brookhaven PDB (*18*). The structure of BPTI has been the subject of several MD studies because it is a small protein which has been fairly well characterized (*8,14,15*). The initial 2-D HBTD generated by STRIPS is given in Figure 1 along with the secondary structure records from the PDB file. There are five original secondary structure strips: 2-7, 47-56, 29-35,18-24 and 45-45. The first strip is a 3_{10} helix, second is a right-handed α-helix and the other three form a single β-bundle. Upon extension, these strips become 1-7, 46-56, 25-35, 8-24 and 36-45 and since there are 58 residues, a new strip was added to the list to include residues 57-58.

The order of the strips are maintained the same as in the PDB file. The two helices are placed in the first two columns and the strand with the zero direction is placed next. The next strand in the bundle is antiparallel to the first, with the hydrogen bond between residues 24 and 29 marking the begining of the sheet. These residues are placed on the same row and subsequent residues in the strips are properly placed so as to maintain the antiparallel direction. The third strand in the bundle is antiparallel to the second and the begining of this sheet is marked by a hydrogen bond between residues 24 and 45. As before, these residues are placed on the same row and other residues of this strand are placed to maintain proper directionality. The last strip is then placed on top of residue 56 to which it is closest in the amino acid sequence. Notice that the residues within each strip are coded using different levels of grey to distinguish various secondary structures specified in the PDB file. At each residue, the shaded hemisphere points towards the N-terminal end of the chain. The line types and thickness reflect the strength of a hydrogen bond.

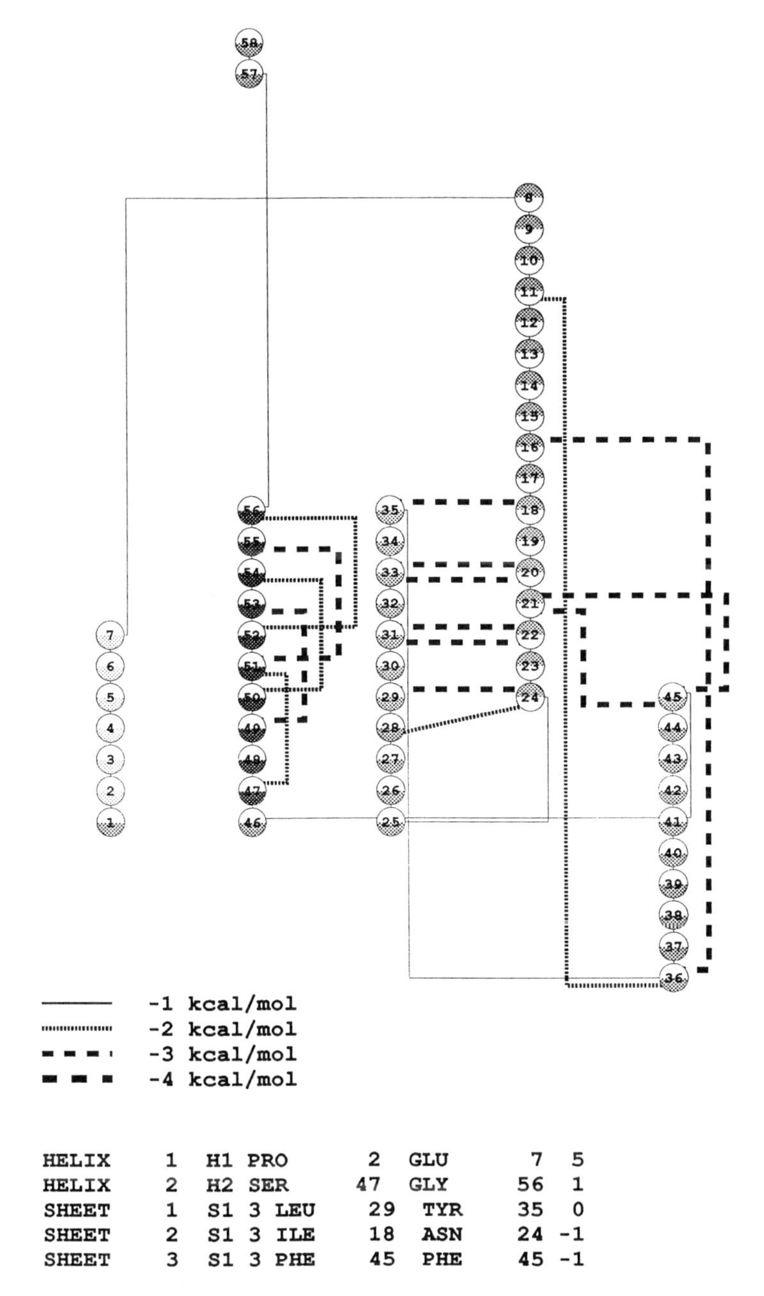

Figure 1. The initial 2-D HBTD of the enzyme BPTI, generated by STRIPS from the HELIX and SHEET records in the PDB file (*18*).

Average hydrogen bond strengths extracted from the first 500 ps of a nanosecond MD run on an aqueous solution of BPTI (*15*) are overlayed on STRIPS-generated 2-D HBTD showing the directionality of each hydrogen bond. The hydrogen bonds extracted from the MD trajectory were chosen using the following criterion: if the distance between the H (from N-H) and O (from C=O) is 2.25 ± 0.75 Å and if the C-O-H angle is 150° ± 30° and if the electrostatic interaction between the N-H and C=O group is -4±3 kcal/mol, we count it as a hydrogen bond. In addition to the geometric criterion, an energy based filter is employed to weed out very weak hydrogen bonds arising from the liberal use of distance and angle tolerance. This criterion did very well in showing hydrogen bonds between the sheets and the right-handed α-helix, but did not pick up any in the N-terminal 3_{10}-helix. However, one can always modify the selection criteria to pick up these hydrogen bonds also. Also, notice that there is a hydrogen bond between the N-H of residue 28 and the C=O of residue 24 that was not present in the secondary structure records of the PDB file. Similarly, the dynamics shows two additional hydrogen bonds, between residues 11 and 36, and between 16 and 36, that were not present in the PDB file.

The 2-D HBTD for BPTI generated by STRIPS captures the hydrogen bonding detail satisfactorily. A manually generated diagram for BPTI already exists (*5*) and is not very different from the STRIPS generated diagram. The major difference is in the length of the strips and in the placement of the right-handed α-helix. The manually generated 2-D HBTD has a total of five strips: 1-15, 27-37, 16-26, 38-46 and 47-58. They appear in the exact order specified above. Since the PDB file usually places all helices first followed by all sheets, the origin of the discrepancy in the position of the right-handed α-helix is obvious. The difference in the length of the strips arise from the fact that the extension algorithm used by STRIPS always terminates a given strip at the end of a secondary structure whereas the manually generated diagram has no such restriction. In Figure 2, we show the modified 2-D HBTD for BPTI that resembles closely, the manually generated diagram along with the new secondary structure records. The difference in the secondary structure records between Figures 1 and 2 are minor, and this example shows how STRIPS can be used to get a good initial guess that can then be easily manipulated to get the desired picture.

The second example is HIV-1 protease, a dimeric protein comprising 99 residues per chain. The enzyme is important in polyprotein processing in the life cycle of the AIDS virus and has been the focus of several recent studies (*9,17*). The 2-D HBTD produced from the PDB file pdb9hvp.ent (*16*) is presented in Figure 3. For simplicity, we show only one of the two chains and the associated hydrogen bonding patterns. The inter-chain hydrogen bonds play an important role in the structure and dynamics of this protein, and the extensions to STRIPS necessary to show these can be found in reference 13. HIV-1 protease has one right-handed α-helix, and β-bundle with a total of eight β-strands. The secondary structure records present in the PDB file have duplication, in that the residues 65-66 appear twice in two different strands within the same bundle. In such cases, STRIPS uses the strip containing the last occurance of residues 65-66 to place the entire strip. This is obvious from Figure 3, where the strip from 50-66 is placed on the sixth column. This causes some of the hydrogen bond lines to go across five columns and can lead to confusion, however, the problem can be fixed by changing the order of secondary structure records in the

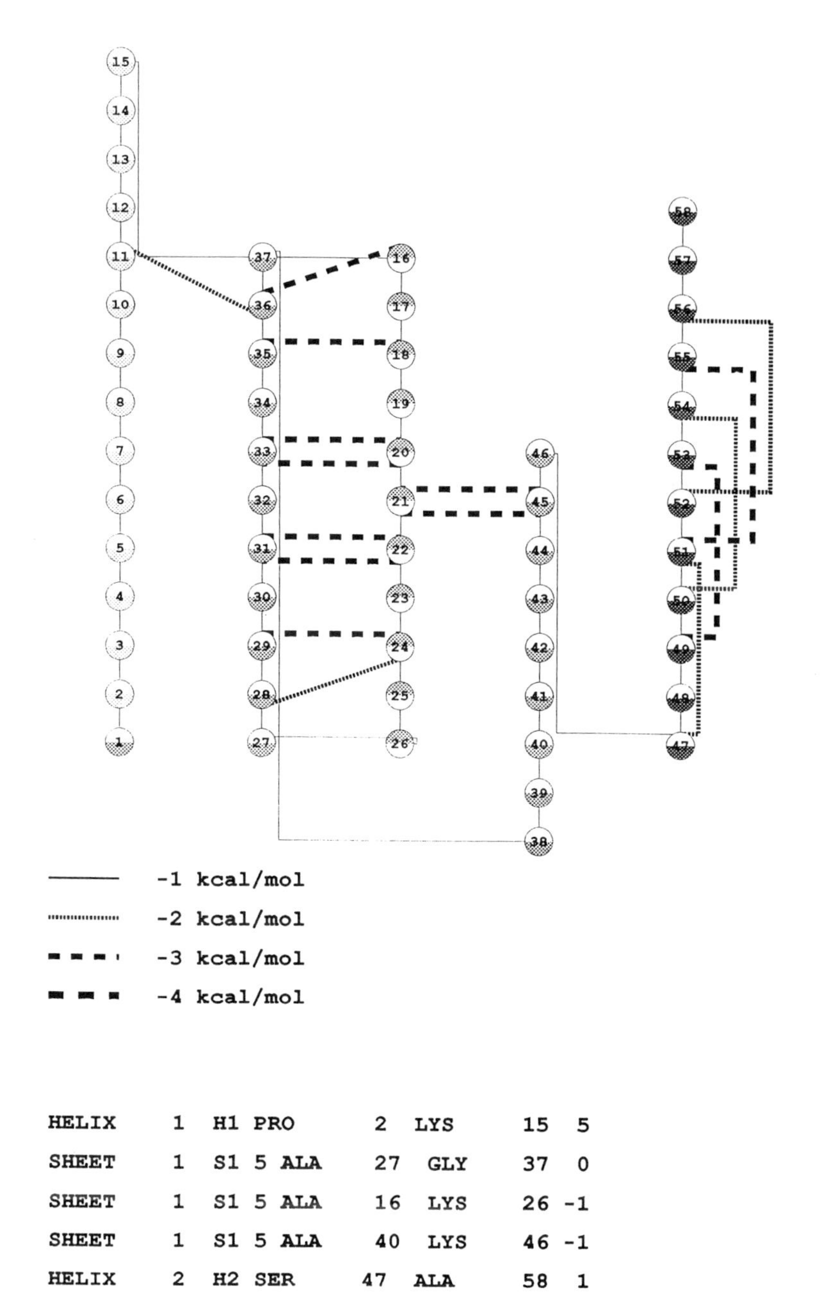

HELIX	1	H1		PRO	2	LYS	15	5
SHEET	1	S1	5	ALA	27	GLY	37	0
SHEET	1	S1	5	ALA	16	LYS	26	-1
SHEET	1	S1	5	ALA	40	LYS	46	-1
HELIX	2	H2		SER	47	ALA	58	1

Figure 2. Customized 2-D HBTD of BPTI obtained after slight modification of the HELIX and SHEET records in the PDB file.

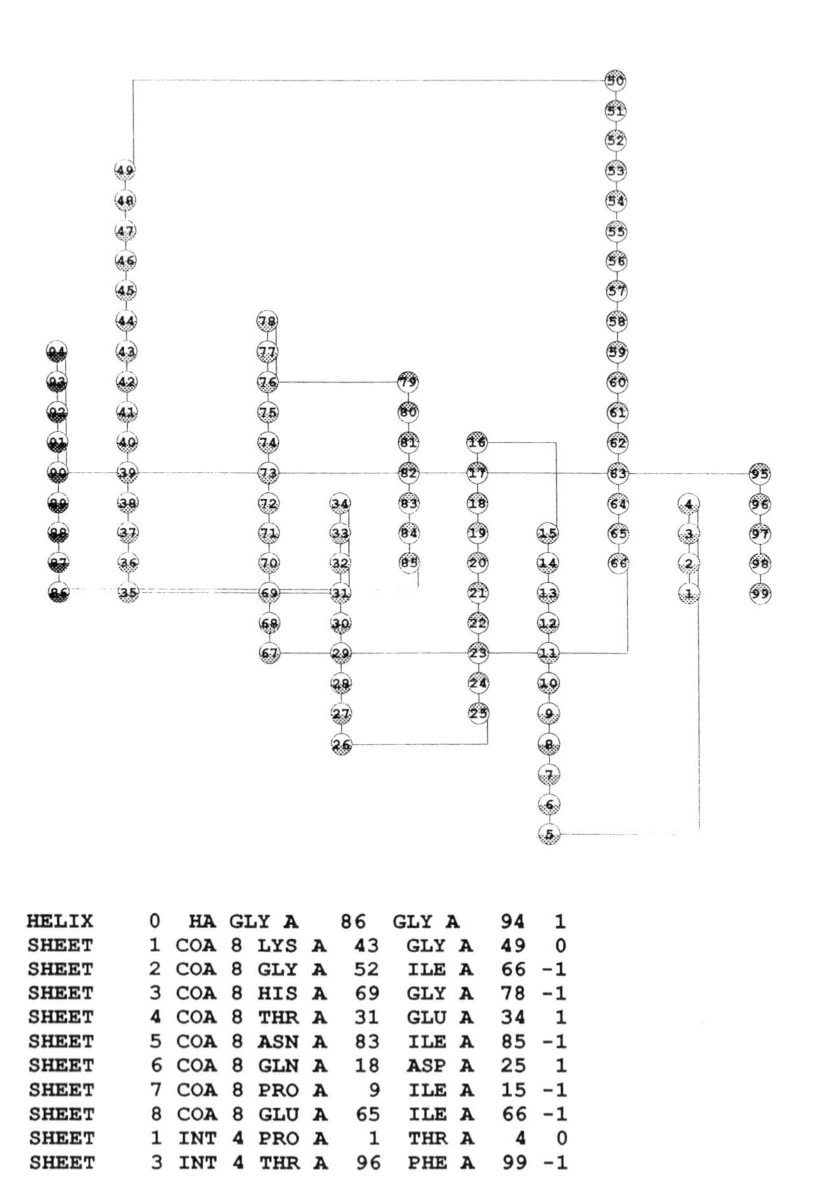

```
HELIX    0  HA GLY A   86   GLY A   94  1
SHEET    1 COA 8 LYS A  43   GLY A  49  0
SHEET    2 COA 8 GLY A  52   ILE A  66 -1
SHEET    3 COA 8 HIS A  69   GLY A  78 -1
SHEET    4 COA 8 THR A  31   GLU A  34  1
SHEET    5 COA 8 ASN A  83   ILE A  85 -1
SHEET    6 COA 8 GLN A  18   ASP A  25  1
SHEET    7 COA 8 PRO A   9   ILE A  15 -1
SHEET    8 COA 8 GLU A  65   ILE A  66 -1
SHEET    1 INT 4 PRO A   1   THR A   4  0
SHEET    3 INT 4 THR A  96   PHE A  99 -1
```

Figure 3. 2-D HBTD of HIV-1 Protease generated by STRIPS from the original HELIX and SHEET records in the PDB file (*16*).

PDB file. The modified diagram is shown in Figure 4. The average hydrogen bond strengths overlayed on the 2-D HBTD are taken from a 100 ps MD simulation on an aqueous solution of the HIV-1 protease dimer, using the GROMOS force field. The ease of use and utility of the STRIPS algorithm is immediately obvious.

The option to manipulate the original diagram easily is essential because the hydrogen bond information appearing in the PDB file does not necessarily define all secondary structural features resulting from an MD simulation. If the MD simulation shows flexibility in a given secondary structure that stretches the original strip by some residues, the user has the ability to manipulate the original diagram to get the desired final diagram. STRIPS on the other hand has an option to take the MD derived secondary structure records as the initial input to produce the 2-D HBTD. We are in the process of testing our implementation of Kabsch and Sander algorithm to produce the secondary structure records from an MD simulation.

The 2-D HBTD can get complicated for large proteins to be of any use. In these cases, STRIPS can be utilized to draw only certain selected secondary structures (*19*). Fluctuations of individual hydrogen bonds within a given α-helix or β-sheet can be analyzed and used to compare with simulations of mutants, complex etc. The strength of the hydrogen bond during the course of the trajectory can be analyzed by drawing the 2-D HBTD separately for different segments of the MD trajectory. It is impossible to list all possible applications of the 2-D HBTD in the analysis of MD simulations, but the ease of use and ability to manipulate the topology diagrams can support many creative applications.

Conclusion

We have shown that the algorithm STRIPS can be used to generate a consistent and useful two-dimensional representation of the complex structure of a protein, on which several hydrogen bond and secondary structural details can be overlayed. The algorithm can utilize the secondary structure records in the Brookhaven PDB files or generate them from atomic coordinates in the PDB file using the kabsch and Sanders algorithm. Alternatively, STRIPS can generate them from a list of hydrogen bonds obtained from the MD trajectory. STRIPS produces graphical output relatively faster, provides an easy mechanism to manipulate and is less cumbersome than manual generation. We are now in the process of extending the algorithm to generate topology diagrams to show hydrogen bonding interactions in protein-DNA complexes.

Acknowledgments

The authors hereby acknowledge funding from the National Institutes of Health, Grant # GM37909 awarded to DLB and Grant # RR07885-01 awarded to DLB and GR and grants of computer time and staff support from the Pittsburgh Supercomputing Center and the Frederick Biomedical Supercomputing Center of the Frederick Cancer Research and Development Center. The authors wish to thank Dr. R. Nirmala for providing the MD trajectory of BPTI and Dr. W. Harte for providing the MD trajectory for HIV-1 protease.

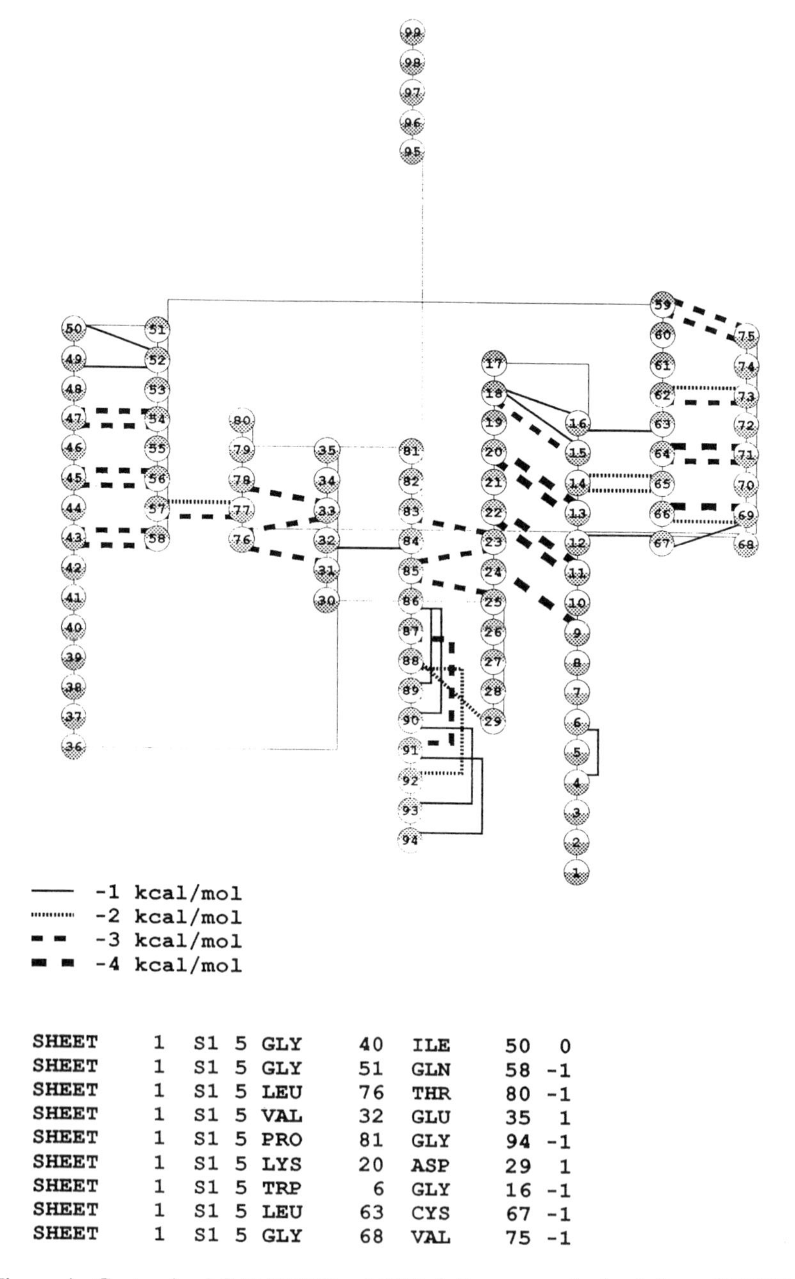

SHEET	1	S1	5	GLY	40	ILE	50	0
SHEET	1	S1	5	GLY	51	GLN	58	-1
SHEET	1	S1	5	LEU	76	THR	80	-1
SHEET	1	S1	5	VAL	32	GLU	35	1
SHEET	1	S1	5	PRO	81	GLY	94	-1
SHEET	1	S1	5	LYS	20	ASP	29	1
SHEET	1	S1	5	TRP	6	GLY	16	-1
SHEET	1	S1	5	LEU	63	CYS	67	-1
SHEET	1	S1	5	GLY	68	VAL	75	-1

Figure 4. Customized 2-D HBTD of HIV-1 Protease, obtained from STRIPS, after slight modification of the HELIX and SHEET records in the PDB file.

Literature Cited

(1) Kabsch, W.; Sander, C. *Biopolymers* **1983**, *22*, 2577.
(2) Saenger, W.; Jeffrey, G. A. *Hydrogen Bonding in Biological Molecules*, Springer-Verlag, 1991.
(3) Richardson, J. S. *Adv. Protein. Chem.* **1981**, *34*, 167.
(4) Carson, M. Ribbons Ver. 2.0, University of Alabama.
(5) Wlodawer, A.; Walter, J.; Huber, R.; Sjolin, L. *J. Mol. Biol.* **1984**, *180*, 301.
(6) Harte, W. E.; Swaminathan, S.; Mansuri, M. M.; Martin, J. C.; Rosenberg, I. E.; Beveridge, D. L. *Proc. Natl. Acad. Sci.* **1990**, *87*, 8864.
(7) van Gunsteren, W. F.; Berendsen, H. J. C. *Angew. Chem. Int. Ed. Engl.* **1990**, *29*, 992.
(8) Swaminathan, S.; Ravishanker, G.; Beveridge, D. L.; Lavery, R.; Etchebest, C.; Sklenar, H. *Proteins: Struct. Func. and Genet.* **1990**, *8*, 179.
(9) Harte, W. E.; Swaminathan, S.; Beveridge, D. L. *Proteins: Struct. Func. & Genet.* **1992**, *13*, 175.
(10) Bernstein, F. C.; Koetzle, T. F.; Williams, G. J. B.; Meyer, E. F.; Brice, M. D.; Rodgers, J. R.; Kennard, O.; Shimanouchi, T.; Tasumi, M. *J. Mol. Biol.* **1977**, *112*, 535.
(11) Atomic Coordinate and Bibliographic Entry Format Description, Protein Data Bank, 1985.
(12) Ravishanker, G.; Beveridge, D. L. MD Toolchest Ver. 2.0, Wesleyan University, Middletown, CT 06459.
(13) Ravishanker, G.; Vijayakumar, S.; Beveridge, D. L. *manuscript in preparation for submission to J. Comp. Chem.*
(14) Levitt, M.; Sharon, R. *Proc. Natl. Acad. Sci.* **1988**, *85*, 7557.
(15) Nirmala, R.; Ravishanker, G.; Beveridge, D. L. *manuscript in Preparation.*
(16) Erickson, J.; Neidhart, D. J.; Van Drie, J.; Kempf, D. J.; Wang, X. C.; Norbeck, D. W.; Plattner, J. J.; Rittenhouse, J. W.; Turon, M.; Wideburg, N.; Kohlbrenner, W. E.; Simmer, R.; Helfrich, R.; Paul, D. A.; Knigge, M. *Science* **1990**, *249*, 527.
(17) Harte, W. E.; Beveridge, D. L. *J. Am. Chem. Soc.* **1993**, *115*, 1231.
(18) Eigenbrot, C.; Randal, M.; Kossiakoff, A. A. PDB entry 'pdb9pti.ent', October 15, **1992**.
(19) Vijayakumar, S.; Ravishanker, G.; Beveridge, D. L. Proceedings of the Symposium on Hydrogen Bonding, Abstract# 27, 206th National Meeting of the American Chemical Society, August 1993.

RECEIVED June 2, 1994

Applications to Molecules and Polymers

Chapter 14

Role of Specific Intermolecular Interactions in Modulating Spectroscopic Properties of Photoreactive Compounds Bound to Proteins

Marshall G. Cory, Jr., Nigel G. J. Richards, and Michael C. Zerner

Department of Chemistry, University of Florida, Gainesville, FL 32611

The rational design of photosubstrates will require a detailed understanding of the effects of protein environments upon the spectroscopic properties of small molecule ligands. We here describe preliminary INDO/S calculations upon two models of the complex between a tripeptide containing a 3-amino-isochromene moiety which represents a potential photocaged amide, and the aspartic proteinase, rhizopuspepsin. Hydrogen bonding interactions appear to play an important role in causing a hypochromic (red) shift of approximately 19 nm (1800 cm^{-1}) in the electronic absorption spectrum of the tripeptide chromophore, a substantial difference compared to the absorption maximum associated with the n-->π* transition of the tripeptide in the gas-phase. In addition, these calculations confirm that charged groups within the protein are likely to modulate low-energy electronic transitions in the photosubstrate. Our results demonstrate that when computational methods are used to predict the spectroscopic and photochemical reactivity of photocaged compounds, specific interactions between the protein environment and the photosubstrate, such as hydrogen bonds, must be included in the calculation.

The demonstration that synchrotron X-ray sources allow sufficiently rapid data collection for protein crystallography has opened the possibility of directly observing the structural features of enzymes as they catalyze reactions in the solid state (*1*). For example, interpretable electron density maps at atomic resolution have been obtained for the enzyme glycogen phosphorylase by exposing the crystal to an X-ray beam for only 0.8 second (*2*). With the solution of the technical problems in data collection, the initiation of reaction in all of the molecules in the crystal at a well-defined instant has become a major hurdle to obtaining snapshots of enzyme-substrate complexes at various stages of reaction (*3*). For simple reactions, experimental strategies involving either rapid jumps in temperature (*4*) or pH (*5*) have been reported as a means for achieving temporal synchronization of reactions throughout the crystal. Indeed, for the enzyme trypsin, identification of the water molecule involved in acylenzyme hydrolysis was achieved using a pH jump to initiate reaction (*6*) although the detailed interpretation of this experiment remains the subject of some debate (*7*). An alternative approach has been to obtain compounds which represent caged substrates (often termed "photosubstrates") for the reaction of interest (*8*), or photolabile enzyme derivatives (*9*,

0097–6156/94/0569–0222$08.00/0

10) which can be released through the absorption of high-intensity light. Such compounds allow the solution of two problems associated with time-resolved X-ray crystallography (*11*). First, the unreactive form of the photosubstrate can be designed to possess high affinity for the enzyme active site, resulting in a large occupancy ratio in crystals of the photosubstrate/enzyme complex. Second, photochemical release of the necessary functional group means that reaction is initiated in all of the protein molecules within a narrow time range, leading to coherence in the chemical reactions occurring in every complex in the crystal. Such a temporal coherence potentially allows the detection of long-lived reaction intermediates and/or the complex between protein and reaction products. While the use of 'caged' enzymes in time-resolved X-ray crystallography has not yet yielded data of sufficient quality to be useful (*12*), the successful use of photosubstrates in allowing direct observation of the complex between Ha-*ras* p21 protein and GTP has been reported (*13*).

The development and application of further examples of photosubstrates to study enzyme-catalyzed reactions involving more complex functional groups might be aided by the use of computational strategies for the evaluation of candidate structures. There are a number of stringent design criteria which must be satisfied by such compounds (*14*). First, any photosubstrate must be closely related, in terms of its gross molecular structure, to the natural substrate so that it has a high affinity for the enzyme active site. Second, the photosubstrate must possess sufficient functionality that the "caged" portion is located correctly within the active site. The photosubstrate must also absorb light at a wavelength at which the protein, the surrounding solvent molecules in the crystal, and the photoproduct are transparent. Finally, the photochemical reaction to unmask the functional group of interest should proceed, in reasonable quantum yield, from the S_1 state to avoid the formation of long-lived radical intermediates which might undergo alternative reactions with the functional groups in the protein.

The inclusion of environment-dependent effects remains a problem of interest in theoretical calculations (*15*). While efforts in this area have focussed upon the influence of environment on structural preferences and energetics (*16, 17*), the modulation of photochemical and spectroscopic properties of ligands by their environment remains much less explored (*18-20*). INDO/S semi-empirical techniques (*21-23*) have been used successfully to study the spectroscopy of a large number of compounds, including, most recently, the photochemical properties of chlorophyll in the photosynthetic reaction center (*24, 25*). In this paper we outline the results of preliminary INDO/S calculations on model complexes between a potential photocaged tripeptide incorporating a 3-amino-isochromene moiety and the aspartic proteinase, rhizopuspepsin (*26*). These studies support the hypothesis that specific, intermolecular hydrogen bonding interactions will be important in modulating the absorption properties of this ligand when bound within the protein.

Isochromenes as Potential Photocaged Amides

Aspartic proteinases represent a particularly important class of enzymes (*27*) which catalyze the hydrolysis of amide bonds. Many of this family of proteases are involved in critical biological roles, including viral assembly (*28*) and the regulation of blood pressure (*29*). Significantly, it is thought that protein loops fold onto the peptide substrate and play a role in mediating catalysis. Support for such a hypothesis has been obtained by site-specific mutagenesis of conserved residues within these regions which affected the rate of the hydrolysis reaction in the mutant enzymes (*30*). The development of photocaged amides allowing the detailed study of the conformational changes occurring within these loops during catalysis might therefore provide new insight into their functional role. We have recently proposed that 3-aminoisochromenes, of general structure **1**, might represent photocaged amides (Figure 1) (*31*) since irradiation of such compounds should effect an electrocyclic reaction to generate an amide bond. Trapping of the reactive intermediate **2** by a water molecule should then yield a stable substrate which can be hydrolyzed by the aspartic proteinase. Our hypothesis is precedented by studies upon the photochemical reactivity of 3-phenyl-isochromenes in which the existence of such a ring-opened intermediate was demonstrated experimentally (*32, 33*). We have also shown that INDO/S calculations can reliably reproduce the spectroscopic and photochemical behavior observed experimentally for this class of compounds (*31*). These previous calculations employing a self-consistent reaction field (SCRF) (*20, 34-37*) to represent solvent, indicated that the inclusion of explicit hydrogen bonds between solvent and the oxygen atom of the isochromene chromophore was necessary to calculate the UV-visible absorption bands correctly (*31*). The requirement that specific interactions must be included in an SCRF model is not too surprising (*36, 37*). At the SCF level, the reaction field stabilizes the ground state of a molecule through its response to any asymmetric distribution of charge, by an amount ΔE_o given, at the dipolar level, by:

$$\Delta E_o = - [\, g(\varepsilon).\langle \mu_o \rangle^2 \,] / 2$$

where $g(\varepsilon)$ is the Onsager reaction factor (*37*), ε is the dielectric constant of the solvent and μ_o is the dipole moment of the ground state structure. In the case of an amino-isochromene, this term would correspond to a dipole-image dipole interaction. Such a description is incomplete when solvent molecules can interact by hydrogen bonding. Not only is the dipole moment associated with the ground state of the molecule diminished, but the orbital energies are also specifically affected by the solvent. Hydrogen-bond interactions must therefore be explicitly introduced into the model through the use of a super-molecule, and it is the optimized ground state structure of the super-molecule which must be used to obtain the UV-visible spectrum in the INDO/S-SRCF-CIS calculation (*36, 38*). Electronic excitation in a molecule occurs in 10^{-15} to 10^{-18} seconds, and solvent nuclei cannot change their positions during the absorption process. However, electronic polarization of the solvent can occur within this timeframe. Inclusion of this effect in the SCRF calculation to first order in perturbation theory (*37*) yields the following expression (model B in reference 37):

$$\Delta E^*_{abs} = - [\, g(\varepsilon).\langle \mu_o \rangle.\langle \mu_* \rangle] / 2 + [\, g(\eta^2).\langle \mu_* \rangle.(\langle \mu_o \rangle - \langle \mu_* \rangle)] / 2$$

Here, ΔE^*_{abs} is change in energy of the excited state of the molecule relative to the gas-phase, immediately after excitation, μ_* is the dipole moment of the excited state, $g(\eta^2)$ is the Onsager reaction factor and η is the refractive index of the solvent. Thus the change in the energy $\Delta\Delta E$ required for the electronic transition in the SCRF model, relative to that in the gas-phase, is given by:

$$\Delta\Delta E = - [\, \{g(\varepsilon).\langle \mu_o \rangle + g(\eta^2).\langle \mu_* \rangle\}.(\langle \mu_o \rangle - \langle \mu_* \rangle)] / 2$$

In general, this implies that if the ground state dipole is greater than that of the excited state, the absorption band will be blue-shifted, and vice versa. In the 3-amino-isochromene structure (**1**, Figure 1) the two lowest lying excitations are n-->π^*, prinicipally from the lone pairs of the nitrogen atom attached to the isochromene moiety to the $e_g(\pi^*)$ molecular orbitals of the benzene ring. The degeneracy of the $e_g(\pi^*)$ is split by the presence of the fused heterocyclic ring. Both of these excitations lead to a large change in the electronic distribution in the chromophore, resulting in the dipole moment of the excited state being larger than that of the ground state structure. By itself, this would lead to a red shift of the associated absorption bands, but specific interactions involving the nitrogen lone pair, were they to be present, might also modulate the energy of the electronic excitations. In order to establish the magnitude of such environmental effects in designing photosubstrates, we have investigated the perturbation of the absorption properties of a model peptide incorporating the 3-amino-isochromene moiety when bound to the active site of rhizopuspepsin. Since it was likely that specific hydrogen bonds would be formed between the protein and the heteroatoms in the peptide, we anticipated that there would be significant shifts in the absorption properties of the heterocyclic moiety relative to those in the gas-phase. Tripeptide **3** (Figure 2) was chosen as a model compound for these calculations, as (1) this represented the smallest unit which might be accommodated within the binding pocket of rhizopuspepsin, and (2) the complex between rhizopuspepsin and the inhibitor, **4**, which incorporated a similar structural fragment, had been determined crystallographically (*39, 40*).

Modeling the Complex between Rhizopuspepsin and the Model Photosubstrate 3

Two initial models of the tripeptide **3** within the proteinase binding pocket were constructed using the X-ray crystal coordinates of the complex between the inhibitor **4** (Figure 2) and rhizopuspepsin (*40*) obtained from the Brookhaven Protein Database (*41*) (2APR). These structures, **5** (Figure 3A) and **9** (Figure 3B) differed in the protonation state of Asp-35 as the binding of tripeptide **3** within rhizpuspepsin might alter the pKa of the ionizable, sidechain carboxylic acid group of this residue. Thus, Asp-35 was ionized in **5** and in its neutral form in **9**. This catalytically active residue is usually presumed to exist in its protonated form so as to be able to act as a general acid in the initial step of the amide hydrolysis reaction (*42*). The inhibitor structure **4** was graphically modified so as to yield the tripeptide structure **3** with minimal modification of all rotatable dihedral angles. The all-atom structures ($\approx$ 3000 atoms) of both complexes **5** and **9** were energy minimized using a truncated Newton-Raphson optimization algorithm (*43*), as implemented in BATCHMIN V3.5 (*44*), until the total residual RMS gradient, representing the forces on the atoms, was less than 0.01 kJ/Å (*45*). Both empirical force field calculations employed the OPLSA potential energy function and parameters (*46*) augmented by the GB/SA continuum solvation potential for water (*47*). Simulated annealing was then used to ensure that tripeptide **3** was optimally docked into the protein in both models of the peptide/protein complex (*48*). The INDO/S calculations employed structures which comprised all atoms in the tripeptide **3** together with the atoms in the following residues: Asp-35, Thr-36, Gly-37; Tyr-77, Gly-78; Asp-218, Thr-219, Gly-220, Thr-221. In addition, each of the three protein segments (35-37; 77-78; 218-221) were modeled as their N-terminal amides and C-terminal methylamides. The location of the additional atoms in the terminal amide groups were defined by the coordinates of the carbonyl group and $C\alpha$ of the preceding (34, 76 and 217), and the NH and $C\alpha$ of the following (37, 79 and 222), residues in the protein structure. Hence, models of a suitable size for the INDO/S calculations ($\approx$ 200 atoms) on the potential photosubstrate **3** when bound to rhizopuspepsin were obtained in which all of the critical intermolecular interactions were present (Figures 3A and 3B). Given the difference in the ionization state of the Asp-35 sidechain in these models, after energy minimization of the complexes,

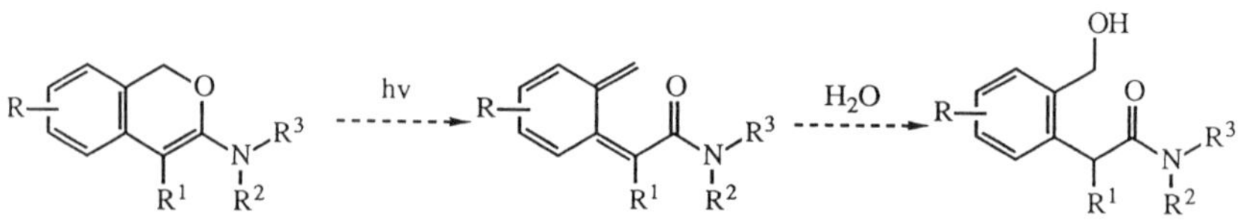

Figure 1. Potential photochemical ring-opening reaction of 3-aminoisochromenes to release a masked amide bond.

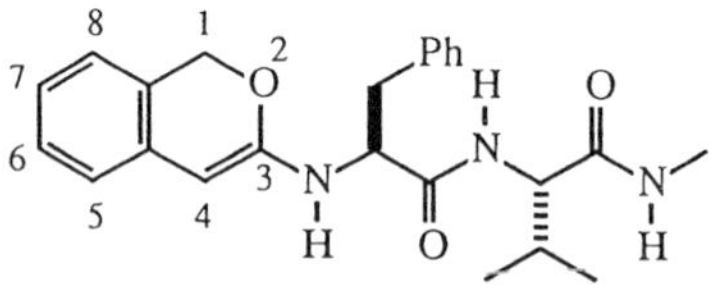

Figure 2. Structures of the tripeptide photosubstrate **3** used in these calculations, and the 'reduced' amide inhibitor **4** used to build the initial model of the complex between **3** and rhizopuspepsin. Small numbers indicate the numbering system for atoms within the isochromene ring.

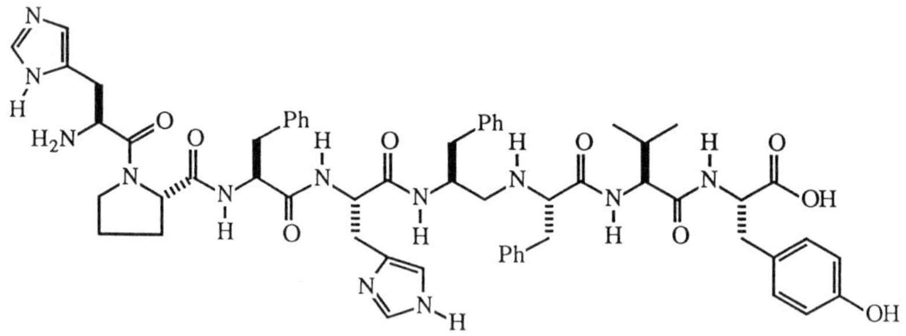

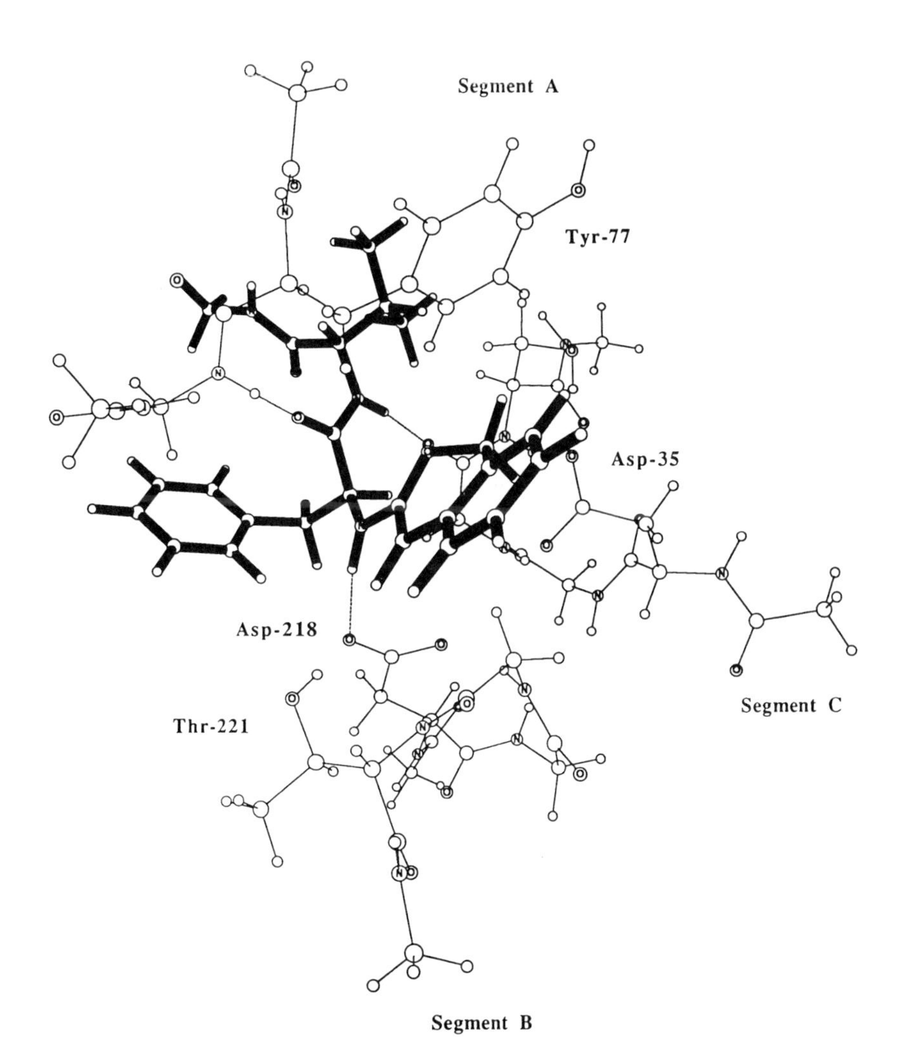

Figure 3. Schematic representations of the subsite models of the complex between tripeptide **3** and the active site residues of rhizopuspepsin. In both figures bonds, intermolecular hydrogen bonding interactions are shown as dotted lines, all bonds in the protein fragments are drawn as single lines irrespective of bond order, and the tripeptide structure is drawn using "thick-lines". Nitrogen and oxygen atoms are represented by the symbols N and O, respectively. (A) Complex **5** in which Asp-35 is in the ionized form.

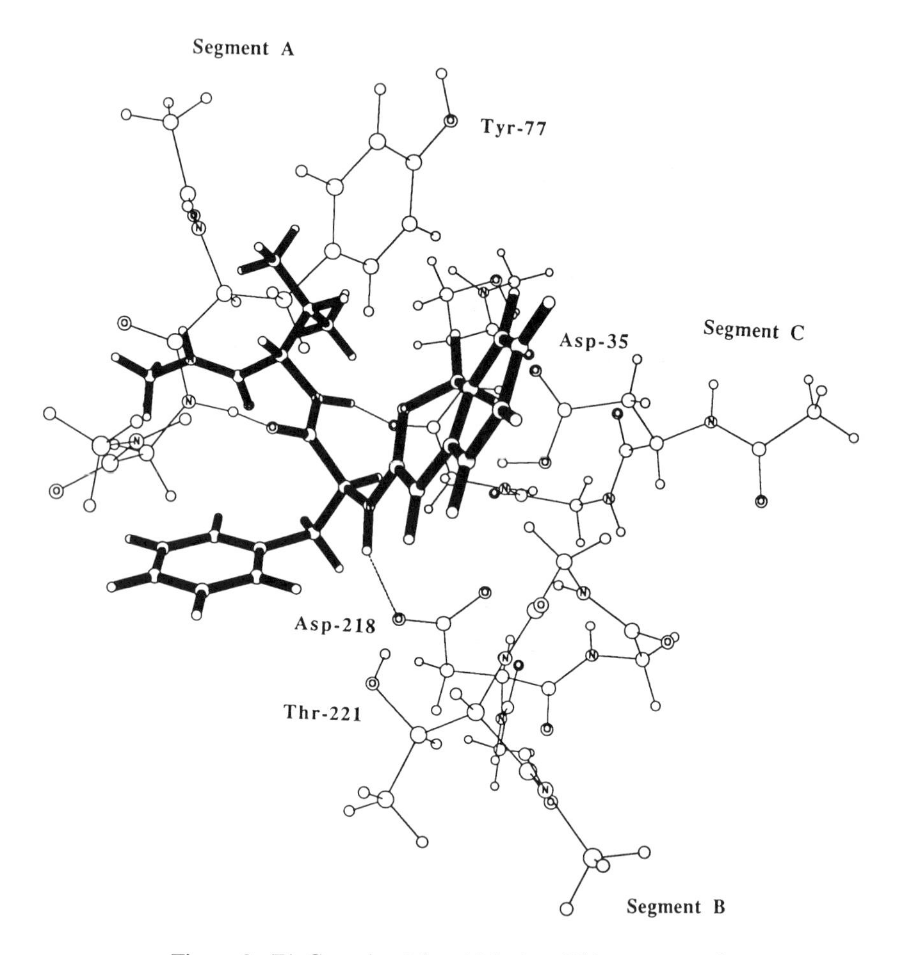

Figure 3. (B) Complex **9** in which Asp-35 is protonated.

tripeptide **3** was bound in two non-identical, although similar, conformations (**3-a** and **3-b**). All subsequent calculations upon the *in vacuo* absorption properties of **3** were therefore carried out upon both **3-a** and **3-b** so that differences in the absorption spectra arising from their specific interactions with the protein were correctly determined.

In the complex **5** in which Asp-35 was modeled as the carboxylate, three hydrogen bonds were detected between the tripeptide in conformation **3-A** and the protein fragments in the optimized structure (Figure 3A). Of particular interest was the hydrogen bond between the NH group of the substituted isochromene and the sidechain of Asp-218, which represented a direct intermolecular interaction that we anticipated would perturb the spectroscopic and photochemical properties of the isochromene group. The aromatic rings of Tyr-77 and the isochromene moiety were in close proximity, raising the possibility that charge transfer between the two π-systems might also be involved in modulating the absorption properties of the isochromene unit when bound to the protein (*49*). For the complex **9** in which Asp-35 was in its protonated, neutral form (Figure 3B), four hydrogen bonds were present in addition to the potential interaction between the aromatic sidechain of Tyr-77 and the isochromene moiety. While three of the hydrogen bonds in **9** were identical to those in **5**, an additional, weak, hydrogen bond appeared to be present between O-2 in the isochromene unit (Figure 2) and the carboxylic acid group of Asp-35 (Figure 3B).

Our initial INDO/S calculations explored the effects of the protonation state of Asp-35 upon the spectroscopic properties of **3** when bound within the protein. Single-point SCF calculations on the two model structures, **5** and **9**, were carried out using ZINDO (*50*) executing on an IBM RS/6000 PowerStation Model 580, in order to obtain the molecular orbitals used to construct the spin-adapted many-electron basis. In these, and all subsequent, calculations a minimal basis set was employed representing only the valence orbitals on each atom. Configuration interaction calculations over singly excited configurations (CIS) were then performed to obtain the UV-visible absorption spectra for these model complexes (Tables I and II). *In vacuo* CIS calculations were also performed for the two conformations of the tripeptide (**3-a** and **3-b**) so as to evaluate the change in the spectroscopic properties of isochromene due to specific interactions with the protein (Tables I and II). For the isolated ligand **3-a**, in the conformation present in peptide/protein complex **5**, there was a strong absorption at 312.4 nm corresponding to an S_1 state, comprising 79% of the HOMO-->LUMO transition (*38*) and 15% of the HOMO-->LUMO+3 transition within the MO picture. For **3-b**, the cognate absorption occured at 313.9 nm. While the difference in energy between these two calculated bands is unimportant, the presence of two low energy bands with significant oscillator strength indicated that the S_1 and S_2 states of the isolated ligand, for both conformations, were close in energy. Variation of the active CIS space, beyond 10-up/10-down, had little effect upon the calculated absorptions or intensities for the model complexes, suggesting that all of the relevant protein and ligand orbitals had been included in our initial computations. This 'control' series of independent CIS calculations employed orbitals from 20-up/20-down to 45-up/45-down centered about the HOMO-LUMO gap. In both model complexes **5** and **9**, the low-energy, electronic transitions cognate to those calculated for the ligand *in vacuo* were shifted to lower energy by about 19 nm (1800 cm^{-1}) and 13 nm (1485 cm^{-1}) for **3-a** and **3-b**, respectively (Tables I and II). The difference in the magnitude of the red shift also appeared to be related to the ionization state of Asp-35. In the S_1 excited state, **3-a** possesses a large dipole due to transfer of charge from the 3-amino substituent to the benzene ring of the isochromene group. Electrostatic stabilization by the charged sidechain should therefore play a direct role in modulating the isochromene absorption spectrum. While it is possible that the shift also results from modification of the MO energies associated with the S_0 state, no evidence for this was observed in

Table I Major absorption bands of tripeptide 3 in a series of models of its complex with rhizopuspepsin

Structure[a] (CI)	State	Absorption λ (nm)	Absorption ν (1000 cm^{-1})	Oscillator Strength	Atoms	Basis Functions
5	S1	331.6	30.2	0.56	200	518
(35x35)	S2	326.8	30.6	0.04		
		.	.	.		
	S26	233.1	42.9	0.23 (x)		
	S27	230.7	43.4	0.19 (y)		
6	S1	332.1	30.1	0.59	189	486
(35x35)	S2	327.6	30.5	0.05		
		.	.	.		
	S17	233.1	42.9	0.25 (x)		
	S19	217.8	45.9	0.09 (y)		
7	S1	332.7	30.0	0.63	160	415
(20x20)	S2	328.8	30.4	0.04		
		.	.	.		
	S13	227.8	43.9	0.36 (x)		
	S19	217.5	45.9	0.09 (y)		
8	S1	327.8	30.5	0.30	111	285
(20x20)	S2	320.5	31.2	0.39		
		.	.	.		
	S14	229.6	43.6	0.20 (x)		
	S18	215.0	46.5	0.07 (y)		
3-a	S1	312.4	30.5	0.51	59	149
(20x20)	S2	308.9	31.2	0.13		
		.	.	.		
	S8	225.8	43.6	0.12		
	S9	224.9	44.5	0.21 (x)		
	S12	213.7	46.8	0.08 (y)		

[a]5; Figure 3A: **6**; Figure 3A, Tyr-77 replaced by alanine: **7**; Figure 3A minus peptide segment A: **8**: Figure 3A minus peptide segments A and C. In all models of the complex, the sidechain of Asp-35 of the protein is modeled as the carboxylate anion. Conformation **3-a** is identical to that present in the enzyme-substrate complexes **5**-**8**, and was used to calculate the gas-phase absorption spectrum. The first excited state is always reported. Numbers in brackets refer to the size of the CI (orbitals up/orbitals down). Only those excited states with significant oscillator strength are given subsequently.

Table II Major absorption bands of tripeptide 3 in a series of models of its complex with rhizopuspepsin

Structure[a] (CI)	State	Absorption λ (nm)	Absorption ν (1000 cm^{-1})	Oscillator Strength	Atoms	Basis Functions
9	S1	327.6	30.5	0.36	201	519
(35x35)	S2	319.2	31.3	0.27		
		.	.	.		
	S24	230.8	43.3	0.22 (x)		
	S25	229.6	43.6	0.11 (y)		
10	S1	327.3	30.5	0.36	190	487
(35x35)	S2	319.3	31.3	0.33		
		.	.	.		
	S20	225.9	44.3	0.29 (x)		
	S24	215.7	46.4	0.08 (y)		
11	S1	328.7	30.4	0.32	161	416
(20x20)	S2	325.9	30.7	0.03		
	S3	320.7	31.2	0.36		
		.	.	.		
	S14	225.9	44.3	0.30 (x)		
	S19	216.8	46.1	0.08 (y)		
12	S1	327.5	30.5	0.30	112	286
(20x20)	S2	320.2	31.2	0.39		
		.	.	.		
	S14	229.2	43.6	0.21 (x)		
	S18	214.8	46.5	0.07 (y)		
3-b	S1	313.9	31.9	0.54	60	150
(20x20)	S2	309.9	32.3	0.11		
		.	.	.		
	S9	225.5	44.3	0.30 (x)		
	S12	213.8	46.8	0.08 (y)		

[a]9; Figure 3B: **10**; Figure 3B, Tyr-77 replaced by alanine: **11**; Figure 3B minus peptide segment A: **12**: Figure 3B minus peptide segments A and C. In all models of the complex, the sidechain of Asp-35 of the protein is modeled as the as the neutral, carboxylic acid. Conformation **3-b** is identical to that present in the enzyme-substrate complexes **9-12**, and was used to calculate the gas-phase absorption spectrum. The first excited state is always reported. Numbers in brackets refer to the size of the CI (orbitals up/orbitals down). Only those excited states with significant oscillator strength are given subsequently.

these calculations. Visualization of the molecular orbitals involved in these electronic transitions (*51*) indicated that the excitation was exclusively located on the isochromene moiety of **3**, and probably corresponded to that already shown to be involved in the desired ring-opening reaction for 3-phenylisochromenes (*31*). The energy splitting of these two states also appeared sensitive to the protein environment. For the two ligand conformations, **3-a** and **3-b**, in the gas-phase, S_1 and S_2 were split by 400 cm^{-1}, while the energy splitting in **5** was smaller than in **9** (400 cm^{-1} and 800 cm^{-1}, respectively). Equally, the partitioning of oscillator strength between S_1 and S_2 was affected by the presence of the protein. For both conformations of the isolated ligand, the ratio of the oscillator strengths was about 4:1. On the other hand, although in both **5** and **9**, the oscillator strengths of S_1 and S_2 summed to approximately 0.65, our calculations indicated that the low-energy region of the absorption spectrum of the 3-amino-isochromene in complex **5** should only consist of comprise a single peak. This contrasts with the case of **9** which consists of two peaks of equal intensity. These results suggest that it might be possible to use the electronic absorption properties of bound ligands to distinguish between the ionization states of groups within proteins.

Modulation of Absorption by Specific Intermolecular Interactions

We reasoned that the stabilization of the S_1 excited state of the isochromene was principally due to the hydrogen bonding interactions with the protein and therefore carried out some simple calculations to determine if this hypothesis was qualitatively correct. For each subsite model of the complex between rhizopuspepsin and tripeptide **3** we therefore explored the effects of specific protein-ligand interactions in contributing to the large shift in the lowest energy absorption band from its gas-phase value (Tables I and II). To do this, a series of model complexes were constructed by the systematic removal of protein fragments participating in specific intermolecular interactions with the tripeptide **3**, and their electronic spectra calculated. Hence, complex **6** was obtained from **5** by removal of the aromatic sidechain of Tyr-77 so as to generate an alanine residue (Figure 3A). In turn, complexes **7** and **8** were obtained by deletion of the peptide fragments corresponding to segments 77-87 (Segment A in Figures 3A and 3B) and 218-221, respectively (Segment C in Figures 3A and 3B). Complexes **10-12** were obtained by an identical procedure to that used for **6-8** except that complex **9**, in which Asp-35 was protonated, was used as the starting structure. For all complexes, the only significant modifications to the spectroscopic properties of **3** occurred when the hydrogen bonding interaction between the NH-substituent on the isochromene and the carboxylate of Asp-218 was broken. This accords with our previous observations upon the importance of modeling specific hydrogen bonding interactions in calculating the properties of heterocyclic chromophores (*31*). We also note that in every system there are two relatively strong bands that we designate (x) and (y) (Tables I and II) which are probably associated with the 3-amino-isochromene chromphore. However, tripeptide conformation **3-a** possesses an additional band arising from the S_8 state which has little intensity in any of the complexes and in **3-b**. Thus, more subtle effects due to chromophore conformation and to the protein environment may be operating. Equally, the importance of dynamic effects has not been evaluated in this study, and more sophisticated calculations employing 'snapshots' from QM/MM simulations (*16*) of peptide/protein complexes will probably be required to address this issue.

Conclusions

While providing only a qualitative picture of the effect of protein environment on the absorption properties of small molecules, these calculations reveal the importance of specific hydrogen bonding interactions in causing significant shifts in absorption bands in the UV-visible region. We therefore conclude that rational discovery of novel photocaged compounds using computational approaches must include such interactions specifically within the model. Current efforts are centered upon the synthesis and characterization of peptides containing the isochromene moiety to verify the results of our calculations. The outcome of these studies will be reported in due course.

Acknowledgments

Partial funding for this work was provided through a Research Development Award from the Division of Sponsored Research at the University of Florida, Proctor and Gamble, and CAChe Scientific, Inc. (NGJR). This work was also supported in part by the Office of Naval Research (MGC, MCZ).

Literature Cited

1. Hajdu, J.; Johnson, L.N. *Biochemistry* **1990**, *29*, 1669.
2. Moffat, K. *Annu. Rev. Biophys. Biophys. Chem.* **1989**, *18*, 309.
3. Hajdu, J.; Acharya, K. R.; Stuart, D. I.; McLaughlin, P. J.; Barford, D.; Oikonomarkos, N. G.; Klein, H.; Johnson, L. N. *EMBO J.* **1987**, *6*, 539.
4. Ringe, D.; Stoddard, B. L.; Bruhnke, J.; Koenigs, P.; Porter, N. *Phil. Trans. R..Soc. Lond. [A]* **1992**, *340*, 273.
5. Rasmussen, B. F.; Stock, A. M.; Ringe, D.; Petsko, G. A. *Nature(Lond.)* **1992**, *357*, 423.
6. Pai, E. F.; Schulz, G. E. *J. Biol. Chem.* **1983**, *258*, 1752.
7. Singer, P. T.; Smalås, A.; Carty, R. P.; Mangel ,W. F.; Sweet, R. M. *Science* **1993**, *259*, 669.
8. Perona, J. J.; Craik, C. S.; Fletterick, R. J. *Science*, **1993**, *261*, 620.
9. McCray, J. A.; Trentham, D. R. *Annu. Rev. Biophys. Biophys. Chem.* **1989**, *18*, 239.
10. Stoddard, B. L.; Koenigs, P.; Porter, N.; Petratos, K.; Petsko, G. A.; Ringe, D. *Proc. natl. Acad. Sci., U. S. A.* **1991**, *88*, 5503.
11. Mendel, D.; Ellman, J. A.; Schultz, P. G. *J. Am. Chem. Soc.* **1991**, *113*, 2758.
12. Moffat, K.; Chen, Y.; Ng, K.; McRee, D.; Getzoff, E. D. *Phil. Trans. R. Soc. Lond. [A]* **1992**, *340*, 175.
13. Stoddard, B. L.; Bruhnke, J.; Koenigs, P.; Porter, N.; Ringe, D.; Petsko, G. A. *Biochemistry* **1990**, *29*, 8042.
14. Schlichtling, I.; Almo, S. C.; Rapp, G.; Wilson, K.; Petratos, K.; Lentfer, A.; Wittinghofer, A.; Kabsch, W.; Pai, E. F.; Petsko, G. A.; Goody, R. S. *Nature(Lond.)* **1990**, *345*, 309.
15. Corrie, J. E. T.; Katayama, Y.; Reid, G. P.; Anson, M.; Trentham, D. R. *Phil. Trans. R. Soc. Lond. [A]* **1992**, *340*, 233.
16. Field, M. J.; Bash, P. A.; Karplus, M. *J. Comput. Chem.* **1990**, *11*, 700.
17. Cramer, C. J.; Truhlar, D. G. *J. Am. Chem. Soc.* **1991**, *113*, 8305.
18. Wong, M. W.; Wiberg, K. B.; Frisch, M. J. *J. Am. Chem. Soc.* **1992**, *114*, 1645.
19. Luzkhov, V.; Warshel, A. *J. Am. Chem. Soc.* **1991**, *113*, 4491.
20. Rauhut, G.; Clark, T.; Steinke, T. *J. Am. Chem. Soc.* **1993**, *115*, 9174.
21. Ridley, J. E.; Zerner, M. C. *Theoret. Chim. Acta* **1973**, *32*, 111.
22. Ridley, J. E.; Zerner, M. C. *J. Mol. Spectrosc.* **1974**, *50*, 457.
23. Pople, J. A.; Beveridge, D.; Dobosh, P. A. *J. Chem. Phys.* **1967**, *47*, 2026.
24. Thompson, M. A.; Zerner, M. C. *J. Am. Chem. Soc.* **1991**, *113*, 8210.

25. Thompson, M. A.; Zerner, M. C. *J. Am. Chem. Soc.* **1990,** *112*, 7828.
26. Chen, Z.; Han, H. P.; Wang, X. J.; Koelsch, G.; Lin, X. L.; Hartsuck, J. A.; Tang, J. *J. Biol. Chem.* **1991,** *266*, 11718.
27. *Structure and Function of Aspartic Proteinases*; Dunn, B. M., Ed.; Advances in Experimental Biology 306; Plenum Press: New York, NY, 1992.
28. Kohl, N. E.; Emini, E. A.; Schleif, W. A.; Davis, L. J.; Heimbach, J. C.; Dixon, R. A. F.; Scolnik, E. M.; Sigal, I. S. *Proc. natl. Acad. Sci., U. S. A.* **1988,** *85*, 4686.
29. Sielecki, A. R.; Hayakawa, K.; Fujinaga, M.; Murphy, M. E. P.; Fraser, M.; Muir, A. K.; Carilli, C. T.; Lewicki, J. A.; Baxter, J. D.; James, M. N. G. *Science* **1989,** *247*, 454.
30. Suzuki, J.; Sasaki, K.; Sasao, Y.; Hamu, A.; Kawasaki, H.; Nishiyama, M.; Horinouchi, S.; Beppu, T. *Prot. Eng.* **1989,** *2*, 563.
31. Richards, N. G. J.; Cory, M. G., Jr. *Int. J. Quantum Chem., Quant. Biol. Symp.* **1992,** *19*, 65.
32. Padwa, A.; Au, A.; Lee, G. A.; Owens, W. *J. Org. Chem.* **1975,** *40*, 1142.
33. Padwa, A.; Au, A.; Owens, W. *J. Chem. Soc. Chem. Commun.* **1974,** 675.
34. Tapia, O.; Goscinski, O. *Mol. Phys.* **1975,** *29*, 1653.
35. Karelson, M. M.; Katritzky, A. R.; Szafran, M.; Zerner, M. C. *J. C. S. Perkin Trans. 2* **1990,** 195.
36. Karelson, M. M.; Zerner, M. C. *J. Am. Chem. Soc.* **1990,** *112*, 9405.
37. Karelson, M. M.; Zerner, M. C. *J. Phys. Chem.* **1992,** *96*, 6949.
38. Abbreviations used in this paper include: SCF, self-consistent field; CIS, configuration interaction single excitations; SCRF, self-consistent reaction field; HOMO, highest occupied molecular orbital; LUMO, lowest unoccupied molecular orbital.
39. Davies, D. R. *Annu. Rev. Biophys. Biophys. Chem.* **1990,** *19*, 189.
40. Suguna, K.; Padlan, E. A.; Smith, C. W.; Carlson, W. D.; Davies, D. R. *Proc. natl. Acad. Sci., U. S. A.* **1977,** *74*, 7009.
41. Bernstein, F. C.; Koetzle, T. F.; Williams, G . J. B.; Meyer, E. F. Brice, M. D.; Rodgers, J. R.; Kennard, O.; Shimanouchi, T.; Tasumi, M.; *J. Mol. Biol.* **1977,** *112*, 535.
42. James, M. N. G.; Sielecki, A. R.; Hayakawa, K.; Gelb, M. H. *Biochemistry* **1992,** *31*, 3872.
43. Schlick, T; Overton, M. *J. Comput. Chem.* **1987,** *8*, 1025.
44. Mohamadi, F. M.; Richards, N. G. J.; Guida, W. C.; Liskamp, R. M. J.; Lipton, M. A.; Caufield, C. E.; Chang, G.; Hendrickson, T. F.; Still, W. C. *J. Comput. Chem.* **1990,** *11*, 440.
45. White, D. N. J. *Comput. Chem.* **1977,** *1*, 225.
46. Jorgenson, W. L.; Tirado-Rives, J. *J. Am. Chem. Soc.* **1988,** *110*, 1657.
47. Still, W. C.; Tempczyk, A.; Hawley, R. C.; Hendrickson, T. F. *J. Am. Chem. Soc.* **1990,** *112*, 6127.
48. Kirkpatrick, S.; Gelatti, C. D. J.; Vecchi, M. P. *Science,* **1983,** *220*, 671.
49. *Photophysics of Aromatic Molecules*; Birks, J. B.; Wiley-Interscience: London, 1970; p. 489.
50. ZINDO, a semi-empirical package written by M. C. Zerner and colleagues, University of Florida.
51. Purvis, G. D., III. *J. Comput.-Aided Mol. Des.* **1991,** *5*, 55.

RECEIVED June 2, 1994

Chapter 15

Comparisons of Hydrogen Bonding in Small Carbohydrate Molecules by Diffraction and MM3(92) Calculations

A. D. French[1] and D. P. Miller[2]

[1]U.S. Department of Agriculture, P.O. Box 19687, New Orleans, LA 70179
[2]D. P. M. Consulting, P.O. Box 423, Waveland, MS 39576

The 1992 version of Allinger's molecular mechanics program has a new description of hydrogen bonding. It was tested on two acyclic 4-carbon sugar alcohols, threitol and erythritol, and on β-D-glucose and maltose. These molecules have a variety of intra- and intermolecular O-H···O hydrogen bonds. Comparisons with diffraction evidence were achieved by construction and optimization of model miniature (630 - 729 atoms) crystals. Despite the absence of long range interactions, such models can be used to rationalize molecular distortion arising from crystal packing, to predict lattice energies, and to determine the optimal dielectric constant. The modeling was judged by the reproduction of observed hydrogen bond geometries, especially the retention of the major and minor components of three-center hydrogen bonds. The relative energies of disordered components were also predicted.

High accuracy in modeling studies is important, philosophically and practically. Philosophically, our caricatures of atomic interactions and molecular structure must be good enough that we can say that the subject is understood. From a practical perspective, many of the important questions in predictive modeling depend on energy differences of about one kcal/mol. An error of three kcal/mol, in the range of the energy of a single hydrogen bond, might cause us to judge an otherwise likely model to be quite improbable.

Hydrogen bonding is especially important in the study of carbohydrates. This is because many carbohydrates have nearly as many hydroxyl groups as carbon atoms. Also, the acetal oxygen atoms in carbohydrate rings and the glycosidic linkages between sugar rings in di-, oligo- and polysaccharides often act as hydrogen bond acceptors. Therefore, determination of the validity of molecular mechanics potential energy functions, including the details of hydrogen bonding, is an important part of our efforts to understand three-dimensional structures of starch and other polysaccharides. Although

0097–6156/94/0569–0235$08.00/0

there are other types of hydrogen bonds, this paper concerns only those involving hydroxyl groups and (hemi) acetal oxygen atoms in anhydrous molecular crystals.

Crystallographic Contributions

The structures of many carbohydrates have been determined by x-ray, and in some cases, neutron diffraction crystallography. A recent book by two crystallographers, Jeffrey and Saenger (*1*) thoroughly covers hydrogen bonding in biological molecules, including especially crystalline carbohydrates, and is a valuable resource in our studies.

Hydrogen atom location. The first step in describing the geometry of hydrogen bonds is to locate the hydrogen atom. Because hydrogen consists of a single proton and electron, this task is not simple. The location of the hydrogen depends on the method of determination. While x-ray diffraction locates the centers of electron density, neutron diffraction locates the proton. Although x-ray diffraction studies typically resolve the single electron on hydrogen with much less precision than other atoms, even the highest-quality x-ray studies often yield bond lengths that are shorter than the internuclear distances. That is because hydrogen's electron is shared with the electronegative oxygen to which it is covalently bonded. Thus, the time-averaged center of electron density for a hydroxyl hydrogen atom is usually closer to the center of the hydroxyl oxygen atom than is the nucleus of the hydrogen atom, leading to short x-ray bond lengths. For example, x-ray diffraction of D-threitol (*2*) gives covalent O-H bond lengths of 0.62, 0.65 and 0.76 Å. On the other hand, internuclear O-H distances determined by neutron diffraction average about 0.97 Å. Distances determined by x-ray diffraction can also be longer.

Therefore, one accepted, but by no means universal, practice in hydrogen bonding studies is to adjust the O-H bond lengths determined with x-rays to match the usual neutron diffraction value (*1*). These corrected covalent lengths give a new position of the hydrogen atom that is then used to calculate the hydrogen bond lengths, H···O. Such normalizations should be applied only in the case of normal or weak hydrogen bonds; strong hydrogen bonds to atoms such as fluorine or ionic oxygen can lengthen the covalent bond by several tenths of an Å, to the point that the hydrogen is equally shared between the two atoms.

Parallel considerations apply in modeling. In this case, the position of the atom during the calculation depends on what property is being calculated. The length used in MM3 for the ideal covalent O-H bond is 0.947 Å, corresponding to the gas-phase electron diffraction result. However, during van der Waals energy calculations in MM3, the lengths of all covalent bonds to hydrogen atoms are reduced by a factor of 0.923 to compensate for the relocation of electrons. Applying this scale factor gives a value of 0.870 Å, 0.1 to 0.2 Å longer than the above x-ray bond lengths for D-threitol. In our MM3-optimized models of maltose, below, the covalently bonded, unscaled internuclear distances range between 0.948 and 0.957 Å.

Low temperature considerations. The resolution and accuracy of diffraction studies is improved by lowering the temperature of the sample. This is especially important in locating the hydrogen atoms. However, some cautions apply. It is important to know whether there are phase changes during cooling, for example. Typically, lattice dimensions contract about 0.5% when cooled to 119° K. The reduced thermal motion of the molecule at low temperature requires less volume, so the lattice can contract. However, bond lengths often show an apparent increase (e.g. 0.007 Å). This is because thermal motion of the atoms at room temperature leads to apparently short bond lengths unless difficult thermal motion corrections are performed. Another effect of cooling may be a loss of static and dynamic disorder. "Flip-flop" hydrogen bonding, wherein hydroxyl groups point first in one direction and then, moments later, point in another direction and form another set of hydrogen bonds (*3*), is a type of dynamic disorder.

Hydrogen bond variety. The variety of O-H···O hydrogen bonds in carbohydrate crystals is extensive. Being much weaker than covalent bonds, their lengths vary greatly. A simple general rule (G. A. Jeffrey, personal communication) is that there will be some lowering of the energy with long hydrogen bonds, but there will be little influence on conformation if the H···O distance is more than 2.0 Å. Because hydroxyl groups can rotate about C-O single bonds, they are able to orient in a way that minimizes the free energy of the system. One consequence is that the hydroxyl groups can function simultaneously as both donors and acceptors. This dual role permits the formation of chains of hydrogen bonds. The cooperativity arising from donor-acceptor-donor-acceptor chains results in stronger (e.g. 12%) hydrogen bonds than an equal number of isolated hydrogen bonds (*4*). The hydroxyl hydrogen is often donated to either hydroxyl or acetal oxygen atoms in neighboring molecules, and in some di- and oligosaccharides, intramolecular bonds are formed. Even in the case of monosaccharides, weak hydrogen bonds sometimes form rings around the molecule. A hydroxyl hydrogen is not limited to a single receptor, but is often shared by two or even more oxygens. Such "bifurcated" hydrogen bonds (the preferred nomenclature (*1*) is "three-center hydrogen bonds") have total energies equal to a single, two-center hydrogen bond (*5*). Sometimes the two components of three-center hydrogen bonds have nearly equal H···O distances. Other times, there are major (shorter) and minor (longer) components. Perhaps the reader is convinced at this point that modeling hydrogen bonding in carbohydrates, and thus determining its role in carbohydrate structures, is far from trivial.

Intramolecular, interresidue hydrogen bonding. One aspect of carbohydrate conformation in which intramolecular hydrogen bonding has been thought to be important is the geometry of the interresidue linkages of disaccharides. The two sugar rings of most disaccharides are linked by two C-O single bonds, as in the 1→4 linkages of maltose. In some other disaccharides, there is a third, C-C, linkage bond, such as the 1→6 linkage of isomaltose. The flexibility of several disaccharides is substantial, as evidenced by ranges as large as 90° in the linkage torsion angles found in crystal structures for families of closely related molecules.

α- and β-maltose give good examples of variation of linkage torsion angles in response to intra- and intermolecular forces. These compounds are composed of two glucose residues with axial-equatorial, 1→4 linkages. Maltose, or its residue after formation of additional bonds, is found in a large number of oligomers, such as the cyclomaltoses, methyl maltotrioside, maltoheptaose, panose and erlose. In the case of hydrated cyclohexaamylose (*6*), there is substantial variation in these linkage torsion angles, despite chemical identity of each residue. Only a subset of the observed maltose linkage torsion angles permits O-2···O-3' intramolecular hydrogen bonding. For other observed conformations, intramolecular, interresidue hydrogen bonding is not geometrically possible.

Compared to maltose moieties, sucrose moieties usually have similar axial-pseudo-equatorial linkages, with a slightly smaller range of observed linkage torsion angles. Crystalline sucrose (*7-8*) has two intramolecular hydrogen bonds that bridge the two residues, but the sucrose linkage in stachyose (*9*), with a nearly identical linkage conformation, has none. From these examples we have concluded that intramolecular hydrogen bonding is just one of the determinants of carbohydrate structure. Other important factors include the energetics corresponding to those represented in the model by the linkage torsional potentials, general crystal-packing considerations, and opportunities for intermolecular hydrogen bonding with geometry more ideal than in intramolecular bonds.

Although Jeffrey and Saenger suggest that hydrogen bonding energies should not be included during modeling studies of isolated disaccharides, our recent experience (*10*) suggests otherwise. We showed that MM3(92) and a dielectric constant of 3.0 account for linkage conformations that are observed in crystals of families of related molecules, whether or not hydrogen bonding occurs in the individual molecules. In the past, electrostatic interactions were often overemphasized, leading to incorrect results (*11*). To quantitatively compare modeling results with the most precise experimental data, however, it is necessary to include the crystalline environment, as we describe below.

Methods

At present, molecular mechanics is the most suitable formalism for routine modeling studies of most carbohydrates. Such programs have been extensively parameterized for the atom types appearing in carbohydrates, and calculations proceed at a very useful rate on readily available workstations or even high-speed personal computers. In many cases, structures are reproduced within limits close to experimental error and our recent studies with the MM3 program suggest that energies are internally consistent (*10*). In contrast, ab initio methods are not yet economically feasible for comprehensive studies of disaccharides or larger molecules. At present, semiempirical quantum mechanics is insufficiently accurate (*12-15*).

Computations. We used the 1992 version of MM3, as available from the QCPE, Technical Utilization Corporation, and Tripos Associates. Both a Digital Equipment Company Vaxstation 4000/90 and Gateway 2000 486/33 and 486/66 personal computers were used. Previously (*10*), we had explored the role of dielectric constant (ϵ) in lattice energy calculations. We obtained a self-consistent calculation of the lattice energy at $\epsilon = 2.5$, while the overall atomic movement minimized at $\epsilon = 3.5$, the value used in the present study. An ϵ of 4.0 was suggested for peptide studies (*16*). MM3 results are much more sensitive to ϵ than are MM2 results.

Standard 1985 and earlier versions of MM2 based hydrogen bonding energies on dipole-dipole attractions, the same as all other electrostatic interactions. Subsequent versions of Allinger's MM series use a "hydrogen-bond aware", modified Buckingham potential in place of the standard van der Waals calculation when a hydrogen-bonding atomic sequence is identified, besides the dipole-dipole energies. The 1992 version of MM3 includes directionality in the modified Buckingham calculations. From the MM3(92) manual, the components of hydrogen bond energy are:

Dipole-dipole energies:

$$EU_{AB} = F\,(\mu_A)\,(\mu_B)\,(\cos X - 3\cos\alpha_A \cos\alpha_B)\,/R^3\,/DE$$

where F is the conversion from ergs/molecule to kcal/mol, μ_A and μ_B are the bond moments of the two (covalent) bonds in question, X is the angle between the two dipoles, α_A and α_B are the angles of the dipole moment between the dipole axes and the line connecting their midpoints, R is the length of the line connecting the midpoints of the two dipoles and DE is the symbol used in the manual for the dielectric constant. In MM3, the bond moment is 1.17 debye units for bonds between sp^3 carbon and acetal or hydroxyl oxygen; the O-H bond moment is -1.67 debye units. As indicated, the strengths of the electrostatic interactions in MM3 depend on the cube of the reciprocal distance. With a point charge model of electrostatic interactions, commonly used in other force fields, the energy depends instead on the reciprocal distance between the atoms.

Additional hydrogen bonding energy:

$$E_{HB} = (\epsilon_{HB}/D) * [184000 * \exp(-12.0/P) - 2.25 * P^6 * \cos\theta * l/l_o]$$

where ϵ_{HB} is the hydrogen bonding energy parameter, D is the dielectric constant symbol, P is r_{HB}/R, where r_{HB} is the equilibrium O···H hydrogen bond distance in Å and R is the current distance, θ (Figure 1) is the angle between H-O and O···O, l is the current bond length of O-H, and l_o is the strainless bond length of the O-H bond. (To prevent collapse during initial optimization of crude structures, a different energy equation is used when H···O distances are less than 0.6 Å.) The parameters for a O-H···O hydrogen-bond are r_{HB} = 1.82 Å and ϵ_{HB} = 3.3 kcal/mol. Thus, with D of 3.5 and l and R at their ideal values, the strongest possible augmentation energy is -1.06 kcal/mol. We have not yet tested the distance-dependent dielectric option available in MM3(92).

The methanol dimer. Four interactions contribute to the dipole-dipole energy for a simple dimer of methanol with a single hydrogen bond (Figure 2). The C-O -- C-O and C-O -- O-H (non-hydrogen bonding) interactions raise the energy, while the O-H -- O-H and hydrogen-bonding O-H -- C-O interactions lower the energy. At $\epsilon = 1.5$, the sum of the dipole-dipole energies (-3.35 kcal/mol) and the additional term (-2.50 kcal/mol) for the methanol dimer is -5.85 kcal/mol. At $\epsilon = 3.5$, the respective values are -1.26, -0.97 and -2.23 kcal/mol. Figure 2 shows various geometric parameters for these models. Another view of the strength of the hydrogen bond in methanol is given by comparing

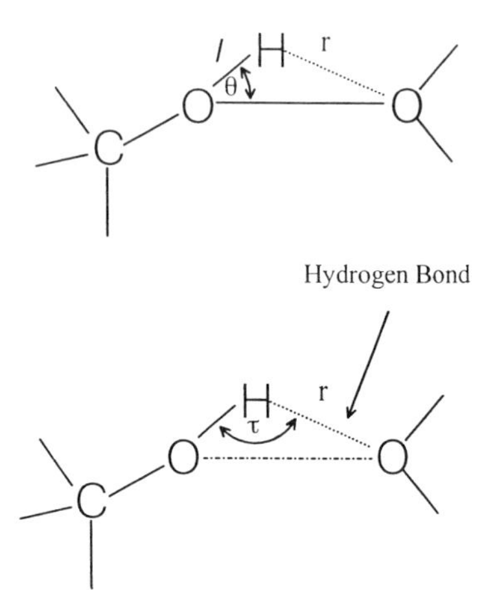

Figure 1. Schematics of hydroxyl-hydroxyl or hydroxyl-acetal oxygen hydrogen bonding. Upper diagram: Terms in the MM3(92) hydrogen bonding expression. Lower: Terms in this paper describing hydrogen bonds.

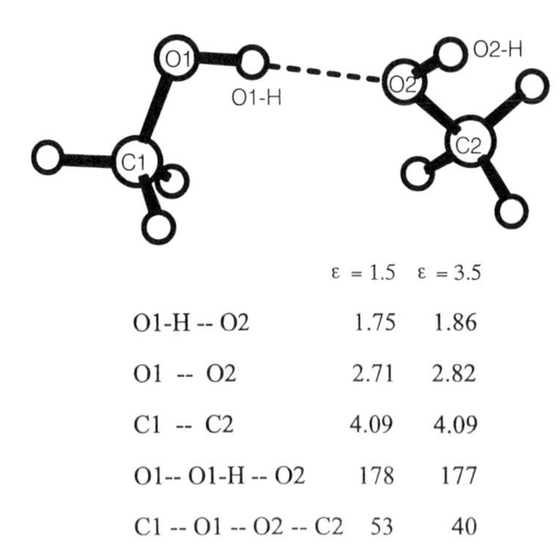

	ε = 1.5	ε = 3.5
O1-H -- O2	1.75	1.86
O1 -- O2	2.71	2.82
C1 -- C2	4.09	4.09
O1-- O1-H -- O2	178	177
C1 -- O1 -- O2 -- C2	53	40

Figure 2. The methanol dimer, drawn at dielectric constant 3.5, with various geometric parameters noted for dielectric constants 1.5 and 3.5.

the total energies of the dimer at equilibrium separation and at infinite separation of the two molecules. The energy at $\epsilon = 1.5$ obtained by holding the two molecules 50Å further apart than the equilibrium distance is 5.32 kcal higher than the equilibrium energy. Because the van der Waals repulsions are included, this definition of the hydrogen bonding energy is 0.53 kcal less stabilizing than when calculated with only the dipole-dipole and modified Buckingham terms.

Minicrystals. The CRSTL program (*17*) delivered with MM3(92) (*18*) is intended for the development of non-bonded potential functions for intermolecular interactions. For accuracy, it uses large (e.g. 15 x 15 x 15 unit cell arrays) assemblies of molecules. On the other hand, the molecules are left in their initial conformation. Inspired by this program, we have been making model miniature crystals (*10,19*) consisting of about 30 monosaccharide residues, and optimizing them with MM3. Despite lacking the important, long-range interactions, our minicrystal models are still informative regarding molecular deformation from crystal packing, lattice energy and optimal dielectric constant. We accepted the lack of long-range interactions (because of the size limits within MM3 on our hardware) in exchange for the ability to include intramolecular interactions. This allows testing of the balance of all terms of the potential energy function. During optimization of the minicrystals, greater movement occurred in the direction of the weaker, van der Waals attractions (*19*).

Miniature crystals for the present study were constructed based on the published atomic coordinates and unit cell dimensions for D-threitol (*2*), D-erythritol (*20*), β-D-glucose (*21*), and α-D-maltose (*22*) shown in Figure 3. Their construction was facilitated by the modeling package CHEM-X (developed and distributed by Chemical Design, Ltd.). The models contained 608 to 729 atoms, with the molecules arrayed in compact arrangements based on the reported space group symmetries. As an example, the minicrystal of β-D-glucose is shown, both before and after optimization, in Figure 4. No constraints, such as crystal symmetry or periodic boundary conditions, were used. The unoptimized miniature crystal of maltose is shown in Ref. 10. In it, six glucose molecules were used in addition to the 13 maltose molecules, to obtain a complete mono-molecular shell surrounding the central molecule without exceeding 750 atoms.

Difficulties regarding the fluctuation of surface hydroxyl hydrogens during MM3 minimization of the miniature crystals are described elsewhere (*19*). Although we used MM3's termination criterion of n * 0.00008 kcal/mol/iteration, where n is the number of atoms, it is also necessary to monitor atomic movement. If the energy criterion had been met (or the energy had increased!) but mean atomic movement had not diminished to less than 0.00010 Å per iteration, fluctuating surface hydroxyl groups were reoriented with CHEM-X and the minimization was restarted.

Experimental values. Adjusted hydrogen bond dimensions for glucose and maltose were taken from Jeffrey and Saenger's book, with corrections described below. The erythritol structure was determined by neutron diffraction; those values were used unchanged. The threitol structure was determined by low-temperature (119° K) x-ray diffraction. We extended its O-H bond lengths to 0.97 Å to mimic neutron diffraction results.

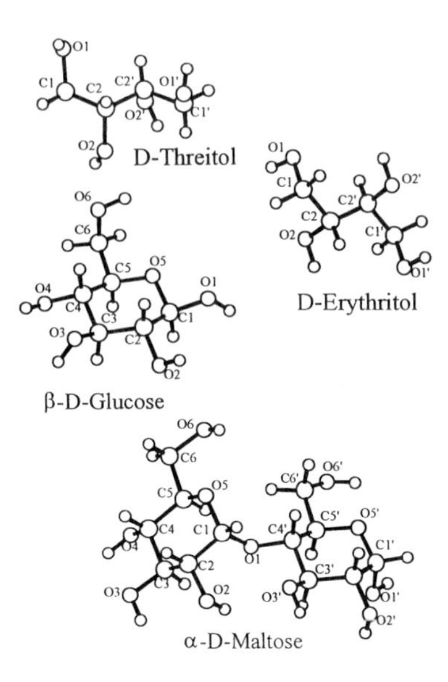

Figure 3. The molecules compared in this paper, drawn from their crystal structure coordinates. Hydrogen atoms are not labeled.

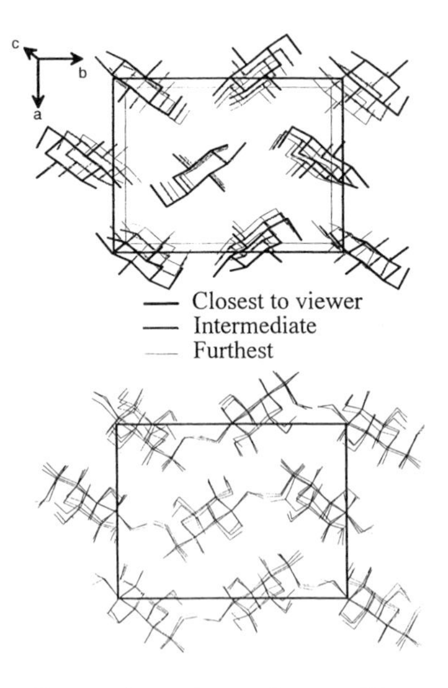

Figure 4. Upper: Perspective view of crystalline β-D-glucose, prior to optimization, hydrogen atoms omitted, showing the 27 molecules in the minicrystal. Lower: Optimized minicrystal, including the hydrogen atoms, projected along the c-axis, no perspective employed. It shows the extent of movement resulting from optimization.

Hydrogen bond lengths refer to the internuclear distance between the donated hydroxyl proton and the receptor oxygen nucleus. We also monitored the O···O distances, important in less accurate single crystal and fiber diffraction work. Jeffrey and Saenger included as hydrogen bonds H···O distances as long as 3.0 Å, and O-H···O angles, τ (Figure 1), as small as 90° (*1*).

The MM3 values for the hydrogen bond lengths in the minicrystals were taken by hydrogen bond scans with CHEM-X as well as visual inspection of the miniature crystals with CHEM-X. All representative hydrogen bonds of each type were included and the standard deviations were calculated. A few of the values deviated more than two standard deviations from the means. They were for hydrogen bonds located near surfaces of the minicrystals and were not included in our reported results. Lattice energies were calculated in a manner similar to the method in CRSTL. The energy of the entire minicrystal is compared with the sum of the energies of the central molecule and the shell of molecules around it. Mean atomic movements caused by minimization (for the oxygen and carbon atoms only) were obtained with CHEM-X for both the entire minicrystal and just the central molecule.

Results

Various results from optimization of the minicrystals are given in Table I. In particular, the statistics on β-D-glucose differ from values we reported earlier (*19*) because of more thorough optimization. Also, our earlier work was based on MM3(90). The lattice energy for maltose is low by perhaps 15 kcal/mol, partly because of the dielectric constant of 3.5; it was 76 kcal/mol when $\epsilon = 2.5$ (*10*). (Note that stabilizing lattice energies are positive.) Similarly, the lattice energy for β-D-glucose seems about 8.6 kcal/mol too small. The ΔH_f calculated from the heat of combustion of glucose (*23*) is -303 kcal/mol. The lattice energy obtained by subtracting the ΔH_f derived from the MM3 bond-energy for the optimized central molecule, -256.2 kcal/mol, is 46.8 kcal/mol. These discrepancies arise from the absence of the long range attractions, which were offset by a slightly low dielectric constant in the earlier work.

Table I. Minicrystal Descriptions

Compound	Threitol	Erythritol A	Erythritol B	Glucose	Maltose
# of Molecules	35	37	37	27	13+6 glucose
Total E (kcal/minicrst)	73.3	18.4	57.9	42.3	186.0
$\Delta E_{lattice}$ (kcal/mol)	30.0	29.1	33.0	38.2	61.3
Mean move (total) Å	0.228	0.193	0.264	0.229	0.199
Mean move (center) Å	0.029	0.014	0.130	0.052	0.067

Threitol. The crystal structure of this linear, four-carbon sugar alcohol was confusing. According to Ref. 2, the space group is $P3_121$, which requires half a molecule in the asymmetric unit. Two-fold axes generate the other halves of the molecules, and the three-fold axis accounts for the three molecules per unit cell. The position of the hydroxyl hydrogen atom on C-2 is described as two-fold disordered, with 50% occupancy in each position. Three minicrystals were examined. In the first (AA), all O-2 hydrogen atoms and their symmetrically-generated O-2' hydrogen atoms were in the first position. This model had high energy because of conflicts between these two hydrogen atoms, and the mean atomic movements were also large. In the second model (BB), all O-2 and O-2' hydroxyl hydrogen atoms were in the second position. Again the miniature crystal energy and movement were high. The third model (AB), shown in Figure 5, placed the O-2 hydroxyl in the first position and the O-2' hydroxyl in the second position. A much lower energy was obtained, and movement was small. This latter structure, in combination with its opposite arrangement (BA), was taken as correct. This structure has infinite chains of hydrogen bonds (Table II). A miniature crystal model based on a prepublication structure of D-threitol had lower symmetry. Optimization of the model showed that higher symmetry was appropriate, agreeing with the final report (2). However, the lower energy of the AB or BA models, relative to the AA or BB models, argues strongly against a random disorder. A random disorder would have a general energy increase due to random occurrences of unformed hydrogen bonds and steric conflict. More likely would be a domain crystal where unrealized hydrogen bonding would occur only at domain boundaries (AB-BA). Thus, the space group does not apply at the molecular level, but only to the average properties. A new determination of the L-threitol crystal structure (*24*) revealed three different molecules having similar (AB) conformations in the asymmetric unit. There is no disorder, but the crystal was twinned, making the determination especially difficult.

Table II. Hydrogen Bond Parameters For the AB Model of D-Threitol[a]

	H···O (Å)		O···O (Å)		τ (°)	
Bond	**obs.**	**calc.**	**obs.**	**calc.**	**obs.**	**calc.**
O-1 O-1	1.73	1.86(2)	2.69	2.79(3)	170	164(5)
O-2 O-2	1.83	1.80(2)	2.68	2.74(3)	144	172(3)
O-2' O-2'	1.73	1.94(6)	2.68	2.80(6)	163	150(4)

[a] *Numbers in parentheses are the standard deviations of the last digit.*

Erythritol. The crystal structure of erythritol, also an acyclic four-carbon sugar alcohol, was determined by neutron diffraction in 1980 at 22.6° K (*20*). It is also disordered, but unlike threitol, the disordered hydroxyl hydrogen is a minor component at 15%. Models were optimized based on pure major component (A) and pure minor component (B). The expected ratio of A to B can be predicted by the relative energies of the two forms, including the lattice energy. In the pure A model, the central molecule has a steric energy of 9.0 kcal/mol, plus the lattice energy of -29.1, giving -20.1 kcal/mol. Pure B has values of 11.8 and -33.0 for a total of -21.2 kcal/mol. At room temperaure, the difference of about 1.1 kcal/mol corresponds to 13% B, close to the amount observed in the crystal. Because an energy difference this large corresponds to a negligible amount of B at 22.6°, the disorder existing at room temperature must have been "frozen" when the temperature was reduced for the diffraction experiment. These hydrogen bonds (Table III) are also in infinite chains.

Table III. Hydrogen Bond Parameters for *meso*-Erythritol

	H···O (Å)		O···O (Å)		τ (°)	
Bond	**exp.**	**calc.**	**exp.**	**calc.**	**exp.**	**calc.**
O-1 O-1A	1.70	1.86(4)	2.68	2.80(3)	173	167(2)
O-2 O-2A	1.79	1.93(2)	2.73	2.84(2)	165	161(5)
O-1 O-1B	1.70	1.85(5)	2.68	2.79(5)	173	170(3)
O-2 O-2B	1.85	1.93(3)	2.73	2.82(3)	164	153(3)

β-Glucose. The results for glucose are in Table IV. This crystal structure was determined by x-ray diffraction in 1970 (*21*). No disorder was detected. Five hydrogen bonds, including one not originally reported, are present. That hydrogen bond, with O-4H donating to O-2, is quite long at 2.32 Å according to Jeffrey and Saenger (*1*). The book's O-4-H···O-2 angle of 100° is apparently in error, with our measurement of the crystallographic result being 161°. O-2 is a double-acceptor and the interaction with O-4H is apparently weak, but no other potential acceptor is in the neighborhood. The chain of hydrogen bonds terminates at O-5.

Table IV. Hydrogen Bond Parameters for β-D-Glucose

	H···O (Å)		O···O (Å)		τ (°)	
Bond	**exp.**	**calc.**	**exp.**	**calc.**	**exp.**	**calc.**
O-1H O-6	1.72	1.82(5)	2.67	2.75(3)	164	168(8)
O-2H O-3	1.60	1.91(4)	2.69	2.82(4)	160	158(6)
O-3H O-5	1.84	1.92(5)	2.77	2.84(5)	160	165(10)
O-4H O-2	2.32	2.51(20)	3.27	3.27(14)	161	139(12)
O-6H O-2	1.80	1.89(6)	2.71	2.78(5)	156	156(7)

α-Maltose. The results for maltose are in Table V. This crystal structure was determined by x-ray diffraction in 1974. There is a disorder (18%) involving the anomeric hydroxyl group and hydrogen at C-1', as found in some other reducing mono- and disaccharides. This disorder was not modeled. The hydrogen bonding system is very complex, with two intramolecular hydrogen bonds and ten intermolecular hydrogen bonds. This complexity led to several errors in Jeffrey and Saenger's summary. A projection of the atoms involved in hydrogen bonding to maltose is given in Figure 6. Figure 7 is a new hydrogen bonding scheme, in the style of Jeffrey and Saenger. There are four three-center bonds in maltose. In all of them, the hydrogen atoms are donated to two oxygen atoms.

Table V. Hydrogen Bond Parameters for α-D-Maltose.

	H···O (Å)		O···O (Å)		τ (°)	
Bond	**exp.**	**calc.**	**exp.**	**calc.**	**exp.**	**calc.**
O-1'H O-5	2.03	1.86(2)	2.93	2.80(3)	140	167(5)
O-1'H O6	2.80	2.85(11)	3.12	3.41(17)	99	118(6)
O-2'H O-5'	1.80	1.93(7)	2.74	2.82(6)	165	157(3)
O-3'H O-2*	1.81	1.84(3)	2.77	2.76(3)	169	163(4)
O-3'H O-4*	2.47	2.55(7)	2.93	2.89(2)	109	103(3)
O-6'H O-2'	2.32	2.44(17)	2.96	2.99(11)	123	118(7)
O-6'H O-3'	2.18	2.07(12)	3.09	2.98(9)	160	163(8)
O-2H O1'	2.35	2.34(12)	3.01	3.10(8)	125	136(4)
O-2H O-2'	1.89	2.11(11)	2.83	2.95(9)	156	147(4)
O-3H O-6	1.86	2.00(18)	2.80	2.89(8)	162	164(6)
O-4H O-6'	1.79	2.02(10)	2.71	2.75(2)	157	134(10)
O-6H O-4	1.81	2.06(10)	2.78	2.97(10)	175	161(2)

* *Intramolecular hydrogen bond*

Discussion

Several previous studies have used molecular mechanics to model hydrogen bonding in carbohydrates. Among the early workers were Jeffrey and Taylor, who modified MM1 for carbohydrates (*25*). Intramolecular interactions were studied in simple molecules such as glycerol, and fairly satisfactory results were obtained with variants of MM2 (*26*). In a study more closely related to the present, the overall packing effects on the molecular conformation of crystalline glucose were simulated with a rigid cage of water molecules, the oxygens of which replaced the hydroxyl groups of neighboring glucose molecules (*27*). The GROMOS molecular dynamics studies of α-cyclodextrin

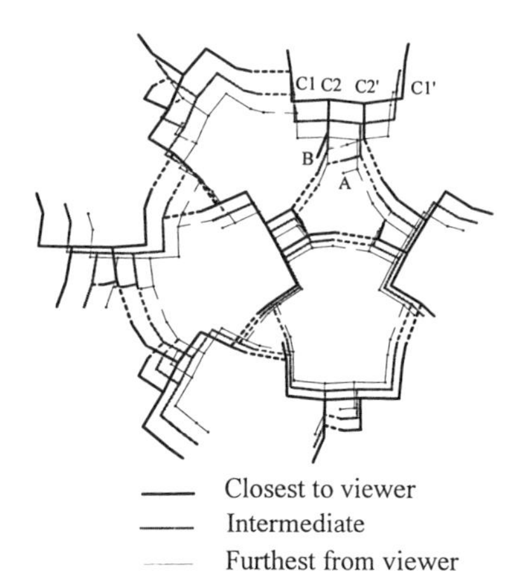

Figure 5. Hydrogen bonding in the unoptimized minicrystal model of D-threitol (an A-B domain). The carbon atoms on one molecule are labeled, as are the two hydroxyl hydrogen atoms (A and B) that correspond to the reported disorder. Hydrogen atoms bonded to carbon atoms are not shown. Each molecule contains hydroxyl hydrogens on the central two oxygen atoms in both the disordered positions. The hydrogen bonds form continuous helices. Only three of the five layers of threitol molecules in the minicrystal are shown.

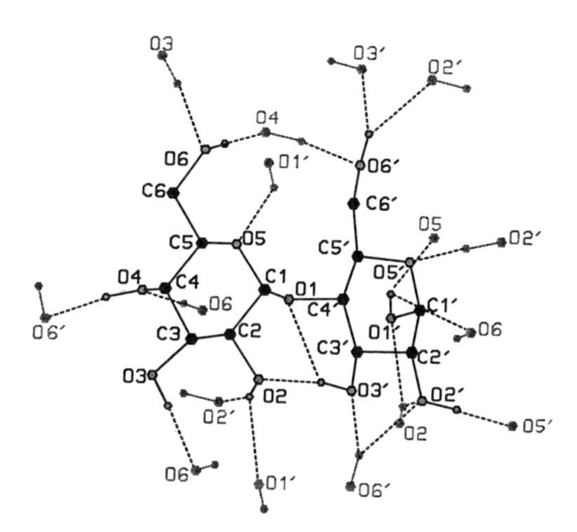

Figure 6. Geometric arrangement of hydrogen bonds in crystalline maltose. All the neighbor donors and acceptors are shown (gray lines and labels). Thus, the interresidue hydrogen bonds are shown twice.

hexahydrate and β-cyclodextrin dodecahydrate (*28-29*) also gave reasonably satisfactory results. They also benefited from a rigid box containing the structure, as well as rigid bond lengths. The present work is, as far as we know, the first detailed study of hydrogen bonding in crystalline arrays of carbohydrate material in which all interactions were fully optimized with MM3(92).

Besides single donor-acceptor bonds, in both finite and infinite chains, some of the crystal structures in this report have three-center bonds. Also, there is disorder of several types, all factors in O-H···O hydrogen bonding systems of carbohydrates. Of all these types of hydrogen bonds, the O-3'H···O-2 interresidue hydrogen bond of maltose was most accurately reproduced. This might have been expected since it is intramolecular, and the general expansion of the minicrystal would not be a factor. However, O-3'H also donates to O-1, and the linkage torsion angles of maltose are not at their minimum energy values for the isolated molecule. Inappropriate strengths of the O3'H···O1 bond or unsuitable torsional parameters might have moved the structure away from the observed conformation during minimization. Instead, the minicrystal field held the model geometries very close to the corrected crystallographic values. The geometries for all the hydrogen bonds are perhaps as close as could be expected, given the numerous approximations in the models and the vagaries in the experimental results and our treatment thereof. Figure 8 plots the combined experimental and modeled distances for all structures. For the most part, the discrepancies in hydrogen bond parameters are roughly comparable to the overall mean atomic movements that occur during optimization.

It is impressive that the strength relationships of the three-center hydrogen bonds, as shown by the differences in the H···O distances, were not extensively altered by the minimizations. The agreement between the predicted and observed amounts of erythritol conformers was another encouraging indicator. Experience with the miniature crystal models of carbohydrates suggests that they would be a useful routine supplement to diffraction studies.

Improvements in the geometric modeling specific for carbohydrates would seem to require improved experimental data, from which averaged data could be developed for several subtypes of hydrogen bonding. To test new refinements it would be useful to have larger crystal models or periodic boundary conditions. In all instances of intermolecular, two-center hydrogen bonds in the current work, short experimental H···O and O···O distances got longer during optimization. The increases may be due to expansion in the minicrystal, and could be lessened with a slightly lower dielectric constant (as shown by the methanol calculations in Methods). Long O···O distances (> 2.90 Å) are either shortened or left alone. These conclusions arise from the slopes (0.81 and 0.80) and intercept values (0.47 and 0.63 Å), respectively, in Figure 8. Given the limitations of our minicrystal models, we are unable to suggest changes to MM3 to attempt to improve its modeling of hydrogen bonding. However, the potential energy function is being revised for MM3(94), based on new ab initio studies (Lii, J.-H.; Allinger, N. L.; personal communication, 1993).

Historically, hydrogen bonding has been assigned a predominant role in molecular structure of carbohydrates. Besides the material offered above on intramolecular, interresidue hydrogen bonding in disaccharides, some perspecitve can be had from some

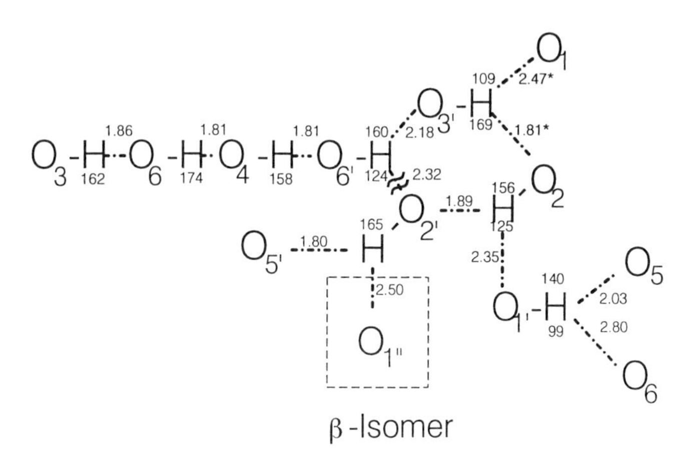

Figure 7. Corrected schematic of hydrogen bonding in α-maltose, also showing the role of the disordered β-anomer. Asterisks indicate intramolecular hydrogen bonds.

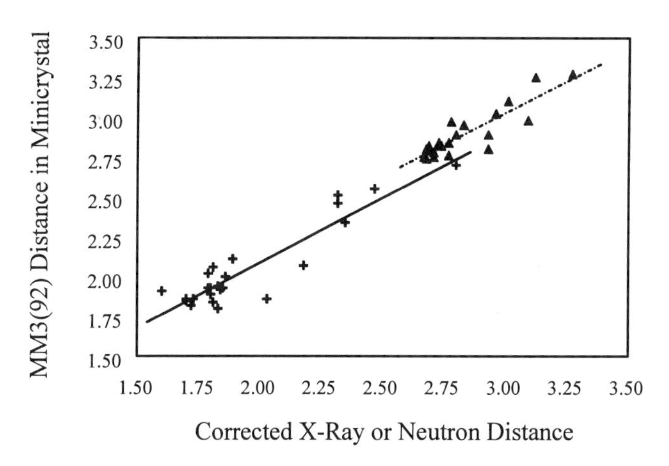

Figure 8. Regression lines of the observed and MM3(92)-predicted H···O (+, solid line) and O···O (Δ, dashed line) distances. Regression R values were 0.92 and 0.88, respectively, and the slopes were 0.81 and 0.80, respectively. Each point represents a single type of a hydrogen bond, observed and calculated. Data were taken from Tables II - V.

simple calculations. We have calculated (*10*) that crystalline maltose has a lattice energy of about -76 kcal/mol. If we arbitrarily (but generously) allow 5 kcal/mol for each of its seven intermolecular donor hydrogen atoms, the total hydrogen bonding energy is -35 kcal/mol, leaving 54% of the lattice energy to be explained by van der Waals forces. According to fiber diffraction studies of native cellulose, there is only one intermolecular hydrogen bond per glucose residue, so van der Waals forces account for nearly 75% of its lattice energy (*19*). Thus, we view hydrogen bonding as only one of the ingredients in a stew of factors that make carbohydrate modeling so interesting.

Acknowledgments

The authors thank G. A. Jeffrey, N. L. Allinger, J.-H. Lii, M. K. Dowd and E. Stevens for helpful discussions.

Literature Cited

1. Jeffrey, G. A.; Saenger, W. *Hydrogen Bonding in Biological Structures*; Springer Verlag: Berlin, 1991.
2. Jeffrey, G. A.; Huang, D.-b. *Carbohydr. Res.*, **1992**, *223*, 11-18.
3. Zabel, V.; Saenger, W.; Mason, S. A. *J. Am. Chem. Soc.*, **1986**, *108*, 3664-3673.
4. Del Bene, J.; Pople, J. A. *J. Chem. Phys.* **1973**, *58*, 3605-3608.
5. Tse, Y. C.; Newton, M. D. *J. Am. Chem. Soc.*, **1978**, *99*, 611-613.
6. Klar, B.; Hingerty, B.; Saenger, W. *Acta Crystallogr. Sect. B* **1980**, *36*, 1154-1165.
7. Hanson, J. C.; Sieker, L. C.; Jensen, L. H. *Acta Crystallogr. Sect B* 1973, *29*, 797-808.
8. Brown, G. M.; Levy, H. A. *Acta Crystallogr. Sect B* **1973**, *29*, 790-797.
9. Jeffrey, G. A.; Huang, D.-b. *Carbohydr. Res.* **1991**, *210*, 89-104.
10. French, A. D.; Dowd, M. K. *J. Mol. Struct. (Theochem)* **1993**, *286*, 183-201.
11. Kroon-Batenburg, L. M. J.; Kroon, J.; Leeflang, B. R.; Vliegenthart, J. F. G. *Carbohydr. Res.* **1993**, *245*, 21-42.
12. Tvaroška, I.; Carver, J. *J. Chemical Research* **1991**, 123-144.
13. Zheng, Y.-J.; Le Grand, S. J.; Merz, Jr., K. M. *J. Comput. Chem.* **1992**, *13*, 772-791.
14. Gundertofte, K.; Palm, J.; Petterson, I.; Stamvik, A. *J. Comput. Chem.* **1991**, *12*, 200-208.
15. Hoffman, R. A.; van Wijk, J.; Leeflang, B. R.; Kamerling, J. P.; Altona, C.; Vliegenthart, J. F. G. *J. Am. Chem. Soc.* **1992**, *114*, 3710-3714.
16. Lii, J.-H.; Allinger, N. L. *J. Comput. Chem.* **1991**, *12*, 186-189.
17. Allinger, N. L.; Yuh, Y. H.; Lii, J.-H. *J. Am. Chem. Soc.* **1989**, *111*, 8551-8566.
18. Allinger, N. L.; Rahman, M.; Lii, J.-H. *J. Am. Chem. Soc.* **1990**, *112*, 8293-8307.
19. French, A. D.; Aabloo, A.; Miller, D. P. *Int. J. Biol. Macromol.* **1993**, *15*, 30-36.
20. Ceccarelli, C.; Jeffrey, G. A.; McMullan, R. K. *Acta Crystallogr. Sect. B* **1980**, *36*, 3079-3083.
21. Chu, S. S. C.; Jeffrey, G. A. *Acta Crystallogr. Sect. B* **1968**, *24*, 830-838.

22. Takusagawa, F.; Jacobson, R. A. *Acta Crystallogr. Sect. B*, **1978**, *34*, 213-218.
23. *Handbook of Chemistry and Physics;* Weast, R. C., Ed.; Chemical Rubber Co.: Cleveland, Ohio, 1967, 48, D-189.
24. Kopf, J.; Morf, M.; Zimmer, B.; Haupt, E. T. K.; Jarchow, O; Köll, P. *Carbohydr. Res.* **1993**, *247*, 119-128.
25. Jeffrey, G. A.; Taylor, R. *J. Comput. Chem.* **1980**, *1*, 99-109.
26. Vázquez, S. A.; Ríos, M. A. *J. Comput. Chem.* **1992**, *13*, 851-859.
27. Kroon-Batenburg, L. M. J.; Kanters, J. A. *Acta Crystallogr. Sect. B* **1983**, *39*, 749-754.
28. Koehler, J. E. H.; Saenger, W.; van Gunsteren, W. F. *Eur. Biophys J.* **1987**, *15*, 197-210.
29. Koehler, J. E. H.; Saenger, W.; van Gunsteren, W. F. *Eur. Biophys J.* **1987**, *15*, 211-224.

RECEIVED June 2, 1994

Chapter 16

Role of Nonbonded Interactions in Determining Solution Conformations of Oligosaccharides

Robert J. Woods[1], Bert Fraser-Reid[2], Raymond A. Dwek[1], and Christopher J. Edge[1]

[1]Glycobiology Institute, Department of Biochemistry, University of Oxford, South Parks Road, Oxford OX1 3QU, England
[2]Department of Chemistry, Paul M. Gross Chemical Laboratory, Duke University, Durham, NC 27706

Molecular dynamics simulations have been performed on disaccharide models for the man-α-(1→2, 1→3, and 1→6)-man-α linkages in oligosaccharides that are found in asparagine-linked glycoproteins. The recently reported GLYCAM_93 parameter set was employed with the AMBER force field. The hydrogen bonding between the sugar residues and between the sugar residues and the solvent was optimised through the use of HF/6-31G* ESP-derived partial atomic charges that are specific for each monosaccharide. In the case of the 1→6 linkage, the *gauche* preference of the ω-angle is found to be a solution phenomenon. The conformations of the disaccharides were determined over 250-500 ps simulations in water and are compared to the conformations of the corresponding linkages in the oligosaccharide $Man_9GlcNAc_2OH$. In all cases the unconstrained dynamics led to conformations that were in agreement with current NMR data.

The carbohydrates that are covalently attached to proteins (glycoproteins) play an important role in the structure and function of biological systems, being involved in, for example, host-pathogen responses, cell-cell recognition, and hormonal control mechanisms (*1,2*). It is therefore essential that the spatial and dynamic properties of these molecules be accurately determined. Experimental techniques, such as nuclear magnetic resonance (NMR) (*3-6*) and fluorescence energy transfer spectroscopy (*7*) have been valuable aids in structural studies of oligosaccharides and glycoproteins. While NMR spectroscopy has been widely applied to oligosaccharides, the conformations of the glycosidic linkages in these systems are particularly difficult to determine. This difficulty arises largely from two factors, namely, the characteristic paucity of nuclear Overhauser effects (*8*) and the uncertain accuracy of Karplus-type relationships derived from heteronuclear J-coupling constants (*9*).

0097–6156/94/0569–0252$08.00/0

Computational approaches to the problem of glycoprotein and oligosaccharide conformational analysis provide an alternative technique. The use of molecular dynamics (MD) simulations in the analysis of protein and nucleic acid structures is a standard technique that has developed over the past decade (*10*). Numerous force fields and parameter sets have been reported for modelling carbohydrates (*11-17*), however, none has as yet gained widespread acceptance for application to MD simulations of oligosaccharides. Although several different macromolecular force fields, such as CHARMm (*18*), AMBER (*19,20*), GROMOS (*21*) and HSEA (*22*) have been used to examine the conformations of oligosaccharides, in one recent report (*23*) the authors concluded that no current molecular mechanical force field for oligosaccharides is sufficiently accurate for unrestrained MD simulations. We report here preliminary results of calculations on three biologically relevant disaccharides, in which we employ the GLYCAM_93 parameter set (*24*) for oligosaccharides and glycoproteins with the AMBER force field and apply no experimental constraints.

Nonbonded Interactions

It is known that the nonbonded terms within the force field play crucial roles in determining the conformations. Each force field may contain a slightly different mathematical expression for the nonbonded terms. In the AMBER force field, the nonbonded terms take the form:

$$\sum_{i>j}\left(\frac{A_{ij}}{R_{ij}^{12}}-\frac{B_{ij}}{R_{ij}^{6}}\right)+\sum_{i>j}\frac{q_i q_j}{\varepsilon R_{ij}}+\sum_{k>l}\left(\frac{C_{kl}}{R_{kl}^{12}}-\frac{D_{kl}}{R_{kl}^{10}}\right) \quad (1),$$

where the first term in eq 1 describes dispersive or the van der Waals interactions, the second electrostatic or coulombic interactions, and the third is reserved for hydrogen bond interactions.

Despite the hydrophilic character of oligosaccharides, dispersive or "stacking" interactions between the monosaccharide residues within a given oligosaccharide and also between carbohydrates and proteins have been reported (*25,26*). These interactions are believed to create a very rigid overall conformation for the oligosaccharide, which is likely to be important for the specific biological recognition of such oligosaccharides by receptor molecules (*27*).

Specificity in carbohydrate-protein interactions may also be conferred through hydrogen bonding (*25,26,28*). Both intra- and intermolecular hydrogen bonds may be present in carbohydrates, a consequence of the abilities of the hydroxyl groups to act simultaneously as proton acceptors and donors (*29-31*). Moreover, the polyhydroxylated character of sugars, and the rotational freedom of the hydroxyl groups, may create a dynamic network of hydrogen bonds. The difficulty of modelling such a network of hydrogen bonds in solution is compounded by the ability of water itself to form as many as five hydrogen bonds per solvent molecule (*32*). In biological systems, consideration must be given to the influence that the carbohydrate may have on the surrounding water molecules. This influence may expand the effective surface of a carbohydrate far beyond that which may be calculated simply from consideration of the van der Waals' radii.

We now report the results of molecular dynamics (MD) simulations on three disaccharides, namely, the glycosides: methyl 2-, 3-, and 6-*O*-α-D-mannopyranosyl-α-D-mannopyranoside (compounds **1**, **2** and **3**, respectively). These disaccharides serve as models for the corresponding linkages in the oligomannose structure $Man_9GlcNAc_2OH$ (**4**), illustrated in Figure 1. This oligosaccharide is present on many *N*-linked glycoproteins and may, *in vivo*, be converted to smaller oligosaccharides through cleavage or "trimming" of the terminal sugar residues. The trimming of oligosaccharide **4** in *N*-linked glycoproteins generally follows a specific sequence beginning with the cleavage of the man-α-(1→2)-man-α linkages (*33,34*). Such an altered oligosaccharide may be then further elaborated through the addition of other sugar residues, mediated by the action of glycosyl transferases, to generate a diverse variety of structures. However, all *N*-linked glycoproteins contain a common pentasaccharide core (residues 1, 2, 3, 4, and 4' in Figure 1). In the discussions below, the following torsion angle definitions are applied: $\phi = H_1$-C_1-O_X'-C_X', $\psi = C_1$-O_X'-C_X'-H_X', where the prime refers to the aglycon, and in the case of the 1→6 linkages, $\omega_o = O_6$-C_6-C_5-O_5 and $\psi = C_1$-O_6'-C_6'-C_5'.

1→2 Linkage

In the case of the man-α-(1→2)-man-α linkages in **4**, it has been proposed that, after the removal of the first mannopyranosyl residue in the endoplasmic reticulum (ER), a conformational change occurs that facilitates the cleavage of the next residue, through the action of α-mannosidases in the Golgi apparatus (*34*). A comparison of the conformations of the disaccharides with those of the corresponding linkages in **4** should provide an indication of the variations induced by the inter- and intramolecular interactions.

There are four man-α-(1→2)-man-α linkages in **4**. Three of these linkages (D1-C, D2-A, D3-B) occur between the penultimate and the terminal residues of each arm of the oligosaccharide. The fourth immediately precedes the D1-C linkage (C-4). *In vivo* the 1→2 linkage between rings D2 and A is cleaved by the action of an α-mannosidase in the ER (*35-37*); the remaining 1→2 linkages are cleaved by the action of α-mannosidases in the Golgi apparatus (*38*). Consequently, the conformations of the man-α-(1→2)-man-α linkages are of direct relevance to an understanding of the action of mannosidases in the biosynthetic processing of *N*-linked glycoproteins. The conformation preferred by the terminal man-α-(1→2)-man-α linkages in **4** have been examined both experimentally, using ^{1}H NMR spectroscopy (*8,39,40*), and theoretically (*17,39*). For the 1→2 linkage between rings D2 and A these conformations have been characterised by ϕ,ψ values (±10°) of -40°,-20° (**4.1**) and -40°,60° (**4.2**) (*8*). Conformation **4.1** has been reported to be close to a minimum on the potential energy surface from both gas-phase semiempirical molecular orbital calculations (*39*), and gas-phase molecular mechanical calculations (*17*).

Molecular Dynamics. To determine the extent to which solvation altered the conformational properties of the man-α-(1→2)-man-α linkage we performed an "all atom" MD simulation in water with the GLYCAM_93 parameter set on **1**. We began the simulation by solvating the energy minimised structure corresponding to conformation **4.1**. After an initial period of approximately 15 ps, during which the conformation showed little change, a slow spontaneous transition in the ψ angle occurred spanning a further 40 ps as illustrated in Figure 2. Thus, at the end of the first 55 ps a new conformation developed that remained throughout a further 213 ps of the simulation (only the first 250 ps are shown in Figure 2). The average values of the glycosidic angles for the period 55-268 ps were $\phi = -40 \pm 9°$ and $\psi = 47 \pm 13°$. This new conformation, which is the equilibrated structure, was in close agreement with experimental conformation **4.2** for the D2-A linkage, and suggests that the terminal residues may adopt conformations that are unaffected by the bulk of the oligosaccharide. An examination of this structure indicated that it formed a sandwich in which the hydrophobic faces of the residues were facing each other. A similar structure for the 1→2 mannobioside was postulated over 10 years ago as a rationalisation for the observed NMR chemical shifts of the anomeric protons (*41*).

In the case of the nonterminal man-α-(1→2)-man-α linkage (C-4) the conformation has been determined experimentally in both **4** and in a related oligosaccharide, in which each of the terminal 1→2 linkages are absent ($Man_6GlcNAc_2Asn$). In the examination of **4**, it was concluded that all of the 1→2 linkages adopted conformation **4.1** (*39*), whereas, in the case of the latter compound it was reported that the 1→2 linkage adopted a slightly altered geometry, namely, with $\phi = -45 \pm 20°$ and $\psi = 20 \pm 20°$ (*42*). It should be noted that the nOe data employed to assign conformation **4.1** (*39*) were later shown to be consistent with either conformation **4.1** or **4.2** (*8*). The value for the ψ angle for the 1→2 linkage in $Man_6GlcNAc_2$ (20 ± 20°) is intermediate between that of conformers **4.1** (-20°) and **4.2** (60°) and may arise either from conformational averaging or from a unique conformation. However, both values for the ψ angle fall within the estimated error limits for our MD-derived value for this angle (47 ± 13°).

Hydrogen Bonding. We then examined the equilibrium conformation to determine the reason for its stability in solution. We focused our examination on the hydrogen bonding network between the sugar and any bound water molecules. The presence of an inter-residue hydrogen bond between the hydroxyl proton of the hydroxymethyl group and the ring oxygen atom (HO-6•••O-5') has been reported for the computed gas-phase structure (*39*). Although the extent of the persistence of inter-residue hydrogen bonding in solution is unknown, such bonds have previously been assumed to be absent (*39,43*).

To address this question, we performed a search of the dynamics trajectory (55-268 ps) for this or any other hydrogen bonds between two sugar oxygen atoms

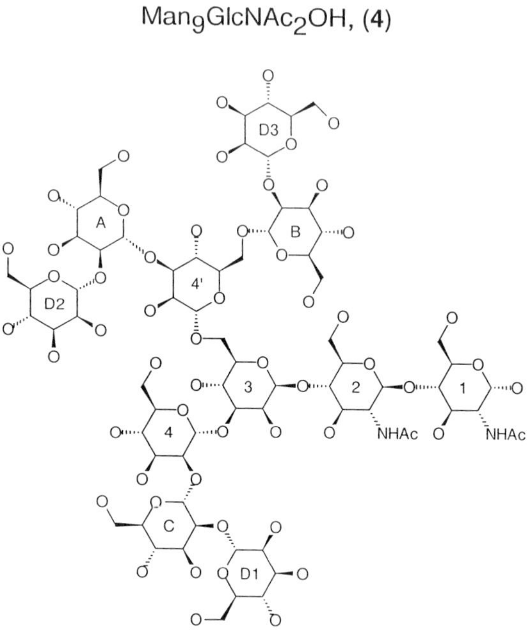

$Man_9GlcNAc_2OH$, (**4**)

Figure 1. Schematic representation of the oligosaccharide $Man_9GlcNAc_2OH$. Hydrogen atoms have been omitted for clarity.

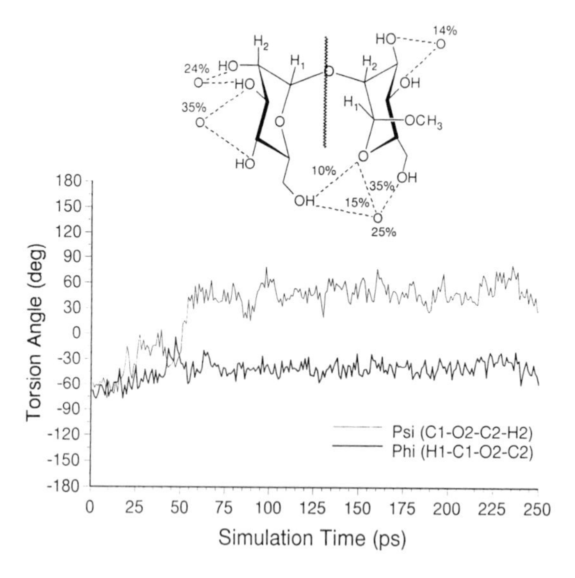

Figure 2. Molecular dynamics trajectory and schematic illustration for the glycosidic angles of the 1→2 linkage in **1**. The percentages refer to hydrogen bond residency times.

either of a direct nature or mediated by a bridging water atom. The search was limited to strong hydrogen bonds by accepting only water molecules whose oxygen atoms were within 3.0 Å of both sugar oxygen atoms. Furthermore, the possibility of a hydrogen bond between HO-6 and O-5' was accommodated in the search by accepting any conformations in which the separation between O-6 and O-5' was less than 3.0 Å. The results of the search indicated that HO-6 was hydrogen bonded to O-5' for approximately 10% of the simulation period from 55-268 ps. Moreover, it was observed that any one of several water molecules formed hydrogen bonds simultaneously to both O-6 (or HO-6) and O-5' for a total of approximately 15% of the same simulation period (see Figure 2). Further, hydrogen bonding between a water molecule and both hydroxymethyl groups existed for approximately 25% of the simulation. Thus, the equilibrated conformation was stabilised by hydrogen bonding for approximately 25 to 40% of the MD simulation. Similar bridging water residency times were found for water molecules coordinated to vicinal hydroxyl groups (intra-residue hydrogen bonds). However, in analogy to DNA base-stacking interactions, in the stacked conformation of **1**, dispersive or van der Waals interactions would be expected to make a large stabilising contribution (*44*). The extent of these interactions is the subject of continuing research employing free energy perturbation and MD simulations.

1→3 Linkage

The branched structure of oligosaccharide **4** arises from the presence of two mannopyranosyl residues to each of which are attached sugars at the 3- (A-4' and 4-3) and 6-positions (B-4' and 4'-3). Once each of the 1→2 linkages has been cleaved during *in vivo* processing, the 1→3 and 1→6 linked residues become susceptible to further enzymatic cleavage. The extent to which this sequential processing depends on the ability of the residues to adopt a conformation favourable to enzymatic hydrolysis is not known. Numerous experimental studies have sought to determine the conformation of man-α-(1→3)-man linkages in mannobiose and larger oligosaccharides (*8,23,39,42,45-48*). Theoretical studies of this linkage have been reported that have employed molecular mechanical (*23,46,49*) and semiempirical quantum mechanical (*50*) approaches. Both experimental and theoretical values for the glycosidic angles have indicated two possible conformations, namely, $\phi = -45 \pm 15°$, $\psi = -15 \pm 15°$ (**4.3**) (*42*) and $\phi = -20 \pm 20°$, $\psi = 30 \pm 20°$ (**4.4**) (*39,50,51*). However, the magnitude of the uncertainties in the NMR-derived structure for this linkage in **4** are sufficiently large that conformations **4.3** and **4.4** may be indistinguishable (*8*). In the case of the 1→3 mannobioside NMR data support conformation **4.3** (*46*).

Molecular Dynamics. Following a protocol analogous to that employed with **1**, we performed a 500 ps simulation on the the 1→3 linked mannobioside **2**. The values for the ϕ (-56 ± 11°) and ψ (-32 ± 16°) angles exhibited only small oscillations throughout the simulation, with the exception of a brief transition at approximately 130-135 ps (see Figure 3). This conformation is in excellent agreement with that derived from NMR data for the 1→3 linkages in both the mannobioside and larger oligomannosides (-45 ± 15°, -15 ± 15°). However, to determine whether the second proposed conformation might be stable, a second simulation was performed with initial values for the torsional angles corresponding to those reported for **4.4**. The variations in the glycosidic torsion angles during the second simulation are illustrated in Figure 4. For approximately 100 ps the trajectories for the ϕ and ψ angles oscillated about mean values of -30° and 65°, respectively, before reverting to the earlier conformation (-56°,-32°). It is noteworthy that the transient conformation existed for a considerable period during the second simulation. Moreover, the exhibited values for the torsional angles were in reasonable agreement with those estimated for **4.4**. However, the spontaneous transition back to the earlier conformation supports calculations that indicated that ϕ = 40°, ψ = 30° corresponds to a shallow local minimum on the potential energy surface for the mannobioside (*46*).

Hydrogen Bonding. A hydrogen bonding analysis of the 500 ps MD trajectory for **2** indicated that a hydrogen bond between HO-2' and the glycosidic oxygen was present (HO-2'•••O-3' separation less than 2.4 Å) for approximately 16% of the simulation. Moreover, a water molecule was found to occupy a bridging position between the two oxygen atoms (O-5•••Ow and Ow•••O-4' less than 3.0 Å) for a further 22%. A network of hydrogen bonds between O-6, O-5, O-1' and O-4' and one or two water molecules was also present for approximately 40% of the same period (see Figure 3). However, an examination of the geometry of this conformation indicated that O-5 and O-4' were never close enough to hydrogen bond directly to each other.

In comparison, the conformation present initially in the second simulation (ϕ = -40°, ψ = 30°) displayed an arrangement of inter-residue hydrogen bonds that has previously been computed to be present in the gas-phase minimum energy conformation of this linkage in a related *N*-linked oligosaccharide (*52*). In this conformation we found hydrogen bonds between O-2' and both O-5 and O-6 for approximately 50% of this short simulation period (90 ps). Interestingly, Carver and Cumming reported that this second conformation was stable only when the computational method (HSEA) included a hydrogen bonding term (*52*), and that in the absence of a hydrogen bond potential the conformation reverted to that shown in Figure 3. Our analysis indicates that both conformations exhibit inter-residue hydrogen bonds, however, the equilibrium conformation also exhibits an extensive water-mediated inter-residue hydrogen bonding network.

1→6 Linkage

The presence of a third rotatable bond in 1→6 linkages, as compared to either 1→2 or 1→3 linkages, is expected to lead to enhanced flexibility in this linkage. Most of the experimental attempts to elucidate the flexibility of this linkage have employed NMR spectroscopy. Based on early NMR studies, it was proposed that there was

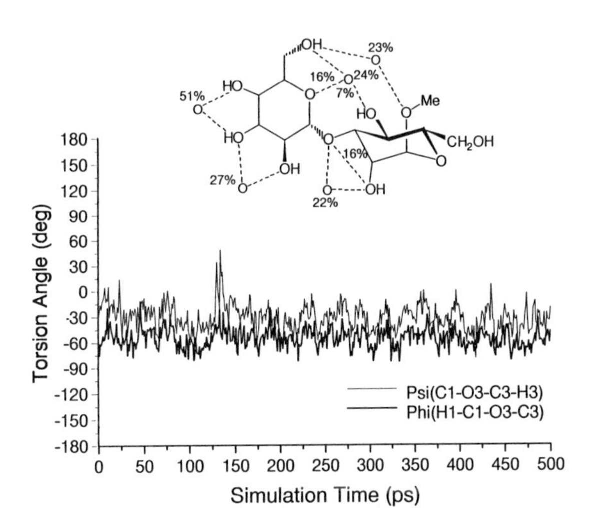

Figure 3. Initial molecular dynamics trajectory for the 1→3 linkage in **2**. See the caption with Figure 2 for further description.

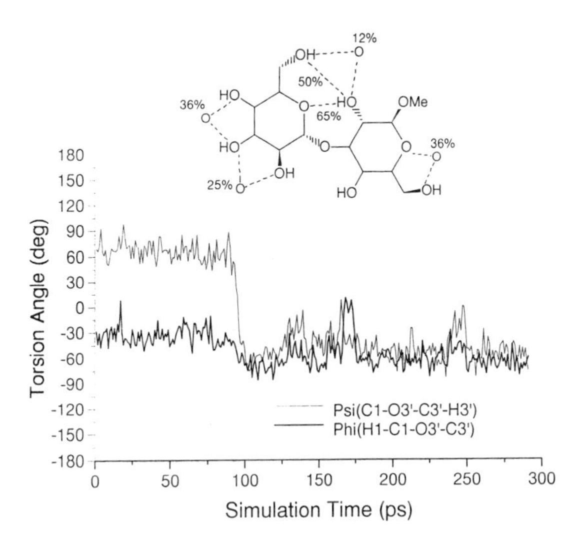

Figure 4. Second molecular dynamics trajectory for the 1→3 linkage in **2**. See the caption with Figure 2 for further description.

free rotation around the C5-C6 bond (*45,47*). However, a considerable body of experimental data has since accumulated that indicates that only two conformations are found (*8,43,46,53-55*). In both of the preferred conformations O-6 adopts a *gauche* orientation with respect to the ring oxygen atom (O-5) and may be either *gauche* to C4 (the *gg* rotamer, $\omega_o = -60 \pm 20°$) or *trans* to C4 (the *gt* rotamer, $\omega_o = 60 \pm 20°$). This linkage, like the 1→3 linkage, is found at branch points in *N*-linked oligosaccharides and variations in the ω-angle could lead to large conformational changes (*55*).

Recent evidence from fluorescence energy transfer experiments suggests that the orientations of these linkages in *N*-linked oligosaccharides depend on the oligosaccharide sequence (*56*). In the case of oligosaccharide **4** there are two 1→6 linkages (4'-3, and B-4') and their orientations appear to depend on the presence or absence of the terminal sugar residues (D3 and D2). In the presence of these terminal residues, both linkages exhibit a preference for the *gg* conformation, whereas, in their absence a mixture of *gg* and *gt* conformations may be present. As the size of the oligosaccharide decreases, there appears to be an increasing preference for the *gt* rotamer (*43,46,54*).

The orientations of the φ and ψ angles in this linkage have received somewhat less attention. It has generally been assumed that the φ angle would adopt an orientation that would be preferred by the *exo*-anomeric effect (*57*) despite the observation that the two commonly observed nOe's are consistent with two values for the φ angle (±50°) (*8*). In the case of mannobioside **3**, values for the φ, ψ and ω_o angles of -50 ± 20°, 90-200° and approximately 60°, respectively are consistent with experimentally determined nOe and T_1 relaxation data (*46*). Based on heteronuclear coupling constant values for a ^{13}C enriched sample of a mannotriose closely related to **3**, a more precise estimate of the value of the ω_o angle (190 ± 20°) has been reported (*46*).

Molecular Dynamics. To explore the effect of the orientation of the ω-angle on the MD simulations we performed three simulations. Each initial configuration employed one of the three staggered orientations for this angle ($\omega_o = 60°$, *gt*; $\omega_o = -60°$, *gg*; $\omega_o = 180°$, *tg*). As expected, the *tg* configuration was unstable (*58,59*) and rapidly converted to the *gt* conformation, remaining there throughout the rest of the simulation. In the case of the *gt* simulation, no transitions in either the φ or the ω angles were observed. The φ angle displayed a very stable trajectory, giving rise to a single orientation (-54 ± 12°), in excellent agreement with experiment, as did the ω_o angle (48.9 ± 11°) (*46*). Despite the fact that this simulation was initiated in the experimentally determined conformation, the ψ angle was found to oscillate between two values, namely 76° and 164°. While the average value of ψ (85 ± 30°) is within the broad experimental estimates for this angle, it is more appropriate to view the trajectory as representing a mixture of two conformations (ψ = 76 ± 14° and ψ = 164 ± 18°). The trajectory for the simulation of the *gg* conformer is illustrated in Figure 5. As in the cases of the *tg* and *gt* rotamers (see Figure 6), the φ angle remained in the orientation preferred by the *exo*-anomeric effect and observed experimentally (-51 ± 12°). However, the ψ and ω angles both displayed transitions. The ω_0 angle remained in the *gg* orientation (-48° ± 13°) for the first 250 ps of the simulation before converting to the *gt* conformation (52° ± 11°). It is noteworthy that within 2 ps of the rotation of the ω angle from *gg* to *gt* the ψ angle underwent a transition from 182 ± 11° to 79 ± 14°. Ongoing simulations on larger oligosaccharides will address the issue of flexibility of the ψ angle in 1→6 linkages.

Each simulation reached the same final conformation (*gt*) regardless of the starting configuration. However, the *gg* conformation was observed for a significant period (~250 ps). If the first 50 ps of each simulation are allowed for equilibration and the remaining times employed for analysis, then the ratio of *gt* to *gg* conformations is approximately 4:1. Both this population ratio, and the individual conformations, are consistent with experimental data for **3** (*46*). The observation of two conformations for the ψ angle indicates that these conformations may be similar in energy, a feature which has been previously reported (*46*).

Based on these simulations it appears that there may be a considerable contribution to the flexibility of the 1→6 linkage through transitions in the ψ angle. The simulations do not indicate that the ω angle is more flexible than either of the other torsion angles. However, increased flexibility of this linkage may arise also from the combined motions of each torsion angle.

Hydrogen Bonding. In the case of the *gg* conformation, the possibility of a hydrogen bond between O-6 and O-4' has been reported (*43,54*). However, it was assumed that rotation of the hydroxymethyl group (O-6) and solvent intervention would severely attenuate such an interaction (*43,54*). An examination of the first 250 ps of the *gg* trajectory revealed that at no time were O-6 and O-4' within direct hydrogen bonding distance (less than 3.2 Å). However, when the search was extended to include bridging water molecules, water was found to coordinate to both O-6 and O-4' for 30% of the simulation. This value is approximately the same as that found for waters that were coordinated to the vicinal oxygen atoms (O-2/O-3, O-3/O-4). Given the geometrically fixed relationship of vicinal hydroxyl groups, the water coordination time of 30% must be approximately the maximum possible value for this linkage.

Unlike the *gg* conformer, no inter-residue hydrogen bonds, either bridged by water or direct, were detected for the *gt* conformer. Moreover, in the *gt* conformer the average duration of interactions between vicinal oxygen atoms and water molecules was decreased by a factor of approximately 2. The infrequent hydrogen bonding between the *gt* conformer and the water indicates that this conformation must be considerably more mobile than the *gg*. It seems reasonable then to expect that the 1→6 linkages in larger, and less mobile, oligosaccharides would prefer the *gg* conformation.

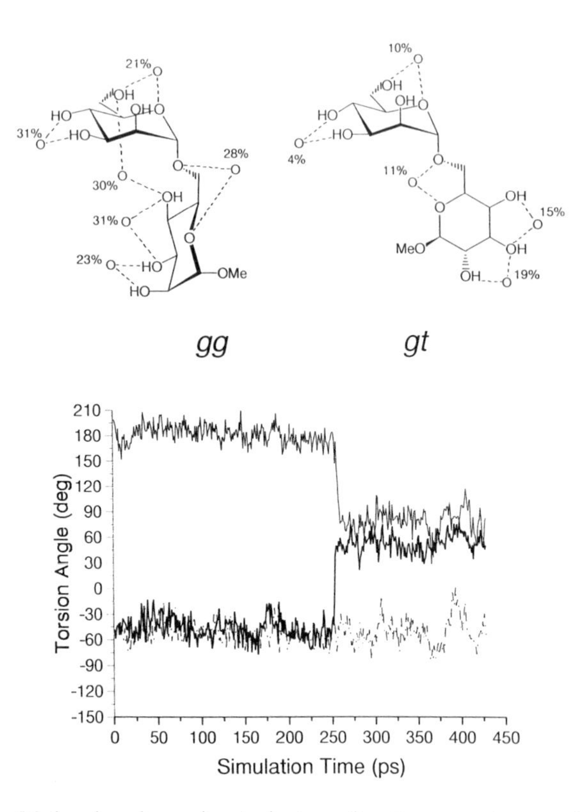

Figure 5. Molecular dynamics trajectory for the *gg* rotamer of the 1→6 linkage in **3**. Legend: ϕ dashed line, ψ thin line, ω_0 thick line. See the caption with Figure 2 for further description.

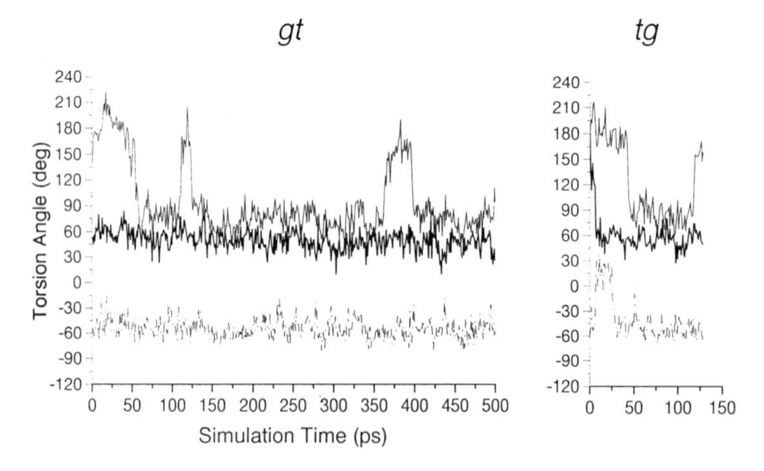

Figure 6. Molecular dynamics trajectory for the *gt* and *tg* rotamers of the 1→6 linkage in **3**. Legend as in Figure 5.

Conclusions

The successful application of a newly derived parameter set for oligosaccharides (GLYCAM_93) in an analysis of the conformational properties of three disaccharides has been reported. In all cases the orientation of the ϕ angles was that preferred by the *exo*-anomeric effect. In contrast, the value of the ψ angle showed marked variation. Van der Waals, hydrogen bond, and bulk solvent interactions each play a role in determining the conformation of the ψ angle. As expected, the 1→6 linkages exhibited only *gg* and *gt* populations, with the latter being strongly preferred. The MD simulations indicate the need for explicit inclusion of water when modelling the solution properties of carbohydrates. Moreover, comparatively long trajectories (250-500 ps or longer) are required to generate a reasonable description of the properties of an oligosaccharide in solution. Had either short simulation times, or nOe constraints been imposed, several of the spontaneous transitions would not have been observed. In each simulation the trajectories led to conformations that were in good agreement with available experimental data. In the case of the 1→2 linkage the final stacked conformation exhibited both van der Waals and hydrogen bonded inter-residue interactions. In contrast the 1→3 linkage appears to be stabilised by a network of interactions involving bound water molecules. Lastly, in the 1→6 linkage, no direct inter-residue hydrogen bonds were observed, and only one water-bridged hydrogen bond was detected. Although this bridging water appears to coordinate strongly in the *gg* conformation, no similar interactions were found in the preferred *gt* conformation. The lack of any discrete nonbonded inter-residue interactions in the 1→6 mannobioside implies that the preference for the *gt* conformation is related to bulk solvation properties.

Computational Method

Molecular dynamics (MD) simulations were performed using the recently reported GLYCAM_93 parameter set (*24*). This parameter set utilises torsion potentials that were derived from ab initio calculations performed at the HF/6-31G* level on a series of tetrahydropyran derivatives that serve as models for the glycosidic linkages. Each ϕ-angle for an α-configuration employed the torsion potential illustrated in Figure 7. The AMBER torsion profile was generated with the following coefficients (kcalmol^{-1}) and phases: $V_1 = 1.39$, $V_2 = 0.70$, $V_3 = 0.91$, $\gamma_1 = 265.77°$, $\gamma_2 = 312.04°$, $\gamma_3 = 347.72°$. As the preference for a *gauche* conformation appears to arise from solvation or crystal matrix effects (*58,59*) neither the "Hassel-Ottar" (*60,61*) nor "*gauche*" (*62,63*) effects were incorporated explicitly into GLYCAM_93. The AMBER torsion profile for O-C-C-O rotation was generated with the following coefficients and phases: $V_1 = 1.34$, $V_2 = -1.15$, $V_3 = 0.77$, $\gamma_1 = 0.0°$, $\gamma_2 = 180.0°$, $\gamma_3 = 0.0°$, and is illustrated in Figure 8. An accurate description of inter-atomic coulombic interactions, such as present in hydrogen bonding, was provided in the parameter set through unique sets of partial atomic charges for each sugar residue, that reproduce the molecular electrostatic potentials (ESPs) at the HF/6-31G* level. The partial atomic charges used in the simulations are presented in Table I. All MD simulations were performed with the

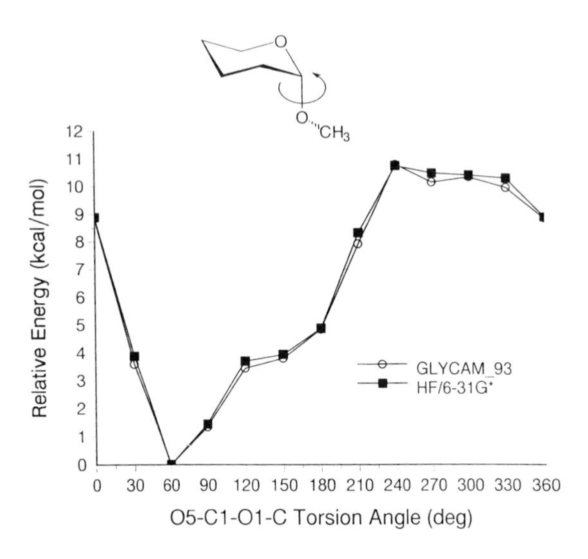

Figure 7. Relative energy profile for rotation of the ϕ angle in axial 2-methoxytetrahydropyran (a model for α-glycosides).

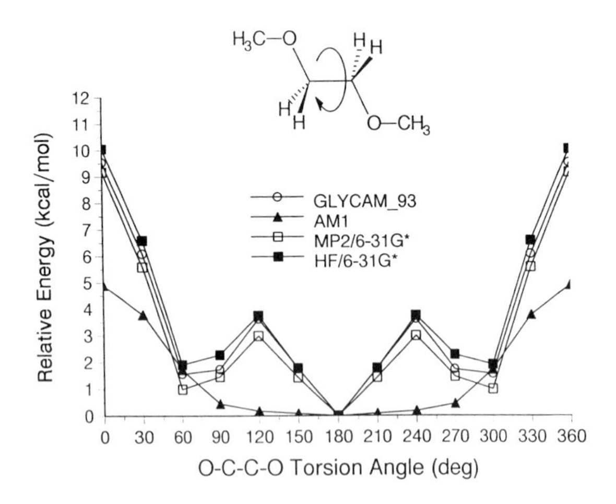

Figure 8. Relative energy profile for rotation of the ω_0 angle in a model for the 1→6 linkage.

Table I: Net Atomic Charges from the HF/6-31G* Electrostatic Potentials

Atom	**1**	**2**	**3**
C-1/C-1'	0.3331	0.3331	0.3331
C-2/C-2'	0.1749	0.1749	0.1749
C-3/C-3'	0.0935	0.0935	0.0935
C-4/C-4'	0.7630	0.7630	0.7630
C-5/C-5'	0.0420	0.0420	0.0420
C-6/C-6'	0.2410	0.2410	0.2410
C-Me	0.1964	0.1964	0.1964
O-2	-0.7216	-0.7216	-0.7216
O-3	-0.7495	-0.7495	-0.7495
O-4/O4'	-0.8245	-0.8245	-0.8245
O-5/O5'	-0.4822	-0.4822	-0.4822
O-6	-0.7085	-0.7085	-0.7085
O-1'	-0.5099	-0.5099	-0.5099
O-2'	-0.4976	-0.7216	-0.7216
O-3'	-0.7495	-0.5301	-0.7495
O-6'	-0.7085	-0.7085	-0.5054
H-1/H-1'	0.1264	0.1264	0.1264
H-2/H-2'	0.0334	0.0334	0.0334
H-3/H-3'	0.0537	0.0537	0.0537
H-4/H-4'	-0.0477	-0.0477	-0.0477
H-5/H-5'	0.0463	0.0463	0.0463
H-6a/H-6a'	0.0209	0.0209	0.0209
H-6b/H-6b'	0.0179	0.0179	0.0179
Ha-Me	0.0032	0.0032	0.0032
Hb-Me	0.0427	0.0427	0.0427
Hc-Me	0.0271	0.0271	0.0271
HO-2	0.4645	0.4645	0.4645
HO-3	0.4599	0.4599	0.4599
HO-4/HO-4'	0.4604	0.4604	0.4604
HO-6	0.4436	0.4436	0.4436
HO-2'	-------	0.4645	0.4645
HO-3'	0.4599	-------	0.4599
HO-6'	0.4436	0.4436	-------

MINMD module of AMBER 4.0 (*64*). Calculations were performed on an IBM RISC 6000/580 or a DEC Alpha AXP 3000/500 computer. All atoms were treated explicitly. The disaccharides were solvated in a box of approximately 570 Monte Carlo TIP3P (*65*) water molecules, with approximate dimensions 30 Å X 26 Å X 25 Å. The initial configurations were then subjected to 1000 cycles of steepest descent energy minimisation. All minimisations and subsequent dynamics were performed with a dielectric constant of unity and a cutoff value for nonbonded pair interactions of 10.0 Å. The standard value of 0.5 was used to scale all 1-4 interactions. Newton's equations of motion were integrated using a Verlet algorithm with a 1 fs time step. Initial atomic velocities were assigned from a Maxwellian distribution at 5 K. During the dynamics a constant pressure of 1 atm was maintained with isotropic position scaling employing a pressure relaxation time of 0.2 ps^{-1}. A constant temperature of 300 K was maintained through weak coupling to an external bath with a coupling constant of 0.25 ps^{-1}. All bond lengths were constrained to their equilibrium values through application of the SHAKE algorithm (*66*), which is consistent with the choice of solvation model. The simulations were performed for at least 250 ps total time, from which an initial period of approximately 50 ps was employed for equilibration. The dynamics simulations were not constrained in any way to reproduce experimental observables, such as nOe intensities.

Acknowledgments. We thank the North Carolina Supercomputing Center and the Oxford Centre for Molecular Sciences for use of their facilities. This work was supported in part by a grant from Monsanto Co., and by the Natural Sciences and Engineering Research Council of Canada, in the form of a Fellowship (to R.J.W.).

Literature Cited

1. Rademacher, T. W.; Parekh, R. B.; Dwek, R. A. *Ann. Rev. Biochem.* **1988**, *57*, 785-838.
2. Varki, A. *Glycobiology,* **1993**, *3*, 97-130.
3. Lemieux, R. U.; Bock, K. *Arch. Biochem. Biophys.* **1983**, *221*, 125-134.
4. Homans, S. W. *Prog. NMR Spect.* **1990**, *22*, 55-81.
5. Tvaroska, I.; Kozár, T.; Hricovíni, M. In *Computer Modeling of Carbohydrate Molecules*; French, A.D. and Brady, J.W., Eds.; ACS Symposium Series 430; American Chemical Society: Washington, D.C., 1990; pp 162-176.
6. Carver, J. P.; Mandel, D.; Michnick, S. W.; Imberty, A.; Brady, J. W. In *Computer Modelling of Carbohydrate Molecules*; French, A.D. and Brady, J.W., Eds.; ACS Symposium Series 430; American Chemical Society: Washington, D.C., 1990; pp 266-280.
7. Rice, K. G.; Wu, P.; Brand, L.; Lee, Y. C. *Biochemistry,* **1991**, *30*, 6646-6655.
8. Wooten, E. W.; Edge, C. J.; Bazzo, R.; Dwek, R. A.; Rademacher, T. W. *Carbohydr. Res.* **1990**, *203*, 13-17.
9. Cano, F. H.; Foces-Foces, C.; Jiménez-Barbero, J.; Alemany, A.; Berbabé, M.; Martín-Lomas, M. *J. Org. Chem.* **1987**, *52*, 3367-3372.
10. van Gunsteren, W. F.; Mark, A. E. *Eur. J. Biochem.* **1992**, *204*, 947-961.
11. Norskov-Lauritsen, L.; Allinger, N. L. *J. Comput. Chem.* **1984**, *5*, 326-335.
12. Tvaroska, I.; Pérez, S. *Carbohydr. Res.* **1986**, *149*, 389-410.

13. Ha, S. N.; Giammona, A.; Field, M.; Brady, J. W. *Carbohydr. Res.* **1988**, *180*, 207-221.
14. Homans, S. W. *Biochemistry,* **1990**, *29*, 9110-9118.
15. Yan, Z.-y.; Bush, A. C. *Biopolymers,* **1990**, *29*, 799-811.
16. Scarsdale, J. N.; Ram, P.; Prestegard, J. H.; Yu, R. K. In *Computer Modeling of Carbohydrate Molecules*; French, A.D. and Brady, J.W., Eds.; ACS Symposium Series 430; American Chemical Society: Washington, D.C., 1990; pp 240-265.
17. Edge, C. J.; Singh, U., C.; Bazzo, R.; Taylor, G. L.; Dwek, R. A.; Rademacher, T. W. *Biochemistry,* **1990**, *29*, 1971-1974.
18. Brooks, B. R.; Bruccoleri, R. E.; Olafson, B. D.; States, D. J.; Swaminathan, S.; Karplus, M. *J. Comput. Chem.* **1983**, *4*, 187-217.
19. Weiner, S. J.; Kollman, P. A.; Case, D. A.; Singh, U. C.; Ghio, C.; Alagona, G.; Profeta, S., Jr.; Weiner, P. *J. Am. Chem. Soc.* **1984**, *106*, 765-784.
20. Weiner, S. J.; Kollman, P. A.; Nguyen, D. T.; Case, D. A. *J. Comput. Chem.* **1986**, *7*, 230-252.
21. Hermans, J.; Berendsen, H. J. C.; Van Gunsteren, W. F.; Postma, J. P. M. *Biopolymers,* **1984**, *23*, 1513-1518.
22. Thogersen, H.; Lemieux, R. U.; Bock, K.; Meyer, B. *Can. J. Chem.* **1982**, *60*, 44-57.
23. Rutherford, T. J.; Partridge, J.; Weller, C. T.; Homans, S. W. *Biochemistry,* **1993**, *32*, 12715-12724.
24. Woods, R. J.; Fraser-Reid, B.; Dwek, R. A.; Edge, C. J. *J. Am. Chem. Soc.* **1993**, submitted for publication.
25. Vyas, N. K. *Curr. Opin. Struct. Biol.* **1991**, *1*, 732-740.
26. Vyas, N. K.; Vyas, M. N.; Quiocho, F. A. *J. Biol. Chem.* **1991**, *266*, 5226-5237.
27. Foxall, C.; Watson, S. R.; Dowbenko, D.; Fennie, C.; Lasky, L. A.; Kiso, M.; Hasegawa, A.; Asa, D.; Brandley, B. K. *J. Cell Biol.* **1992**, *117*, 895-902.
28. Drickamer, K. *Nature,* **1992**, *360*, 183-186.
29. Jeffrey, G. A.; Maluszynska, H. *Int. J. Quantum Chem., Quantum Biol. Symposium,* **1981**, *8*, 231-239.
30. Jeffrey, G. A. In *Molecular Structure and Biological Activity*; Griffin, J.F. and Duax, W.L., Eds.; Elsevier Science Publishing Co., Inc.: New York, 1982; pp 135-150.
31. Jeffrey, G. A.; Mitra, J. *Acta Cryst.* **1983**, *39*, 469-480.
32. Sciortino, F.; Geiger, A.; Stanley, H. E. *Nature,* **1991**, *354*, 218-221.
33. Kornfeld, R.; Kornfeld, S. *Ann. Rev. Biochem.* **1985**, *54*, 631-664.
34. Cummings, R. D. In *Glycoconjugates. Composition, Structure, and Function*; Allen, H. J. and Kisailus, E. C., Eds.; Marcel Dekker, Inc.: New York, 1992; pp 331-360.
35. Bischoff, J.; Kornfeld, R. *J. Biol. Chem.* **1983**, *258*, 7907-7910.
36. Byrd, J. C.; Tarentino, A. L.; Maley, F.; Atkinson, P. H.; Trimble, R. B. *J. Biol. Chem.* **1982**, *257*, 14657-14666.
37. Trimble, R. B.; Atkinson, P. H. *J. Biol. Chem.* **1986**, *261*, 9815-9824.
38. Bischoff, J.; Moremen, K.; Lodish, H. F. *J. Biol. Chem.* **1990**, *265*, 17110-17117.

39. Homans, S. W.; Pastore, A.; Dwek, R. A.; Rademacher, T. W. *Biochemistry,* **1987**, *26*, 6649-6655.
40. Woods, R. J.; Edge, C. J.; Wormald, M. R.; Dwek, R. A. In *Complex Carbohydrates in Drug Research*; Bock, K., Clausen, H., Krogsgaard-Larsen, P., and Kofod, H., Eds.; In press; Munksgaard: Copenhagen, Denmark, 1993.
41. Ogawa, T.; Sasajima, K. *Carbohydr. Res.* **1981**, *97*, 205-227.
42. Brisson, J.-R.; Carver, J. P. *Biochemistry,* **1983**, *22*, 3671-3680.
43. Homans, S. W.; Dwek, R. A.; Boyd, J.; Mahmoudian, M.; Richards, W. G.; Rademacher, T. W. *Biochemistry,* **1986**, *25*, 6342-6350.
44. Friedman, R. A.; Honig, B. *Biopolymers,* **1992**, *32*, 145-159.
45. Homans, S. W.; Dwek, R. A.; Fernandes, D. L.; Rademacher, T. W. *FEBS Lett.* **1982**, *150*, 503-506.
46. Brisson, J.-R.; Carver, J. P. *Biochemistry,* **1983**, *22*, 1362-1368.
47. Homans, S. W.; Dwek, R. A.; Fernandes, D. L.; Rademacher, T. W. *FEBS Lett.* **1983**, *164*, 231-235.
48. Cumming, D. A.; Dime, D. S.; Grey, A. A.; Krepinsky, J. J.; Carver, J. P. *J. Biol. Chem.* **1986**, *261*, 3208-3213.
49. Imberty, A.; Tran, V.; Pérez, S. *J. Comput. Chem.* **1989**, *11*, 205-216.
50. Homans, S. W.; Dwek, R. A.; Rademacher, T. W. *Biochemistry,* **1987**, *26*, 6571-6578.
51. Homans, S. W.; Dwek, R. A.; Rademacher, T. W. *Biochemistry,* **1987**, *26*, 6553-6560.
52. Carver, J. P.; Cumming, D. A. *Pure Appl. Chem.* **1987**, *59*, 1465-1476.
53. Brisson, J.-R.; Carver, J. P. *Biochemistry,* **1983**, *22*, 3680-3686.
54. Cumming, D. A.; Carver, J. P. *Biochemistry,* **1987**, *26*, 6676-6683.
55. Wooten, E. W.; Bazzo, R.; Edge, C. J.; Zamze, S.; Dwek, R. A.; Rademacher, T. W. *Eur. Biophys. J.* **1990**, *18*, 139-148.
56. Rice, K. G.; Wu, P.; Brand, L.; Lee, Y. C. *Biochemistry,* **1993**, *32*, 7264-7270.
57. Lemieux, R. U.; Koto, S.; Voisin, D. In *Anomeric Effect. Origin and Consequences*; Szarek, W.A. and Horton, D., Eds.; ACS Symposium Series 87; American Chemical Society: Washington, D.C., 1979; Chapter 2.
58. Ha, S.; Gao, J.; Tidor, B.; Brady, J. W.; Karplus, M. *J. Am. Chem. Soc.* **1991**, *113*, 1553-1557.
59. Kroon-Batenburg, I. M. J.; Kroon, J. *Biopolymers,* **1990**, *29*, 1243-1248.
60. Hassel, O.; Ottar, B. *Acta Chem. Scand.* **1947**, *1*, 929-942.
61. Marchessault, R. H.; Perez, S. *Biopolymers,* **1979**, *18*, 2369-2374.
62. Wolfe, S. *Acc. Chem. Res.* **1972**, *5*, 102-111.
63. Kirby, A. J. *The Anomeric Effect and Related Stereoelectronic Effects at Oxygen*; Springer-Verlag: New York, 1983.
64. Pearlman, D. A.; Case, D. A.; Caldwell, J. C.; Seibel, G. L.; Singh, U. C.; Weiner, P.; Kollman, P. A. *AMBER 4.0*; University of California: San Francisco, 1991.
65. Jorgensen, W. L.; Chandrasekhar, J.; Madura, J. D.; Impey, R. W.; Klein, M. L. *J. Phys. Chem.* **1983**, *79*, 926-935.
66. Ryckaert, J.-P.; Ciccotti, G.; Berendsen, H. J. *J. Comput. Phys.* **1977**, *23*, 327-341.

RECEIVED June 2, 1994

Chapter 17

Modeling Hydrogen Bond Energies in Pseudoladder Polymers Based on the Polybenzobisazole Backbone

S. Trohalaki[1], A. T. Yeates[2,3], and D. S. Dudis[2,3]

[1]AdTech Systems Research Inc., 1342 North Fairfield Road, Dayton, OH 45432–2698
[2]Wright Laboratory, Materials Directorate, Wright-Patterson Air Force Base, OH 45433–7750

Semiempirical and *ab initio* studies of the hydrogen bond strength in 2-hydroxy PBO were performed using the model compound 2-(2-hydroxyphenyl)benzoxazole. The calculations were carried out at the AM1, RHF and MP2 levels. The *ab initio* calculations were performed with the 6-31G* basis set and then repeated placing a 6-31+G(2d,2p) basis on the atoms able to participate in the hydrogen bond. The results indicate that the lowest energy conformation in this system contains a hydrogen bond of about 11 Kcal/mole between the OH and the N. This bond increases the phenyl rotational barrier, thereby inducing a higher degree of planarity in the system.

Rigid-rod polymers are prime candidates for applications that require high tensile strength and high modulus. The molecules are naturally aligned with their long axes parallel, which implies that tension in this direction stretches covalent bonds in the molecule, resulting in their high tensile properties. The rigid-rod material *cis* poly(*p*-phenylene benzobisoxazole) (PBO) has the molecular structure shown in Figure 1a with X=O. This material has demonstrated tensile properties that are among the highest for rigid-rod polymers to date.[*1*] The deficiency of rigid-rod polymers is their compressive strength. Fibers made from these materials buckle under compressive stresses that are orders of magnitude lower than can be sustained under tension.[*2*]

Various strategies have been advanced to alleviate the problem of low compressive strengths. One of the most promising is the concept of "hairy rod" molecular composites.[*3*] This is a rigid-rod polymeric system in which coil-like

[3]Corresponding authors

0097–6156/94/0569–0269$08.00/0

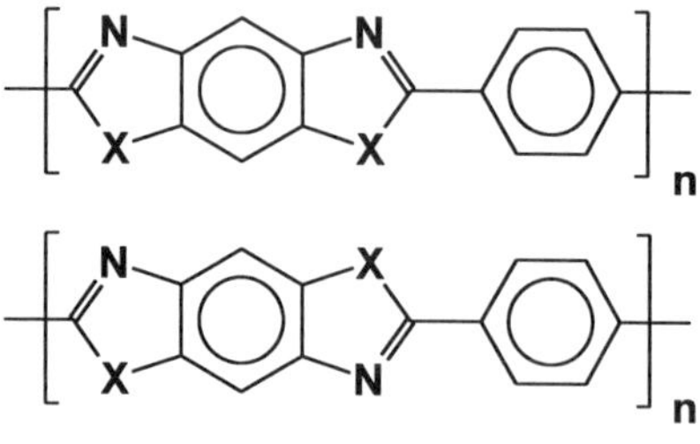

segments are grafted to the rod backbone. In principle, the coil segments can interact with each other and effectively cement the rigid-rod backbones so that they do not easily bend under compression. Attempts to achieve a material with both good tensile and compressive properties *via* this mechanism have so far failed. A molecular dynamics simulation of hairy-rods[4] found that grafting of the coil segments to the rigid-rod increased the flexibility of the rod, which would presumably reduce the tensile modulus. The mechanism for this flexibility comes from a combination of the out-of-plane bending and rotation between the heterocyclic and phenyl rings. Both of these motions have very small spring constants in these systems.[5-8] A possible route to overcome this difficulty may be to modify the structure to include a hydroxyl group at the 2 position of the phenyl ring. It is envisaged that the hydroxyl will form a hydrogen bond with the nitrogen of the benzoxazole ring and increase the rotational spring constant sufficiently to decouple the bending and rotational motions. It is hoped that this will lead to an effective stiffening of the molecule. This intramolecular hydrogen bond is the subject of this paper.

Figure 1: Molecular structure of PBX polymers. (a) *cis*-PBX; (b) *trans*-PBX

The molecules represented in Figure 1 also demonstrate unusually high nonlinear optical (NLO) properties. The bulk third order NLO susceptibility, $\chi^{(3)}$, is partially a consequence of the extended conjugation along the polymer backbone. The polymer *trans* poly(*p*-phenylene benzobisthiazole) (PBT or PBZT), which has the molecular structure represented in Figure 1b with X=S, has a measured $\chi^{(3)}$ value of 4.5 x 10^{-10} esu.[9] This is among the highest non-resonant $\chi^{(3)}$ values of any organic polymer, though still several orders of magnitude below what is needed for most device applications.

One method to increase $\chi^{(3)}$ response is to increase the effective conjugation length of the molecules. Studies have shown that the second hyperpolarizability γ - a molecular quantity related to $\chi^{(3)}$ - has a power law dependence on the conjugation length. For the smaller oligomers of polyacetylene, the exponent of the power law is somewhere between 3 and 4.[10] It is known from x-ray studies[11] that the phenyl rings in PBT are significantly nonplanar with respect to the benzthiazole rings. This leads to a decrease in the effective conjugation length of the molecule. Quantum mechanical modelling[5-8] of PBT model compounds predict a planar gas phase structure with a small rotational barrier. Thus, crystal packing forces could significantly affect the planarity of the molecule. Intramolecular hydrogen bonding can be used in the PBT systems as well to make the molecules more planar, increasing their conjugation length and thus their $\chi^{(3)}$ value.

In this paper we will discuss the strength of the hydrogen bond in the hydroxy PBO system and whether it can lead to a significant increase in the planarity and rotation barrier of the polymer molecule. No attempt will be made to estimate the intermolecular hydrogen bond. However, with the experimental observations available[*21*], there is no indication of this type of bonding. While hydroxy PBO is only one of many possible hydroxy PBX systems, the hydrogen bond between the OH of the phenyl ring and the nitrogen in the benzoxazole ring should be representative of the other members in the family. This presupposition will be tested in future papers.

Computational Model and Semiempirical Results

The hydrogen bond in PBO was modeled using the molecule 2-(2-hydroxyphenyl)benzoxazole, shown in Figure 2. The dashed line indicates the hydrogen bond between the nitrogen of the benzoxazole ring and the OH attached to the phenyl ring. For the AM1 calculations, the molecule was completely optimized subject to the constraints that: (a) the atoms of the benzoxazole ring are coplanar and (b) the atoms of the phenyl ring and oxygen are coplanar. The hydrogen bond strength was estimated by rotating the OH group about the CO bond so that the hydrogen pointed away from the nitrogen and reoptimizing the geometry. Additional constraints were placed on the molecule in this second optimization, namely, the phenyl torsion angle was kept fixed at its value in the first optimization[*12*] and the C-C-O-H torsion was kept at 180°. Since there is no energy change upon rotating the OH group in phenol by 180°, any energy change obtained by this procedure must be due to the interaction of the OH with the benzoxazole ring. The difference in energy between these two configurations is ascribed to the hydrogen bond.

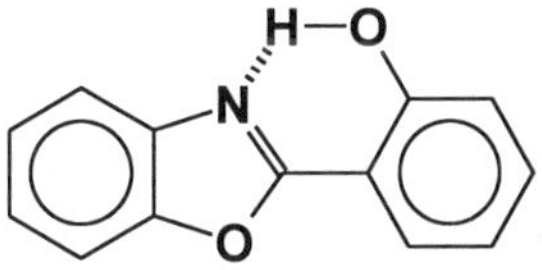

H—O
N
O

Figure 2: 2-(2-hydroxyphenyl)-benzoxazole used as the hydroxy-PBO model compound. The dashed line indicates the hydrogen bond.

It is well known that the semiempirical methods underestimate the strength of the hydrogen bond. However, it will be instructive to compare the hydrogen bond from an AM1 calculation to that of the *ab initio* calculations shown later. The AM1 optimized geometry of 2-(2-hydroxyphenyl)-benzoxazole is completely planar for both the hydrogen bonded and non-hydrogen bonded forms, even if no constraint is placed on the dihedral angle between the phenyl ring and the benzoxazole ring.[*12*] The presence of a hydrogen bond does not lead to gross geometrical changes in the molecule. The detailed geometrical structure of this compound will be published elsewhere. The heat of formation for the hydrogen bonded conformation - with the hydrogen of the OH pointing toward the nitrogen of the benzoxazole - was 16.00 Kcal/mole. Rotating the OH group 180° around the C-O bond increases the heat of formation to 19.43 Kcal/mole. Hence AM1 predicts the hydrogen bond strength in the hydroxy-PBO model compound to be 3.44 Kcal/mole.

Since oxygen can also form hydrogen bonds, an AM1 calculation was performed in the conformation in which a hydrogen bond will form between the oxygen in the heterocycle and OH. The geometry in this situation was again optimized subject to the same constraints as were used for the previous calculations. The heat of formation of the hydrogen bonded conformation is 16.94 Kcal/mole while that with the OH rotated is 18.82, leading to a hydrogen bond strength of 1.88 Kcal/mole.

***Ab Initio* Results**

For all *ab initio* RHF calculations the GAUSSIAN 90[*13*] suite of programs was used on the Cray X-MP at Wright-Patterson AFB. A 6-31G* basis was used to optimize the geometries of all the conformations within the Hartree-Fock SCF method. The hydrogen bond energies were computed using a 6-31G* basis on all atoms in an analogous manner to the AM1 calculation. These optimizations were then repeated placing a 6-31+G(2d,2p) basis set[*14*] on the nitrogen, hydrogen and oxygen that were involved in the hydrogen bond in order to test the completeness of the calculations. MP2 single point corrections were applied to some of the 6-31+G(2d,2p) calculations. These were performed on a 4 Gw Cray Y-MP M90 using GAUSSIAN 92.[*15*] The results are shown in Table 1.

As with the AM1 calculation, the *ab initio* geometry is planar in both the hydrogen bonded and the non-hydrogen bonded conformations. The largest geometric difference between the two conformations is in the C-C-O angle formed by the OH group with the phenyl group, which decreases by 3.8° from 123.4° in the hydrogen bonded conformation to 119.6° in the non-hydrogen bonded conformation (moving the oxygen closer to the nitrogen). This indicates the presence of a small amount of steric repulsion between the hydrogen and the nitrogen. The total energy change involved in the geometry relaxation upon breaking the hydrogen bond amounts to about 1.6 Kcal/mole with the extended basis set, which is about 15% of the total energy change.

As seen in the first two rows of Table 1, the hydrogen bond in the hydroxy-PBO model is predicted to be about 11-12 Kcal/mole. This is an extremely strong hydrogen bond, being nearly twice the strength of hydrogen bonds observed in water,[*16,17*] ammonia/water,[*18-20*] or amides.[*17*] Changing from a 6-31G* to an extended basis set in which the N, O, and H involved in the hydrogen bond have a 6-31+G(2d,2p) basis has little effect on the strength of the hydrogen bond in this conformation. Even MP2 corrections for post Hartree-Fock correlation do not greatly affect the hydrogen bond strength. Hence, the hydrogen bond strength appears converged with respect to basis set and correlation.

The basis set superposition error (BSSE) was estimated by adding hydrogen basis functions to the molecule in row 2 at the position of the hydrogen on the OH in row 1. This should overestimate the correction, because it will improve the description of the OH as well as the nitrogen. The observed energy change was less than 0.5 Kcal/mole, which indicates that BSSE is not a significant factor in the calculations.

The previous calculations were repeated after rotating the phenyl ring 180° making the hydrogen bond to the oxygen of the benzoxazole ring instead of the nitrogen. As with the calculations for the hydrogen bond to the nitrogen, these showed no significant basis set effects. The 6-31G* basis indicates a hydrogen bond strength of 5.1 Kcal/mole. Extending the basis set on the oxygens and the hydrogen

Table 1: Computed energy differences (Kcal/mole) between conformations of the hydroxy-PBO model compound relative to the lowest energy conformation for a given basis set with a hydrogen bond to the nitrogen.

Molecule	RHF/6-31G*	RHF/extended basis	MP2/extended basis
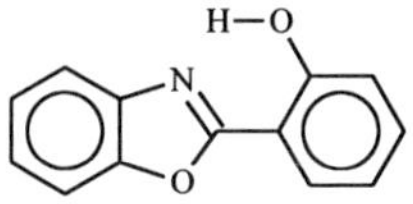 H—O / N / O	0	0	0
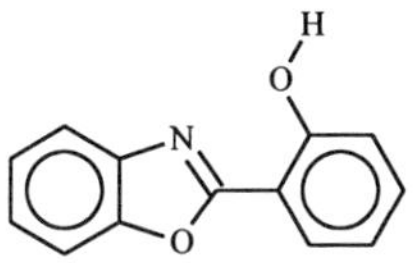 H / O / N / O	11.1	10.4	11.7
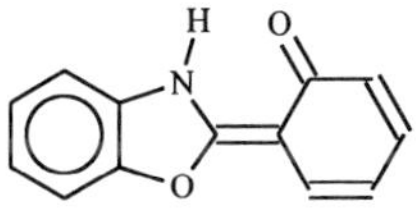 H O / N / O	–	16.5	18.5
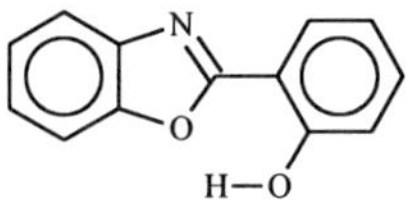 N / O / H—O	1.7	4.1	–
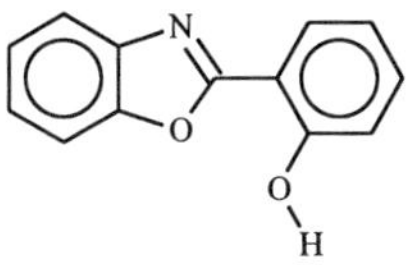 N / O / O / H	6.8	9.4	–
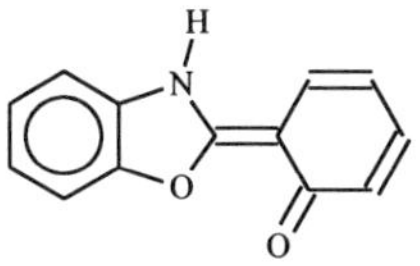 H / N / O / O	–	32.3	–

of the OH to 6-31+G(2d,2p) increases the hydrogen bond strength slightly to 5.3 Kcal/mole. These bond strengths are comparable to those observed in other similar molecules.[*16-20*] No MP2 corrections were calculated for this conformation.

One other set of conformations was studied. It is known that excited states of molecules like 2-(2-hydroxyphenyl)benzoxazole will undergo hydrogen transfer reactions to a quinoidal structure in which the hydrogen of the OH completely transfers to the nitrogen of the benzoxazole (row 3 of Table 1).[*21-23*] These structures have been optimized with the extended basis set, i.e., 6-31+G(2d,2p) on N, O, and OH. The calculations indicate that the energy required to make this transformation is higher than that required to break the hydrogen bond by about 60%.

However, the predicted energy change is still much smaller than the 47 Kcal/mole difference observed spectroscopically by Ernsting. [*21*] This discrepancy is under further study. Since the hydrogen bond is still capable of being formed in this conformation, the molecule in row 6 of Table 1 was studied to estimate the strength of the hydrogen bond. In this instance, rotation of the phenyl ring does not lead to an energetically equivalent situation in the absence of the hydrogen bond. Despite this concern, it is not expected that the carbonyl oxygen will interact very differently with the nitrogen or the oxygen of the benzoxazole, so the energy difference between rows 3 and 6 of Table 1 should be largely due to the breaking of the hydrogen bond. However, due to the nonequivalence of the two conformations, all that can be said is that the energy difference of 15.8 Kcal/mole is comparable to the hydrogen bond energy in the ground state molecule.

The rotational potential for the phenyl ring in 2-(2-hydroxyphenyl)-benzoxazole was computed with a basis set consisting of 6-31+G(2d,2p) on the N, O, and OH and 6-31G* on all other atoms. Each of the calculations involved a geometry optimization in which the phenyl and benzoxazole rings were kept individually planar and the torsion angle between them was kept fixed at the values used for the potential. Figure 3 is a plot of the results compared to the PBO model compound (without the hydroxyl), for which the 6-31G* basis was used for all atoms. The minimum energy in both molecules was considered as the zero point of energy for the graph. Both graphs peak at about 90° torsion angle. The rotational barrier for the hydroxy-PBO model compound is about 3.5 Kcal/mole higher than the non-hydrogen bonded compound. This can be interpreted as being due to the breaking of the hydrogen bond during rotation. The change in rotation barrier is considerably less than the hydrogen bond strength. This is reasonable since as the phenyl ring rotates, the increase in energy caused by the hydrogen bond breaking is partially compensated for by the lower repulsion between the hydroxyl and the nitrogen. However, the 3.5 Kcal/mole of energy difference will reduce the room temperature Boltz-mann factor for right angle twists to about 0.003 of its value for the non-hydrogen bonded material. Even considering collective

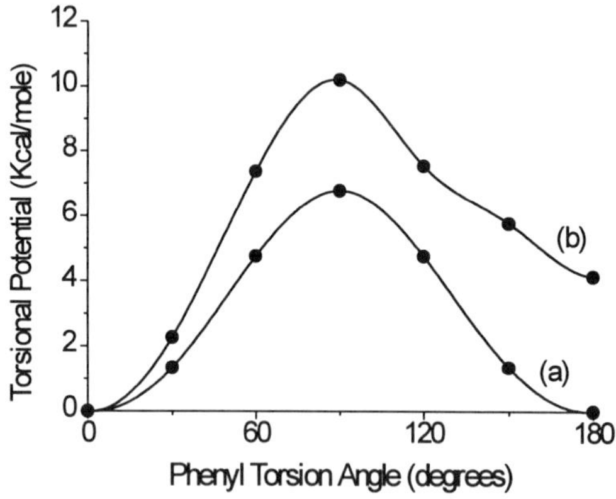

Figure 3: Torsion potential for PBO model compounds. (a) 2-phenylbenzoxazole; (b) 2-(2-hydroxyphenyl)-benzoxazole, 0° corresponds to the planar conformation with the hydrogen bond to the nitrogen.

torsions[*24*], the increased slope of the rotational barrier will cause the length required for significant rotation to increase greatly, leading to a much stiffer polymer.

Conclusions

To the extent that 2-(2-hydroxyphenyl)benzoxazole accurately models the hydroxy-PBO polymer, it has been shown that the intramolecular hydrogen bond in this molecule is among the strongest hydrogen bonds seen to date. AM1 severely underestimates the strength of this bond compared to *ab initio* calculations, which appear to be well converged with respect to both basis set and correlation. The presence of the hydrogen bond leads to a significant increase in the energy required to rotate the phenyl ring, which should lead to a much stiffer polymer. This property may be able to be used to improve many of the already impressive characteristics exhibited by the PBX polymer systems.

Acknowledgments

The authors would like to express their appreciation to David Slowinski of Cray Research for performing the MP2 calculations on their Y-MP M90 computer.

References

1. E. W. Choe and S. N. Kim, *Macromol.*, **14**, 920(1981); D. R. Ulrich, *Polymer*, **28**, 533(1987)
2. S. B. Allen, A. G. Filippov, B. J. Farris, E. L. Thomas, C.-P. Wong, G. C. Berry and E. C. Chenevey, *Macromol.*, **14**, 1139(1981)
3. T. E. Helminiak, et. al., U. S. Patent 4,207,407 (1980)
4. S. Trohalaki and D. S. Dudis, *Makromol. Chem., Macromol. Symp.*, **65**, 163(1993)
5. B. L. Farmer, S. G. Wierschke and W. W. Adams, *Polymer*, **31**, 1637(1990)
6. J. W. Connolly and D. S. Dudis, *Polymer*, **34**, 1477(1993)
7. Y. Yang and W. J. Welsh, *Macromol.*, **23**, 2410(1990)
8. B. L. Farmer, B. R. Chapman, S. S. Dudis and W. W. Adams, *Polymer*, **34**, 1588(1993)
9. C. Y.-C. Lee, J. Swiatkiewicz, P. N. Prasad, R. Mehta and S. J. Bai, *Polymer*, **32**, 1195(1991)
10. G. J. B. Hurst, M. Dupuis and E. Clementi, *J. Chem. Phys.*, **89**, 385(1988)
11. M. W. Wellman, W. W. Adams, R. A. Wolfe, D. S. Dudis and A. V. Fratini, *Macromol.*, **14**, 935(1981)
12. For all AM1 calculations no effect was seen on the energy of the non-hydrogen bonded conformation by releasing the constraint on the phenyl torsion. In subsequent *ab initio* calculations, this constraint was kept in effect.
13. *Gaussian 90*, M. J. Frisch, M. Head-Gordon, G. W. Trucks, J. B. Foresman, H. B. Schlegel, K. Raghavachari, M. A. Robb, J. S. Binkley, C. Gonzalez, D. J. Defrees,D. J. Fox, R. A. Whiteside, R. Seeger, C. F. Melius, J. Baker, R. L.

Martin, L. R. Kahn, J. J. P. Stewart, S. Topiol, and J. A. Pople, Gaussian, Inc., Pittsburgh PA., 1990
14. J. del Bene, *J. Comp. Chem.*, **10**, 603(1989)
15. *Gaussian 92*, Revision A, M. J. Frisch, G. W. Trucks, M. Head-Gordon, P. M. W. Gill, M. W. Wong, J. B. Foresman, B. G. Johnson, H. B. Schlegel, M. A. Robb, E. S. Replogle, R. Gomperts, J. L. Andres, K. Raghavachari, J. S. Binkley, C. Gonzalez, R. L. Martin, D. J. Fox, D. J. Defrees, J. Baker, J. J. P. Stewart and J. A. Pople, Gaussian, Inc., Pittsburgh PA, 1992
16. D. Feller, *J. Chem. Phys.*, **96**, 6104(1992)
17. J. B. O. Mitchell, and S. L. Price, *Chem. Phys. Lett.*, **180**, 517(1991)
18. J. D. Dill, L. C. Allen, W. C. Topp, and J. A. Pople, *J. Am. Chem. Soc.*, **97**, 7220(1975)
19. W. A. Lathan, L. A. Curtiss, W. J. Hehre, J. B. Lisle, and J. A. Pople, *Prog. Phys. Org. Chem.*, **11**, 175(1974)
20. W. C. Topp, and L. C. Allen, *J. Am. Chem. Soc.*, **96**, 5291(1974)
21. N. P. Ernsting, *J. Phys. Chem.*, **89**, 4932(1985)
22. A. Mordzinksi and W. Kuhnle, *J. Phys. Chem.*, **90**, 1455(1986)
23. S.-I. Nagaoka and U. Nagashima, *J. Phys. Chem.*, **94**, 1425(1990)
24. J. W. Connolly and D. S. Dudis, to be published

RECEIVED June 2, 1994

Chapter 18

Borohydrides as Novel Hydrogen-Bond Acceptors

M. A. Zottola[1], L. G. Pedersen[2], P. Singh[3], and B. Ramsay-Shaw[1]

[1]Department of Chemistry, Duke University, Durham, NC 27705
[2]Department of Chemistry, University of North Carolina—Chapel Hill, Chapel Hill, NC 27599
[3]Department of Chemistry, North Carolina State University, Raleigh, NC 27695

An X-ray crystallographic structure of N1-boronated cytosine has provided experimental evidence for stabilizing light atom to light atom contact. This work uses high level ab initio calculations on model borane systems to substantiate the claim that there is a stabilizing light atom to light atom contact. By comparison of these model systems to the water dimer it can be concluded that these light atom interactions are unusual hydrogen bond interactions.

The research focus of our lab has been the development of boronated nucleotides as potential anti-cancer agents and therapeutic agents for use in Boron Neutron Capture Therapy (BNCT)[1,2]. This anti-cancer therapy relies upon incorporating boronated nucleic acids into cells via specific oligonucleotide sequences. Part of our work has been to model the effect on oligonucleotide structure and dynamics induced by the presence of a boronated nucleotide.

The X-Ray crystal structure of N1-boronated cytosine (Figure 1) provided a starting point for the modeling effort. An intriguing aspect of the crystal structure shown here is the apparent close contact between the amino protons and the protons on the cyanoborane unit.

Refinement of the X-Ray crystal structure included the hydrogens. The positions of all the hydrogens were able to be determined from a Fourier difference map. The crystallographically-determined interproton distance was 2.05 Å. This distance implies an overlap of these two protons van derWaals' radii. The resolution of the structure (0.71 Å) afforded a measure of confidence for the apparent close contact between the amino protons and the protons on the cyanoborane unit.

0097–6156/94/0569–0277$08.00/0

We reasoned that there could be at least two possible explanations for this close contact between the two hydrogens. The first possibility was that packing in the unit cell forced a close contact between the protons. Unit cell analysis revealed no intermolecular contacts within 4 Å that would enforce such a conformation. Thus there were no heavy atom to heavy atom contacts which could impose this conformation which forces an overlap of the hydrogen van der Waal's radii.

The second possibility was that a significant attraction could exist between the amino and cyanoboranyl hydrogens. Questions then arise about the strength and type of interaction existing between these hydrogens. However, attraction between two sets of non-bonded hydrogens has not documented, to our knowledge. The remainder of this paper is devoted to determining whether such hydrogen to hydrogen contacts as seen in this crystal structure are real.

Our work began with the premise that the crystal structure contains evidence for the attraction between non-bonded hydrogens. To substantiate that claim, we must first define what we mean by a hydrogen bond. Figure 2 shows the water dimer[3].

The characteristics of this dimer include :

• A bonding motif characterized by an electronegative heavy atom, electropositive light atom, electronegative heavy atom triad. In this case the triad consists of oxygen, hydrogen and oxygen. This exemplifies the electrostatic nature of the interaction[4].

• The O-H···O triad of atoms deviates slightly from linearity due to overlap between the 1s orbital of hydrogen and the 2p lone pair orbital on the acceptor oxygen. Despite the electrostatic component of the interaction, the geometry of the hydrogen bonded complex reflects that some covalent interaction exists between the light and heavy atoms.

A third characteristic of a hydrogen-bonded systems is that the interaction energy for a series of complexes holding the hydrogen bond acceptor (donor) constant is directly related to the Brønsted acidity (basicity) of the donor (acceptor)[5]. The criteria to decide whether or not the interaction that we observe (if real) is a hydrogen bond interaction are :

1) Does the dimer have an electrostatic component?
2) Does the geometry of the complex reflect orbital overlap?
3) Does the interaction energy reflect Brønsted acidities/ basicities?

Our approach to examine the potential hydrogen-bond acceptor behavior of borohydrides follows :

1) We used the molecules HF, H_2O and H_3N to probe the hydrogen bonding potential of borazane.

2) We repeated the series using imineborane (the complex of borane with methylene imine) in place of borazane, to assess the effect of the nitrogen ligand on the potential hydrogen - bonding behavior.

First, we produced an electrostatic map for the borazane molecule (ammoniaborane)[6]. This map reveals a region of negative charge around the borane fragment, a clearly welcome result and an expected one based on the chemical reactivity of boron hydrides. This result implies the possibility for borohydrides to act as hydrogen-bond acceptors.

We examined these complexes at three basis sets[7], hf/6-31g, hf/6-31g** and mp2/6-31g**. We also explored extended basis sets for the hydrogen fluoride complexes (mp2/6-311g** and mp2/6-31g(3d,2p)). In both cases minimal (~0.005Å) geometry and energy (ΔE <~0.5Kcals) changes occurred. Tables I and II summarize the data generated from points one and two above.

For the ammonia complexes, no unusual B-H to NH contacts occurred within the geometry of the complex. This is represented by the value of 0 in the tables. Although unexpected, the lack of interaction of the B-H bond can be rationalized by the weak Brønsted acidity of the ammonia hydrogens.

Interaction energies are expressed in two ways - a ΔE for the total interaction energy of the complex corrected for basis set superposition error (BSSE)[8] and a ΔE for the hydrogen bonding interactions, again corrected for BSSE. The interaction energy for the complex (ΔE(cplx)) represents the sum of the interactions for the borane unit interacting with the amine and the hydrogen bond donor interacting with the borane amine complex. The ΔE of hydrogen bonding (ΔE(hbnd)) represents the interactions of the hydrogen bond donor with the borane amine complex.

To establish ΔE(cplx) and ΔE(hbnd), counterpoise calculations were done for each component of the ternary complex. In Figure 3, Ea' represents the Hartree-Fock energy for component a while ghost orbitals are placed on components b and c. The terms Eb' and Ec' are similarly defined. The three values are summed and that value is subtracted from the Hartree-Fock energy of the complex, to give ΔE(cplx), the total interaction energy for the complex. The interaction energy for the borane-amine complex was calculated using the counterpoise method (ΔEb in Figure 3). This value was subtracted from ΔE(cplx) to yield ΔE(hbnd). This approach is reasonable because the geometry of the borane-amine complex was not greatly altered when optimized with a hydrogen bond donor. The bond length changes between the binary amine-borane complex and the borane-amine unit within the ternary complex were approximately 0.01Å. A similar comparison of valence angle changes showed differences of only 2 - 3 degrees.

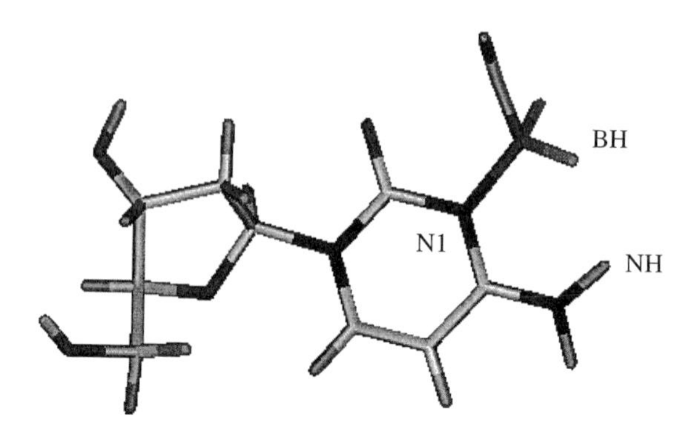

Figure 1. A representation of the X-ray crystal structure of N1-cyanoborane cytidine. Note the close contact between the boron hydride and the amino proton (see text).

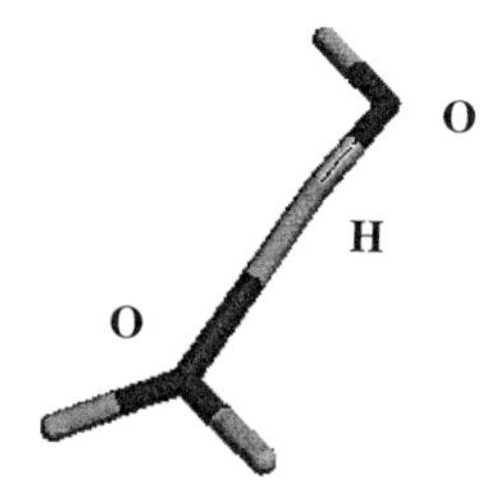

Figure 2. The geometry of the water dimer was optimized using 6-31G basis set. An electrostatic isodensity map of the water dimer reveals the regions of negative charge are over the oxygens while more positively charged regions are over the hydrogens.

1) The putative hydrogen bond complex is considered to be a ternary complex

$$\Delta E(cplx) = Et - (Ea' + Eb' + Ec')$$

2) As the structure of the BN compound does not change during optimization

$$\Delta Eb = Et' - (Ea'' + Eb'')$$

3) Hence the ΔE for the interaction is

$$\Delta E(hbnd) = \Delta Et - \Delta Eb$$

Figure 3. Calculation of ΔE(Hbnd)

TABLE I

H-Bond Complexes with H3BNH3

Complex	Basis Set	r(B-N)	r(X-HN)	r(BH-HX)	dE (cplx)	dE (hbnd)	Dipole momen
HF/BNH6	6-31g	1.690	2.106	2.142	45.40	8.46	3.758
	6-31g**	1.666	2.142	2.045	42.66	8.25	4.023
	mp2/631g**	1.643	2.011	1.725	50.84	9.64	4.169
H2O/BNH6	6-31g	1.687	2.270	1.921	47.89	10.95	4.604
	6-31g**	1.665	2.276	2.092	43.13	8.72	3.798
	mp2/631g**	1.642	2.070	2.000	52.05	10.85	3.693
H3N/BNH6	6-31g	1.687	2.012	0.000	47.46	10.52	5.663
	6-31g**	1.664	2.084	0.000	44.23	9.82	5.434
	mp2/631g**	1.641	1.945	0.000	52.86	11.66	5.240

TABLE II

H-Bond Complexes with H3B-Imine

CPLX	Basis Set	r(B-N)	r(BH-HX)	r(HX-HN)	dE (cplx)	dE (Hbnd)	Dipole Moment
HF/ImineBH3	6-31g	1.630	2.003	2.110	49.640	9.250	9.250
	6-31g**	1.622	1.974	2.156	44.890	8.270	8.270
	mp2/631g**	1.593	1.739	2.036	53.170	9.260	9.260
H2O/ImineBH3	6-31g	1.628	2.288	1.930	51.790	11.400	4.604
	6-31g**	1.665	2.276	2.092	45.380	8.760	4.479
	mp2/631g**	1.591	2.131	1.982	53.980	10.070	3.837
H3N/ImineBH3	6-31g	1.627	0.000	1.982	51.770	11.380	6.259
	6-31g**	1.620	0.000	2.312	46.390	9.770	5.954
	mp2/631g**	1.589	0.000	2.190	55.880	11.970	5.443

The geometries of these potentially hydrogen bonded complexes show a slight but interesting basis set dependence. At the 6-31g basis set, the important feature is that the proton in HF nearly bisects the H-B-H bond angle (Figure 4). Also the line describing the boron-nitrogen bond is nearly parallel to the hydrogen-fluorine bond. The geometry for the 6-31g** basis set, however, shows an anisotropic interaction between the hydrogen of HF and one B-H (Figure 5). The line describing the B-N bond is now skewed to the H-F bond. The 6-31g** basis set provides a more complete description of valence space which provides for the direct interaction (i.e. a single hydride interacting with a single electropositive proton). Applying second order Møller-Plesset[9] corrections to the Hartree-Fock wave function increases the extent of this direct hydrogen to hydrogen interaction.

As expected, these ternary complexes also have present the "traditional" hydrogen bond contact between the heavy atom (F, O or N) and the proton on the nitrogen in the borane amine complex. Thus, ΔE(hbnd) represents a sum of both interactions.

The conclusion one may draw from this set of results is that orbital overlap does play a role in the geometry of these complexes. Although the effect is subtle and only apparent at higher levels of theory, it is significant. This result meets one of the criteria for a hydrogen bonded complex, that the geometry must depend on orbital overlap.

A normal criterion for a given series of hydrogen bond complexes, is that the strength of the interaction (ΔE(hbnd)) should follow the Brønsted acidity/basicity of the donor/acceptor. The dependence ΔE(hbnd) on basis set for the set of ternary complexes with borazane and imineborane shows two distinct features. As noted earlier, these energies are a sum of two contact points in the cases of HF and water.

The first feature is the type of amine ligand on the boron does not significantly alter the ΔE(hbnd) for the complex. The second feature is the trend in the interaction energy as measured by ΔE(hbnd) opposes the trend in Brønsted acidity of the hydrogen bond donors.

Although this trend is contrary to expectation, one would expect hydrogen to hydrogen contact energies partitioned from ΔE(hbnd) to follow the expected trend. Thus, how does one partition the hydrogen to hydrogen energy from ΔE(hbnd)? We attempted to establish an upper bound for the pure hydrogen to hydrogen contact by optimizing the complexes of N - methylimineborane with either hydrogen fluoride or water. The presence of the methyl group on the imine nitrogen precludes any "normal" hydrogen bond interactions. After subtracting the ΔE_b of the parent borane amine complex from the ΔE(cplx) of the ternary complex (each value corrected for BSSE), the residual ΔE (ΔE(hbnd)) estimates an upper bound to the strength of the hydrogen to hydrogen interaction. The upper bound values reflect the expected trend of ΔE vs. Brønsted acidity. This data is summarized in Table III.

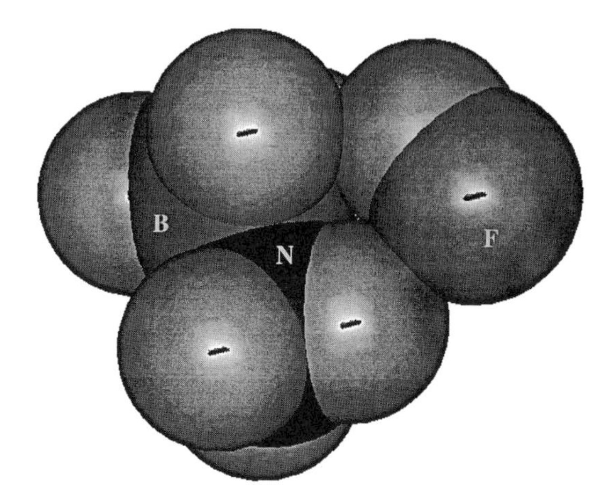

Figure 4. The borazane - HF complex optimized at the 6-31g level. The heavy atoms (B, N and F) are labeled. Note that the H-F bond is nearly parallel to the B-N bond (see text).

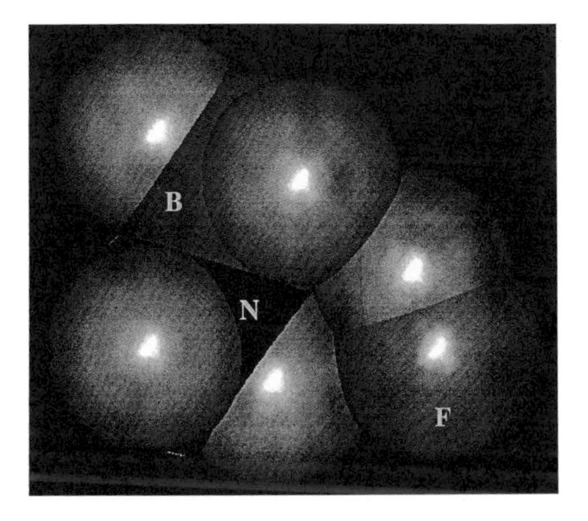

Figure 5. The borazane HF complex optimized using the 6-31g** basis set. The heavy atoms (B,N and F) are labeled. Note that the line describing the H-F bond is skewed relative to the B-N bond.

TABLE III

Upper Bounds for the Hydrogen to Hydrogen Interaction

ΔE (hbnd) for HF = -7.6 Kcal

ΔE (hbnd) for $H2O$ = -6.3 Kcal

These numbers were calculated using the 6-31g basis set with N-methylimineborane instead of imineborane as the hydride source.

Why is the order of ΔE(hbnd) reversed from expectations? As stated before, the ΔE's reflect a contribution from the "traditional" hydrogen bond interaction. As the series progresses from HF to water to ammonia, the Brønsted base strength of the heavy atom increases. Thus the ΔE(hbnd) for these systems reflects a synergy of the two interactions.

In summary, we have shown that stable ternary complexes can be formed between borane amine complexes and hydrogen bond donors. These ternary complexes have a stabilizing hydrogen to hydrogen interaction. Further, this interaction, in analogy to the "normal" hydrogen bond, reveals a strong electrostatic component and a geometry which relies upon orbital overlap.

We are currently exploring the effect of incorporating boronated nucleotides on oligonucleotide structure and dynamics and will provide a report at a later date.

Acknowledgments

We are very grateful to the North Carolina Supercomputing Center for the generous grants of Cray time. MAZ is also indebted to Glaxo Research Institute for both a research fellowship that has funded him while pursuing this work and free reign to use their facilities . Finally, we would like to thank Mrs. Janet Layko for the reading of this manuscript.

Literature Cited

1) Sood, A.; Spielvogel,B. F.; Shaw, B. R. *J. Am. Chem. Soc.* **1989**, *111*, 9234-9235.
2) Sood, A.; Shaw, B. R.; Spielvogel, B. F. *J. Am. Chem. Soc.* **1990**, *112*, 9000-9001.
3) Generated using the Spartan program available from Wavefunction, Inc. The geometry of the dimer was optimized at HF/6-31g.
4) Scheiner, S. *Rev. Comput. Chem.* **1991**, *2*, 165-218.
5) Hehre, W.J.; Radom, L.; Schleyer, P.v.R. and Pople, J.A. **Ab Initio Molecular Orbital Theory,** Wiley Interscience, New York, 1986, pg. 222.
6) Generated using the Spartan program available from Wavefunction, Inc. The geometry of the complex was optimized at hf/6-31g.
7) All the ab initio calculations were carried out using the Gaussian 92 package. Gaussian 92, Revision C, M. J. Frisch, G. W. Trucks, M. Head-Gordon, P. M. W. Gill, M. W. Wong, J. B. Foresman, B. G. Johnson, H. B. Schlegel, M. A. Robb, E. S. Replogle, R. Gomperts, J. L. Andres, K. Raghavachari, J. S. Binkley, C. Gonzalez, R. L. Martin, D. J. Fox, D. J. Defrees, J. Baker, J. J. P. Stewart, and J. A. Pople, Gaussian, Inc., Pittsburgh, PA, 1992.
8) Boys, S.F.; Bernardi, F. **Mol. Phys. 1970, 19,** 553.
9) Moller, C.; Plesset, M. S. *Physical Review* **1934**, *46*, 618.

RECEIVED July 29, 1994

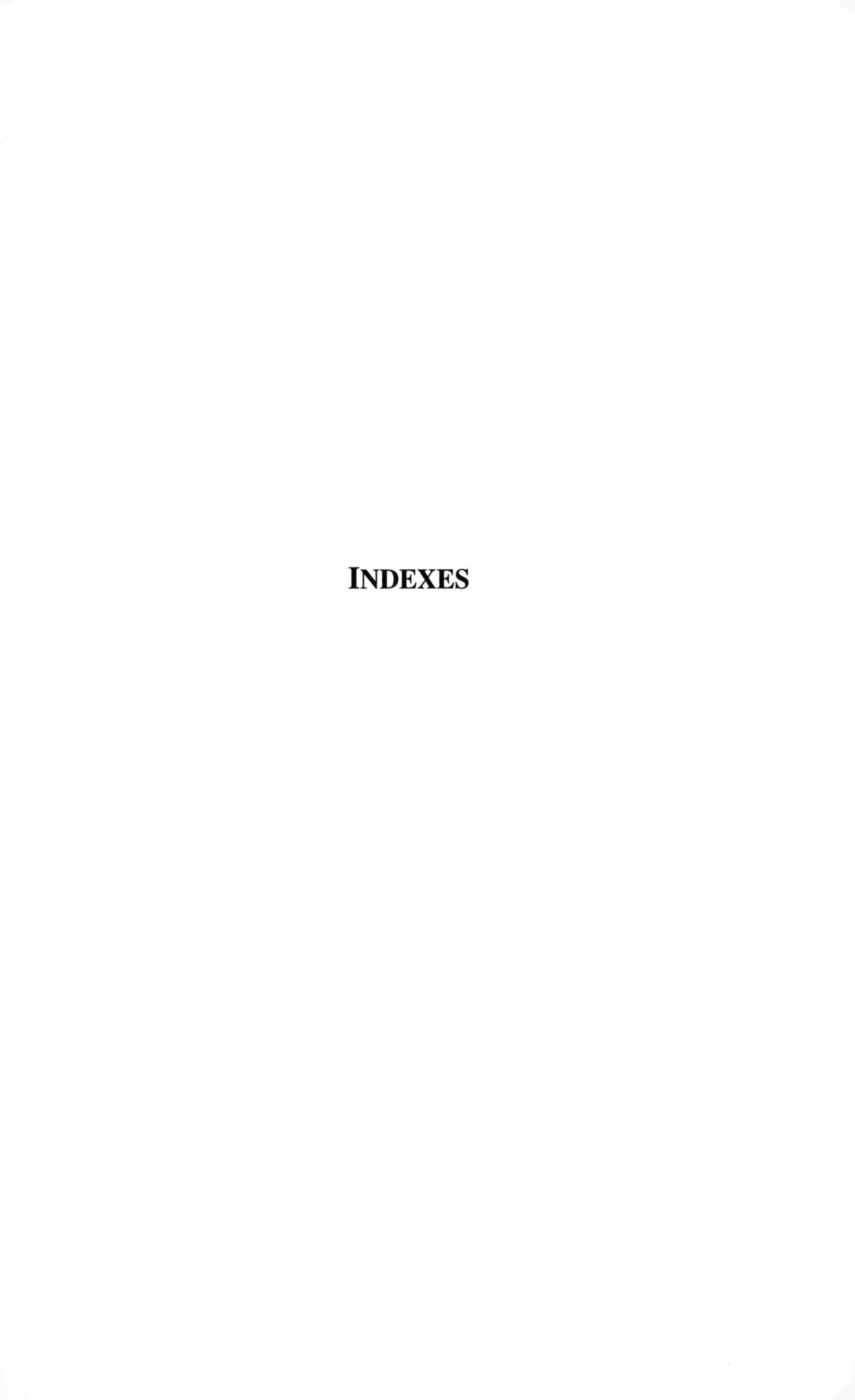

INDEXES

Author Index

Affiliation Index

Subject Index

E

F

G

N

T

Production: Susan Antigone & Charlotte McNaughton
Indexing: Deborah H. Steiner
Acquisition: Barbara Pralle & Rhonda Bitterli
Cover design: Alan Kahan

Printed and bound by Maple Press, York, PA

Highlights from ACS Books

Good Laboratory Practice Standards: Applications for Field and Laboratory Studies
Edited by Willa Y. Garner, Maureen S. Barge, and James P. Ussary
ACS Professional Reference Book; 572 pp; clothbound ISBN 0–8412–2192–8

Silent Spring Revisited
Edited by Gino J. Marco, Robert M. Hollingworth, and William Durham
214 pp; clothbound ISBN 0–8412–0980–4; paperback ISBN 0–8412–0981–2

The Microkinetics of Heterogeneous Catalysis
By James A. Dumesic, Dale F. Rudd, Luis M. Aparicio, James E. Rekoske, and Andrés A. Treviño
ACS Professional Reference Book; 316 pp; clothbound ISBN 0–8412–2214–2

Helping Your Child Learn Science
By Nancy Paulu with Margery Martin; Illustrated by Margaret Scott
58 pp; paperback ISBN 0–8412–2626–1

Handbook of Chemical Property Estimation Methods
By Warren J. Lyman, William F. Reehl, and David H. Rosenblatt
960 pp; clothbound ISBN 0–8412–1761–0

Understanding Chemical Patents: A Guide for the Inventor
By John T. Maynard and Howard M. Peters
184 pp; clothbound ISBN 0–8412–1997–4; paperback ISBN 0–8412–1998–2

Spectroscopy of Polymers
By Jack L. Koenig
ACS Professional Reference Book; 328 pp;
clothbound ISBN 0–8412–1904–4; paperback ISBN 0–8412–1924–9

Harnessing Biotechnology for the 21st Century
Edited by Michael R. Ladisch and Arindam Bose
Conference Proceedings Series; 612 pp;
clothbound ISBN 0–8412–2477–3

From Caveman to Chemist: Circumstances and Achievements
By Hugh W. Salzberg
300 pp; clothbound ISBN 0–8412–1786–6; paperback ISBN 0–8412–1787–4

The Green Flame: Surviving Government Secrecy
By Andrew Dequasie
300 pp; clothbound ISBN 0–8412–1857–9

For further information and a free catalog of ACS books, contact:
American Chemical Society
Distribution Office, Department 225
1155 16th Street, NW, Washington, DC 20036
Telephone 800–227–5558